PERGAMON INTERNATIONAL LIBRARY
of Science, Technology, Engineering and Social Studies
The 1000-volume original paperback library in aid of education, industrial training and the enjoyment of leisure
Publisher: Robert Maxwell, M.C.

INORGANIC CHEMISTRY AND THE EARTH

Chemical Resources, Their Extraction, Use and Environmental Impact

PERGAMON SERIES ON ENVIRONMENTAL SCIENCE

Series Editors: O. HUTZINGER and S. SAFE

Other Books in the Series

Volume 1
*HUTZINGER *et al:* Aquatic Pollutants — Transformation and Biological Effects

Volume 2
ZOETEMAN: Sensory Assessment of Water Quality

Volume 3
*ALBAIGES: Analytical Techniques in Environmental Chemistry I

Volume 4
VOWLES & CONNELL: Experiments in Environmental Chemistry

Volume 5
HUTZINGER *et al:* Chlorinated Dioxins and Related Compounds

Volume 7
*ALBAIGES: Analytical Techniques in Environmental Chemistry II

Also of Interest

HENDERSON: Inorganic Geochemistry

Journals

ATMOSPHERIC ENVIRONMENT
CHEMOSPHERE (Chemistry, Biology and Toxicology as Related to Environmental Problems)
ENVIRONMENT INTERNATIONAL
THE ENVIRONMENTAL PROFESSIONAL
TALANTA
WATER RESEARCH
WATER SCIENCE AND TECHNOLOGY

INORGANIC CHEMISTRY AND THE EARTH

Chemical Resources, Their Extraction, Use and Environmental Impact

by

J. E. FERGUSSON

Department of Chemistry, University of Canterbury
Christchurch, New Zealand

PERGAMON PRESS

OXFORD · NEW YORK · TORONTO · SYDNEY · PARIS · FRANKFURT

U.K.	Pergamon Press Ltd., Headington Hill Hall, Oxford OX3 0BW, England
U.S.A.	Pergamon Press Inc., Maxwell House, Fairview Park, Elmsford, New York 10523, U.S.A.
CANADA	Pergamon Press Canada Ltd., Suite 104, 150 Consumers Road, Willowdale, Ontario M2J 1P9, Canada
AUSTRALIA	Pergamon Press (Aust.) Pty. Ltd., P.O. Box 544, Potts Point, N.S.W. 2011, Australia
FRANCE	Pergamon Press SARL, 24 rue des Ecoles, 75240 Paris, Cedex 05, France
FEDERAL REPUBLIC OF GERMANY	Pergamon Press GmbH, Hammerweg 6, D-6242 Kronberg-Taunus, Federal Republic of Germany

First edition 1982
Reprinted with corrections 1985

British Library Cataloguing in Publication Data

Fergusson, J.E.
Inorganic chemistry and the earth.—(Pergamon series on environmental science; 6)—(Pergamon international library)
1. Environmental chemistry
I. Title
574.5 QD31.2

ISBN 0-08-023995-1 (Hardcover)
ISBN 0-08-023994-3 (Flexicover)

Printed in Great Britain by A. Wheaton & Co. Ltd., Exeter

Preface

Growth in interest in, and the study of, the chemistry of natural systems of the earth has become apparent over the last two decades. This is clear from the growing number of research papers and monographs on environmental chemistry published, and development of courses at Universities. Reasons for this growth are an increasing concern for the future of the world's resources and the effect of pollution on the environment and people. Also the rapid development in methods of fast instrumental analysis, enabling trace amounts of material in a complex matrix to be determined, has been a significant factor.

Much of the chemistry encountered in a study of the earth and environmental pollution is basic chemistry. For example oxidation-reduction processes feature significantly and solubility is an important aspect. The chemistry of the gaseous oxides of the p-block elements feature prominently in air-chemistry. As a consequence there has been a rebirth in interest in this type of chemistry, as well as in analysis; two areas which have been over-shadowed in the last 30 years by the developments in the chemistry of the transition metals. However, while a large proportion of the chemistry is basic and well understood in relation to simple and pure systems, this is not the case for the more complex systems found in nature. For example, soil is a heterogenous mixture of solids, gases and aqueous solutions, and even rain water, with dissolved and suspended species, is not simple. As a result there are many gaps in our knowledge and understanding of the chemistry of natural systems.

The chemistry of the earth cannot be discussed without involving other disciplines, including biology, engineering, geology, mathematics, medicine and physics. The study is truly interdisciplinary, of which chemistry is of major importance. There is a need for people with a chemical training to join interdisciplinary groups in the study of the earth. It is a hope of the author, that this book may stimulate some students to actively particip-ate in such ventures.

The book is divided into four parts which, broadly speaking, cover the areas; the origin and distribution of the chemical elements in the earth, the world's resources, their extraction and some of the chemical processes involved in producing useful commodities, and finally the impact the latter area has on the earth. The subject has not been treated exhaustively as it is an aim to demonstrate the breadth of the involvement of chemistry in our understanding of the earth. The book is intended for use by undergraduates and graduates, and a reasonably large bibliography is given at the end for those who wish to pursue aspects of the subject further.

The author wishes to acknowledge the accurate and expert typing by Chris Haughey, and to Pergamon Press for preparation of the diagrams. The author is grateful to colleagues and students who often, unbeknown to them, have been a sounding board for some of the concepts written in this book. Finally grateful thanks are expressed to my family, who have been very patient over the months of writing.

Christchurch
New Zealand
June 1982

J.E. Fergusson

Contents

PART I

The Chemical Elements in the Environment

CHAPTER 1

Origins: the Chemical Elements and the Earth

Concepts and ideas, on the origin of the universe range over the fields of science, philosophy and religion, and also in the realm of fanciful guess work. More and more evidence is being obtained which supports the contention that the chemical elements are formed by stellar nuclear synthesis, and that the earth's formation can be explained by chemical processes. In this chapter we will discuss both of these aspects.

THE ORIGIN OF THE CHEMICAL ELEMENTS

The Chemical Elements

Chemistry, which is the study of matter, and how it may be modified by physical and chemical processes, is concerned with the chemical elements. There are 90 naturally occurring elements on the earth, 81 of these (from H to Bi, excluding Tc and Pm) have at least one, or more, stable isotope, while the 9 other elements exist as unstable radioactive isotopes (Po to U). The remaining elements of those listed in Table 1.1 (Tc, Pm, and Np to element 106) do not occur naturally on earth today, but have been synthesised by man. However, it is probable that some of these elements did exist in the past, but owing to short half-lives have since decayed. Recent evidence, still being debated, indicates that some elements with atomic numbers in the region 116 to 126 may have existed on the earth in minute amounts. This group of elements are predicted to have relatively stable nuclei compared with their neighbours.

Origin of the Elements

Both the developments in nuclear chemistry over recent years, and the results of space exploration, make it possible to answer questions on the origin and distribution of the chemical elements. The elemental composition of various parts of the universe differ, as indicated by the data in Table 1.2. The relative abundance of the elements, for the universe as a whole, is depicted in Fig. 1.1. Certain features in the table and the figure lead one to raise questions such as; why the variation throughout the universe?

TABLE 1.1 The Occurrence of Elements in Nature

														6		H	He
Li	Be	B											C	N	O	F	Ne
Na	Mg	Al	Si										Si	P	S	Cl	A
K	Ca	Sc	Ti	V	Cr	Mn	Fe	Co	Ni	Cu	Zn	Ga	Ge	As	Se	Br	Kr
Rb	Sr	Y	Zr	Nb	Mo	(Tc)	Ru	Rh	Pd	Ag	Cd	In	Sn	Sb	Te	I	Xe
Cs	Ba	La*	Hf	Ta	W	Re	Os	Ir	Pt	Au	Hg	Tl	Pb	Bi	Po	(At)	(Rn)
Fr	Ra	Ac	Th	Pa	U	(Np**)											
1	2		3					4				5					6

1 Soluble simple salts (Na, K, Rb chlorides, sulphates); cationic constituents in aluminosilicates (Li, Be, Cs).

2 Insoluble carbonates, sulphates.

3 Oxides.

4 Occur as free elements.

5 Sulphides

6 Atmosphere.

○ Does not occur naturally on the earth

* Lanthanides La to Lu

** Actinides Ac to Lr

why does iron feature so prominently? why are elements Li, Be and B of such low abundance relative to their neighbours?

Before pursuing these questions it is necessary to consider, briefly, the structure of the nucleus.

TABLE 1.2 Elemental Composition (% of total atoms) for Selected Systems in the Universe

Universe		Sun (% by vol.)		Bulk Earth (% by mass)		Earth's Crust		Human Body	
H	91	H	92.5	Fe	35.4	O	46.6	H	63
He	9.1	He	7.3	O	27.8	Si	27.2	O	25.5
O	0.057	O	0.068	Mg	17.0	Al	8.13	C	9.5
N	0.042	C	0.037	Si	12.6	Fe	5.00	N	1.4
C	0.021	N	0.0093	S	2.7	Ca	3.63	Ca	0.31
Ne	0.003	Ne	0.0074	Ni	2.7	Na	2.83	P	0.22
Si	0.003	Mg	0.0044	Ca	0.6	K	2.59	K	0.06
Mg	0.002	Si	0.0029	Al	0.44	Mg	2.09	Cl	0.03
Fe	0.002	Fe	0.00185	Co>Na>Mn>K >Ti>P>Cr		Ti>H>C		S>Na>Mg	

(Source: Selbin, J., *J. Chem. Ed.*, 1973, 50, 306.)

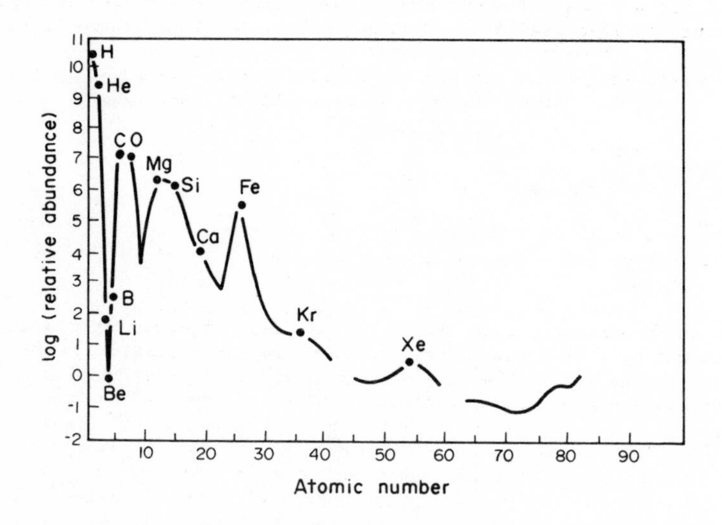

Fig. 1.1. Relative abundances of the elements in the universe, smoothed curve (based on log(abundance of Si) = 6).

Structure of the Nucleus

In broad detail an atom consists of two principal parts, the nucleus and the electron shell or cloud. The nucleus is very small relative to the size of an atom, being about 10^{-12}cm in diameter, whereas an atom, such as sodium, has a diameter of 3.80×10^{-8}cm. The fundamental difference between atoms of one element with those of another, as well as different atoms of the same element lies in the composition of the nucleus.

The two principal particles in a nucleus, the proton and the neutron, account for most of the mass of an atom (Table 1.3). An atom is characterised by its nucleus and by two quantities, the atomic number Z (number of protons) and the atomic weight A (number of protons and neutrons). The symbol $^{A}_{Z}M$ is used where M is the chemical symbol for the element. For any one element Z is constant (e.g. 6 for carbon, 8 for oxygen) but the element may have a number

TABLE 1.3 Atomic Particles

	Electron	Proton	Neutron
Charge	-1	+1	0
Absolute Value	1.6021×10^{-19} coulomb	1.6021×10^{-19} coulomb	0
Mass	9.11×10^{-31}kg	1.67×10^{-27}kg	1.67×10^{-27}kg
Relative Mass	1/1840	1	1

of values of A. The different atoms of an element are called isotopes, which are either stable or radioactive, some occur naturally, others do not.

The two particles in the nucleus are closely inter-related and may inter-convert. This can happen in atoms in order to increase nuclear stability.

$$p^+ \rightarrow n + e^+ + \nu^{(+)} \quad (1.1)$$

$$n \rightarrow p^+ + e^- + \nu^{(-)} \quad (1.2)$$

where e^+ is a positron (+ve electron), e^- is the electron, p^+ is the proton, n is the neutron and ν is a neutrino, (+) and (-) refer to the spin alignment of the neutrino.

Protons and neutrons, called neucleons, are held together by cohesive forces which cannot be just electrostatic, as proton-proton repulsions would destabilize the nucleus. The very strong attractive forces between nucleons are operative over distances less than 1.4×10^{-13}cm. At greater distances the attraction falls rapidly to zero, and the coulomb repulsive force dominates. The intense short range forces are said to be due to a messenger, called a pion, with finite rest mass (274 x mass of the electron). Nucleons are not tightly packed, but occupy approximately four times the volume of a close-packed arrangement.

Origin of the Universe

A question, which many people at some time in their lives ponder, is "how did it all begin?". One fact is that the Universe is expanding, the evidence being that light from the furthermost galaxies is of longer wavelength (red end) than light from nearer galaxies. This is explained by the Doppler effect, i.e. the galaxies are moving outwards, away from each other, and so the light appears to the observer, who is behind the moving object, to be of longer wavelength. If the Universe is expanding it suggests there may have been a starting point.

It is proposed in the *steady state theory* of the origin of the Universe that matter, viz. hydrogen, is being created continuously to fill in the gaps created by the expanding Universe, and all other elements come from the hydrogen by nuclear reactions. The more favoured theory, the *big bang theory* states that at some definite time the Universe came into being, and that the extremely densely packed matter (elementary particles) as a result of high pressure and temperatures eventually "exploded" dispersing matter through space. The material then regrouped under gravitational influences into stars and other bodies as we know them today. An extension of this theory called the *oscillating theory* suggests that at some time in the future the expansion of the Universe will stop and contraction will then occur until a dense mass is reformed, followed by another "big bang".

During the big bang, temperatures in the region of 10^9K would have been produced and a number of nuclear reactions occur, such as:

$$^{1}_{1}H + ^{1}_{0}n \rightarrow ^{2}_{1}H \qquad (1.3)$$

$$^{2}_{1}H + ^{1}_{1}H \rightarrow ^{3}_{2}He \qquad (1.4)$$

$$^{3}_{2}He + ^{1}_{0}n \rightarrow ^{4}_{2}He \quad (\alpha \text{ particle}) \qquad (1.5)$$

$$^{4}_{2}He + ^{1}_{0}n \rightarrow ^{5}_{2}He \qquad (1.6)$$

All this would have happened in the first 30 to 60 minutes, as beyond that time the temperature would have dropped so no further reactions could occur. Also the "chain-type" reaction stops at $^{5}_{2}He$, which has a half-life of 2×10^{-21}s, and decays to $^{4}_{2}He$. Heavier chemical elements are produced in subsequent reactions in the stars.

Nuclear Synthesis

Irrespective of the origin of the Universe it is generally accepted that synthesis of the elements that go to make up our world have been, and still are, carried out in the stars. We will turn our attention to this, after first discussing in more detail nuclear stability.

Nuclear stability. A number of impirical rules concerning the stability of nuclei, and their relative abundance have been formulated. For example, nuclei with even atomic numbers (Z) are more abundant than those with odd atomic numbers. Nuclei containing even numbers of protons and neutrons (even-even nuclei) are both more abundant and more stable than odd-odd nuclei, while even-odd nuclei are of intermediate stability. Often the most abundant and stable nuclei, up to atomic number 20, have a 1:1

neutron:proton ratio, e.g. $^{4}_{2}He$ (99.99%), $^{14}_{7}N$ (99.63%), $^{24}_{12}Mg$ (78.70%) and $^{40}_{20}Ca$ (96.97%). However, some light isotopes with odd atomic numbers, and which do not possess the 1:1 n:p ratio, are the most stable and abundant for the particular element, eg. $^{20}_{9}F$, $^{23}_{11}Na$, $^{27}_{13}Al$. They possess odd-even nuclei which is more stable than odd-odd nuclei (even with a 1:1 n:p ratio). Nuclei with atomic numbers 2, 8, 20, 50, 82, 126 have a special stability, and their atomic numbers are termed "magic numbers". Explanations for the above observations are in terms of inter-nuclear like-like pairing forces and like-unlike interactions and will not be discussed here.

When light nuclei are formed energy is emitted, and the nuclear mass becomes less than the sum of the masses of the component neutrons and protons. The mass of $^{16}_{8}O$ is 15.9949 atomic mass units (amu) whereas the sum of 8 protons and 8 neutrons (8 x 1.0078 + 8 x 1.0087) is 16.1320 amu. The difference, 0.1371 amu, is the mass loss on forming the oxygen nucleus, which corresponds to 128 MeV (using $E = mc^2$). The energy per nucleon (128/16 = 8.0 MeV) is termed the binding energy of the nucleus and this varies with mass number as shown in Fig. 1.2. Since the binding energy is proportional to the number of nucleons, one could expect a regular increase. However, for mass number >56 proton-proton repulsions become more significant destabilizing the

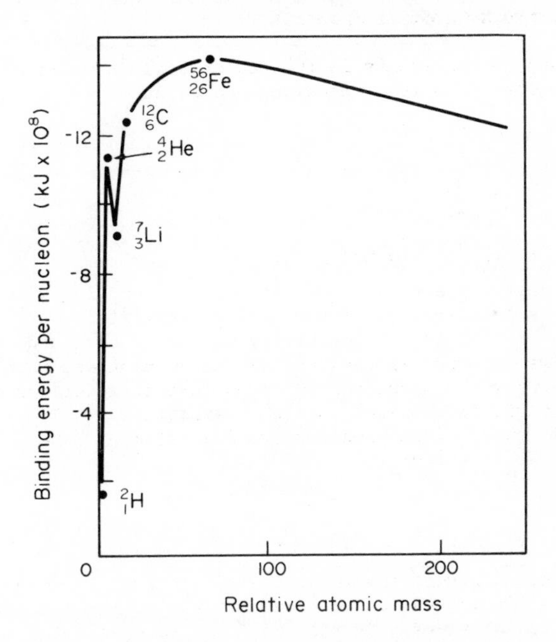

Fig. 1.2. Stabilities of atomic nuclei relative to separated protons and neutrons.

nucleus, which is offset somewhat, by dilution with more neutrons. Therefore heavier nuclei have more neutrons than protons. A plot of atomic number against number of neutrons (Fig. 1.3) indicates this. Nuclei that lie off the stability area undergo radioactive decay in order to achieve stability.

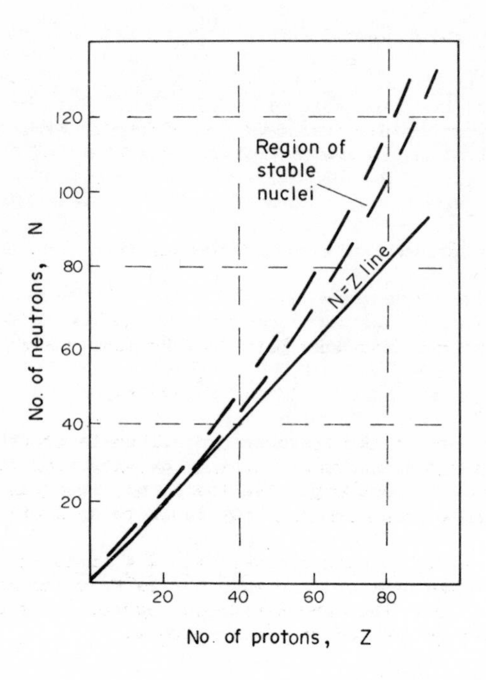

Fig. 1.3. Neutrons versus protons for stable nuclei.

Hydrogen burning. The initial process in stellar nuclear synthesis is *hydrogen burning*, producing helium. Gravitational forces acting upon hydrogen produce a density around 10^5kg m^{-3} and the gravitational energy, converted into heat, raises the temperature to around 1-3 x 10^7K. At this temperature the kinetic energy of the hydrogen nuclei is sufficient to overcome the strong barrier to two positively charged nuclei undergoing nuclear fusion.

$$^{1}_{1}H + ^{1}_{1}H \rightarrow ^{2}_{1}H + \beta^{+} + \nu^{(+)} \text{(neutrino)} \quad (1.7)$$

$$^{2}_{1}H + ^{1}_{1}H \rightarrow ^{3}_{2}He + \gamma \quad (1.8)$$

$$^{3}_{2}He + ^{3}_{2}He \rightarrow ^{4}_{2}He + 2^{1}_{1}H \quad (1.9)$$

$$4^{1}_{1}H \rightarrow ^{4}_{2}He + 2\beta^{+} + 2\gamma + 2\nu^{(+)} \quad (1.10)$$

The process is exothermic (or exoergic) since more stable nuclei are produced. This corresponds to moving down the binding energy curve (Fig. 1.2). The time scale for hydrogen burning is greater than 5 billion years, as our sun, which is around 5 billion years old, is in the early stages of hydrogen burning (around 90% hydrogen remains).

Helium burning. As hydrogen is used up and helium produced, the temperature of the core drops, and the star exterior expands to conserve heat. The star, with a larger surface area and a lower temperature, becomes reddish in colour and is call a *red giant*. However, the core, now mainly helium, continues to collapse under the gravitational forces driving the temperatures up to 1-3 x 10^8K with a density of around 10^8kg m^{-3}. The shell at this stage becomes hotter and hydrogen in it begins to burn, while helium burning begins in the core. The higher temperatures are now sufficient to overcome the even greater energy barrier to the fusion of helium nuclei together. The first product of helium burning is ${}^{8}_{4}Be$ which has a half-life of 2 x 10^{-16}s.

$$2\,{}^{4}_{2}He \rightarrow {}^{8}_{4}Be + \gamma. \qquad (1.11)$$

However, sufficient ${}^{8}_{4}Be$ accumulates for further reactions to occur.

$$ {}^{8}_{4}Be + {}^{4}_{2}He \rightarrow {}^{12}_{6}C + \gamma. \qquad (1.12)$$

Neutrons and protons can also take part in reactions during ${}^{4}_{2}He$ burning, e.g.

$$ {}^{12}_{6}C + {}^{1}_{1}H \rightarrow {}^{13}_{7}N \rightarrow {}^{13}_{6}C + \beta^{+} + \nu. \qquad (1.13)$$

Small stars. When much of the hydrogen and helium is spent in small stars, with a mass less than 1.4 x mass of our sun, gravity produces further contraction to give a *white dwarf*. Eventually both nuclear reactions and gravitational collapse stops and the body cools to an inert dense mass.

Big stars. In stars with a mass greater than 1.4 times the mass of our sun, higher temperatures are produced and when 6 x 10^8K is reached an alternative hydrogen burning process, the *carbon-nitrogen cycle*, takes place, provided some ${}^{12}_{6}C$, which acts as a catalyst, is available.

$$ {}^{12}_{6}C + {}^{1}_{1}H \rightarrow {}^{13}_{7}N + \gamma \qquad (1.14)$$

$$ {}^{13}_{7}N \rightarrow {}^{13}_{6}C + \beta^{+} + \nu \qquad (1.15)$$

$$ {}^{13}_{6}C + {}^{1}_{1}H \rightarrow {}^{14}_{7}N + \gamma \qquad (1.16)$$

$$ {}^{14}_{7}N + {}^{1}_{1}H \rightarrow {}^{15}_{8}O + \gamma \qquad (1.17)$$

$$ {}^{15}_{8}O \rightarrow {}^{15}_{7}N + \beta^{+} + \nu \qquad (1.18)$$

$$ {}^{15}_{7}N + {}^{1}_{1}H \rightarrow {}^{4}_{2}He + {}^{12}_{6}C \qquad (1.19)$$

This adds up to

$$4\,{}^{1}_{1}H \rightarrow {}^{4}_{2}He + 3\gamma + 2\beta^{+} + 2\nu \qquad (1.20)$$

Carbon and oxygen burning also occurs at these temperatures and produce nuclei in the region of ${}^{28}_{14}Si$, for example,

$$ {}^{12}_{6}C + {}^{12}_{6}C \rightarrow {}^{20}_{10}Ne + {}^{4}_{2}He \qquad (1.21)$$

$$ {}^{16}_{8}O + {}^{16}_{8}O \rightarrow {}^{28}_{14}Si + {}^{4}_{2}He \qquad (1.22)$$

$$ {}^{16}_{8}O + {}^{16}_{8}O \rightarrow {}^{31}_{16}S + {}^{1}_{0}n \qquad (1.23)$$

Gamma radiation produced in many of these exoergic reactions can cause disintegration of nuclei, for example $^{20}_{10}Ne$ gives protons, neutrons and particles, which are available for further reactions. The most abundant element produced from $^{16}_{8}O$ burning in $^{28}_{14}Si$, but at higher temperatures (~3 x 10^9K) γ-radiation can lead to the breakdown of $^{28}_{14}Si$ opening the way for synthesis of even heavier nuclei. This is called silicon burning,

$$\gamma + {}^{28}_{14}Si \rightleftharpoons {}^{24}_{12}Mg + {}^{4}_{2}He \quad (1.24)$$

$$^{28}_{14}Si + {}^{4}_{2}He \rightleftharpoons {}^{32}_{16}S + \gamma \quad (1.25)$$

$$^{32}_{16}S + {}^{4}_{2}He \rightleftharpoons {}^{36}_{18}Ar + \gamma \quad (1.26)$$

By this stage an equilibrium between nuclei formation and disintegration occurs, with a continuing tendency towards the formation of heavier elements. The fusion process is still exoergic up to the formation of $^{56}_{26}Fe$ but then stops at $^{56}_{26}Fe$, which is the nucleus with maximum stability (Fig. 1.2). One could expect an accumulation of $^{56}_{26}Fe$ in the Universe compared with other elements. Iron is in fact a relatively abundant element, but there are lighter elements, which are more abundant, indicating that the processes described above are still in progress. The above description helps explain why in the relative abundance curve (Fig. 1.1) there are elements such as $^{12}_{6}C$, $^{16}_{8}O$, $^{20}_{10}Ne$ which are more abundant than their neighbours. We note that each of these abundant nuclei differ from the next one by a $^{4}_{2}He$ nucleus.

Eventually the temperature of big stars may reach 7-8 x 10^9K at which point many endoergic reactions will occur, such as the breakdown of nuclei giving neutrons, protons and α particles, which can be used in further reactions. Neutrinos are also produced (e.g. $e^+ + e^- \rightarrow \nu^{(+)} + \nu^{(-)}$) which carry energy away from the stars. While this is in progress considerable contraction of the core occurs up to a density of around 10^{14}kg m^{-3}. Finally an implosion of the core and explosion of the shell will result. This corresponds to a *supernova*, which flings material out into space.

Heavy elements. Since $^{56}_{26}Fe$ is the most stable nucleus any heavier nuclei can only be produced with an expenditure of energy (endoergic). Also, it becomes increasingly more difficult for highly charged nuclei to get close enough for fusion to occur. It is therefore perhaps surprising that any elements heavier than iron exist. However, one nuclear reaction that circumvents these problems is neutron capture, since the neutron has little difficulty penetrating a positively charged nucleus. Heavy elements are obtained by neutron capture beginning with the "seed" nucleus $^{56}_{26}Fe$.

Neutrons are produced in stars either as part of the normal stellar nuclear evolutionary processes in "red giants" and second generation stars (e.g. $^{13}_{6}C + {}^{4}_{2}He \rightarrow {}^{16}_{8}O + {}^{1}_{0}n$), or in considerable quantities just prior to, and during a supernova period. Stars which emit periodic bursts of radio waves (pulsars) are mainly composed of neutrons (neutron stars). The different sources of neutrons give rise to two distinct neutron capture processes.

(a) "s" process. In the "s" process, or slow neutron capture, one neutron is captured by a nucleus which becomes, in most cases, unstable and decays usually with β^- emission in order to increase the p/n ratio before a second neutron is captured.

$$^{58}_{26}Fe + ^{1}_{0}n \rightarrow ^{59}_{26}Fe \rightarrow ^{59}_{27}Co + ^{0}_{-1}e + \nu^{(-)} \qquad (1.27)$$

(i.e. in the $^{59}_{26}Fe$ nucleus reaction (1.2) has occurred.) Single neutron capture occurs in periods from 1 to 10^5 years which generally gives sufficient time for β^- decay to occur. The "s" process tends to produce the lighter isotopes of elements, i.e. the "proton-rich" isotopes. The heaviest element produced by the "s" process is $^{209}_{83}Bi$. The range of isotopes of tin (see below) and the existence of technetium (longest ½ life 2.6 x 10^6 yrs) in the stars are evidence for the "s" process.

(b) "r" process. Rapid neutron capture occurs in environments of high neutron density, many neutrons may be captured by a nucleus in a period 10^{-2} to 10 seconds before decay occurs. The "r" process gives "neutron-rich" isotopes,

$$^{56}_{26}Fe + 13^{1}_{0}n \rightarrow ^{69}_{26}Fe \rightarrow ^{69}_{27}Co + ^{0}_{-1}e \qquad (1.28)$$

and the heavy elements. For example ^{254}Cf is produced in nuclear explosions and is present in stars.

A summary of the "r" and "s" processes which lead to the isotopes of tin is given in Fig. 1.4. The method by which the isotopes are produced is indicated by the letters "r" and "s" and b (bypassed). The black line corresponds to the "s" process path and clearly the progress from $^{114}_{48}Cd$ to $^{115}_{49}In$ to $^{116}_{50}Sn$

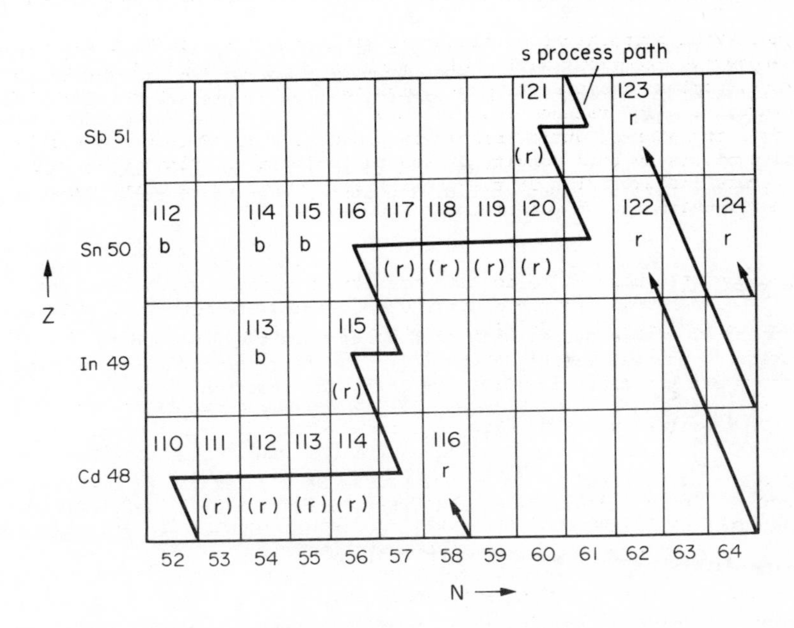

Fig. 1.4. Neutron capture process in the neighbourhood of tin. Isotopes marked by r are produced by the "r" process, (r) by both the "r" and "s" processes, b if they are bypassed, and those unmarked by the "s" process.

is by single neutron captures followed by β^- emission. The three isotopes ^{112}Sn, ^{114}Sn and ^{115}Sn are not produced by the "s" or "r" processes, and it is suggested that they are obtained by rapid proton capture. Isotopes obtained this way are rich in protons and have very low abundance, as proton capture is less favourable than neutron capture.

Deuterium, Lithium, Beryllium and Boron. The nuclei ^{2}D, ^{6}Li, ^{7}Li, ^{9}Be, ^{10}B and ^{11}B are not end products in any of the processes mentioned. The small amounts produced during hydrogen and helium burning are removed as they burn at lower temperatures giving helium, i.e.

$$^{2}_{1}D + ^{1}_{1}H \rightarrow ^{3}_{2}He + \gamma \quad (1.29)$$

$$^{6}_{3}Li + ^{1}_{1}H \rightarrow ^{3}_{2}He + ^{4}_{2}He \quad (1.30)$$

$$^{7}_{3}Li + ^{1}_{1}H \rightarrow \gamma + ^{8}_{4}Be \rightarrow 2^{4}_{2}He \quad (1.31)$$

$$^{9}_{4}Be + ^{1}_{1}H \rightarrow ^{4}_{2}He + ^{6}_{3}Li \xrightarrow{^{1}_{1}H} ^{3}_{2}He + 2^{4}_{2}He \quad (1.32)$$

$$^{10}_{5}B + ^{1}_{1}H \rightarrow ^{4}_{2}He + ^{7}_{4}Be \quad (1.33)$$

$$^{11}_{5}B + ^{1}_{1}H \rightarrow 3^{4}_{2}He + \gamma \quad (1.34)$$

However, the nuclei do exist, but in low abundance, and are probably produced by *spallation* nuclear reactions, when heavy cosmic ray particles collide with elements, such as carbon, nitrogen and oxygen, and cause them to break up to give lighter nuclei.

The details of the above processes will be modified as more data comes to light, but the details given in Fig. 1.5 summarises the present state of our knowledge.

Manufacture of Elements

Bombardment of stable nuclei with high speed particles has led to the production of a great variety of nuclei. The first particle, to be used by Rutherford and his group in 1919, was $^{4}_{2}He$. The reaction;

$$^{14}_{7}N + ^{4}_{2}He \rightarrow ^{17}_{8}O + ^{1}_{1}H, \quad (1.35)$$

produced the stable nucleus $^{17}_{8}O$. However, in many cases the product is radioactive, for example radioactive $^{32}_{15}P$ (½ life 14.3 days) is produced by the reaction;

$$^{31}_{15}P + ^{2}_{1}H \rightarrow ^{32}_{15}P + ^{1}_{1}H. \quad (1.36)$$

Like many radioactive isotopes it is useful in tracer studies in biology agriculture, medicine and analytical investigations.

Neutrons are obtained from nuclear reactions such as:

$$^{9}_{4}Be + ^{4}_{2}He \rightarrow ^{12}_{6}C + ^{1}_{0}n, \quad (1.37)$$

or reactions in a cyclotron, or in nuclear reactors and are useful for producing isotopes.

It is now possible to produce elements heavier than Pu (as well as Tc and Pm), which do not occur on the earth. The following reactions are just a few that can be achieved(page 14).

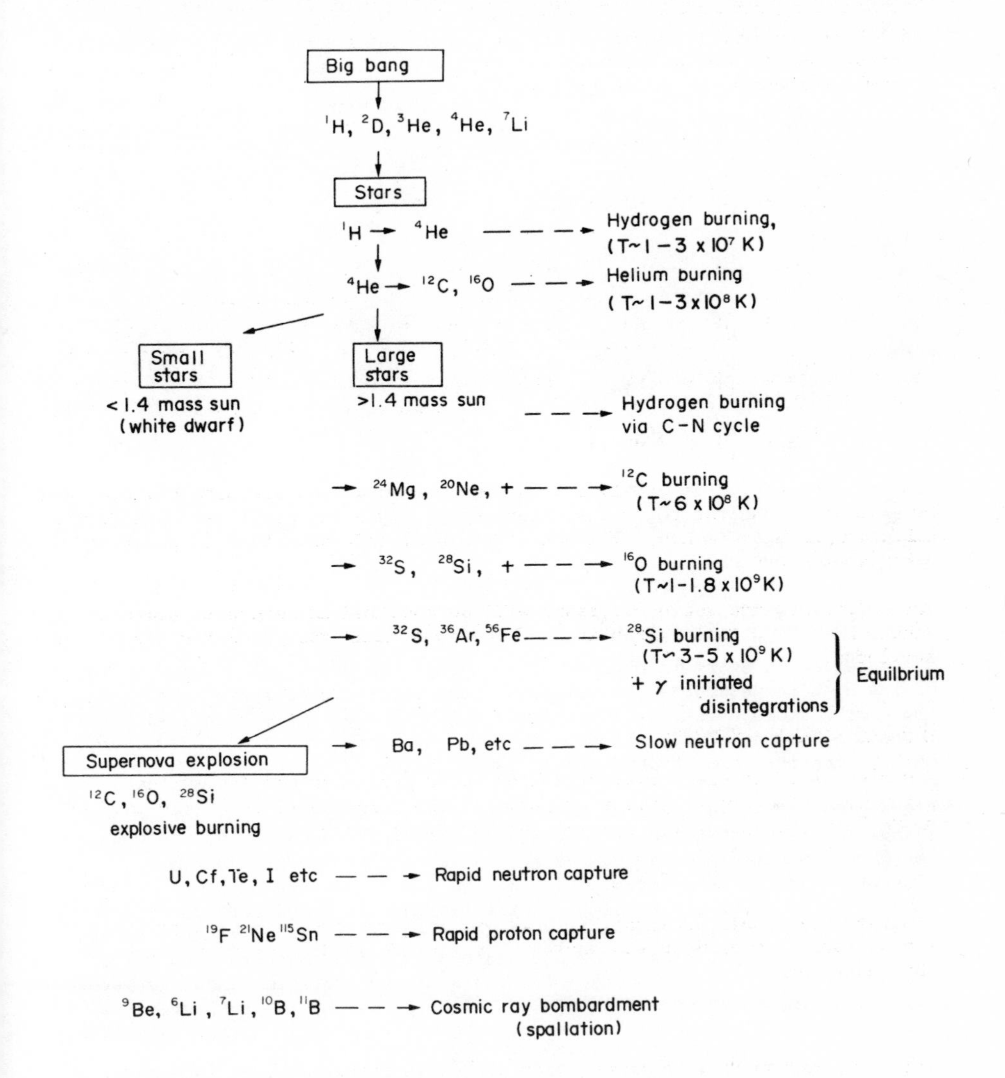

Fig. 1.5. A brief summary of stellar nucleosynthesis. Source: Selbin, J., *J. Chem. Ed.*, 1973, 50, 306, 380.

$$^{238}_{92}U + ^{1}_{0}n \rightarrow ^{239}_{92}U + \gamma \rightarrow ^{239}_{93}Np + ^{0}_{-1}e \qquad (1.38)$$

$$^{239}_{93}Np \rightarrow ^{239}_{94}Pu + ^{0}_{-1}e$$

$$(t_{1/2} = 24{,}360\ yr)$$

$$^{239}_{94}Pu + ^{1}_{0}n \rightarrow ^{240}_{94}Pu + \gamma \xrightarrow{^{1}_{0}n} ^{241}_{94}Pu + \gamma \qquad (1.39)$$

$$^{241}_{94}Pu \rightarrow ^{241}_{95}Am + ^{0}_{-1}e$$

$$(t_{1/2} = 458\ yr)$$

$$^{241}_{95}Am + ^{4}_{2}He \rightarrow ^{243}_{97}Bk + 2^{1}_{0}n \qquad (1.40)$$

$$(t_{1/2} = 4.5\ hr)$$

$$^{252}_{98}Cf + ^{11}_{5}B \rightarrow ^{257}_{103}Lw + 6^{1}_{0}n \qquad (1.41)$$

$$(t_{1/2} = 8\ sec)$$

The amount of material produced is often small, 10^{-9}g, and special chemical techniques have been developed in order to study the chemistry of the transuranic elements.

Molecules in Interstellar Space

The material in stars is nuclear, i.e. atoms stripped of electrons, because of the high temperatures. However, at temperatures comparable with those on the earth, atoms exist, as for example in the space between stars and in certain "cloud-type" formations. This environment is a very good vacuum, much better than can be obtained in laboratories.

Since atoms exist in space it should not be too surprising to find molecules also. The first evidence of molecular species was found in 1937-1939 when the three diatomic species CN, CH and CH^+ were discovered by detecting their ultraviolet emission spectra. Little else was discovered until the mid-sixties, because of the problems of interference in the detection of molecular radiation. For example, detection of infrared radiation is restricted by the heavy absorption of infrared by H_2O and CO_2 in our atmosphere. However, microwave radiation, the energy of which activates rotational energy levels of a molecule, is not so affected and the detection of it from interstellar molecules is now possible. A list of some of the molecules that have been detected is given in Table 1.4.

The molecules form from atoms and other molecular species on the surface of interstellar dust, such as particles of Fe and SiO_2. Two observations may be made concerning the type of compounds observed: (a) the large number of hydrogen containing species probably reflects the high abundance of hydrogen in the universe, and (b) many of the species have multiple bonds, hence strong

TABLE 1.4 Some Molecules Observed in Interstellar Space (Abundance given by (x) = 10^x)

Molecule	Abundance
Diatomic	
H_2	(0)
OH	(-7)
CH	(-8)
CO	(-4)
CN	(-8)
CS	(-7)
NS	(-8)
SO	(-7)
SiO	(-7)
SiS	(-7)
Triatomic	
HCO	(-8)
HCO^+	(-7)
HCN	(-6)
HNC	(-6)
N_2H^+	(-7)
C_2H	(-7)
H_2S	(-8)
SO_2	(-7)
OCS	(-8)
Tetraatomic	
NH_3	(-6)
H_2CO	(-8)
H_2CS	(-10)
HNCO	(-9)
Pentaatomic	
CH_2NH	(-10)
HCOOH	(-10)
HC≡CCN	(-8)
Hexaatomic	
CH_3OH	(-7)
NH_2CHO	(-10)
Heptaatomic	
CH_3CHO	(-10)
CH_3NH_2	(-10)
$H_2C=CHCN$	(-10)
$CH_3C≡CN$	(-9)
HC≡C-C≡CCN	(-10)
Octaatomic	
$CH_3C≡CCN$	(-10)
$HCOOCH_3$	(-10)
Nonaatomic	
CH_3CH_2OH	(-10)
CH_3COCH_3	(-10)

Source: Huntress, W.T., *Chem. Soc. Revs.*, 1977, 16, 295.

bonds between atoms may be an important factor. The "absence" of species such as O_2 and N_2, in which the bonding is strong, may be because they have not yet been detected, being molecules of high molecular symmetry and with no dipole moment. The reactivity of NO may prevent its accumulation.

The existence of species such as $CH_2=NH$ and $HCONH_2$ in interstellar space, (precursors to amino acids?), has led to the suggestion, that before too long amino acids will be discovered.

THE FORMATION AND STRUCTURE OF THE EARTH

We will now consider the nature and formation, and the composition of the earth. The discussion will be directed at the distribution of the elements and chemical compounds in the earth.

Formation of the Solar System

One explanation of the formation of the solar system, is that cosmic dust clouds and interstellar atoms were attracted to each other by gravitational forces. This process, which probably began around 6 x 10^9 years ago, produced a massive ball containing over 99.8% of the total mass, the remaining material being in a disc rotating around it. As the proto-sun undergoes gravitational collapse the heat generated raises the temperature to 1-3 x 10^7K sufficient to allow hydrogen burning to occur, while the disc must have split into zones within which local accretion occurred giving rise to the planets. Heating in the planets was much less owing to the small mass.

Structure of the Earth

The earth has three separate areas, the atmosphere, the hydrosphere and the solid material, a fourth category spanning all three, called the biosphere can also be considered. The atmosphere is principally gaseous, the main constituents being dinitrogen and dioxygen, the hydrosphere is liquid water containing dissolved and particulate material. The solid earth is the most complex of the three, and is heterogeneous in composition (Table 1.5).

TABLE 1.5 Structure of the Earth

Zone	Major Chemical Constituents	State of Matter
Atmosphere	N_2, O_2, H_2O, CO_2, noble gases, particles.	Gas, liquid, solid.
Hydrosphere	H_2O (liquid), ice, snow, dissolved minerals, Na^+, Cl^-, Mg^{2+}, Br^-, etc., particles.	Liquid (solution) suspended solids.
Biosphere	Organic material, mineral skeleton, H_2O, trace elements.	Solid, liquid, colloidal, gas.
Crust	Silicate rocks, oxide and sulphide minerals.	Solid (heterogenous) (intermixed with water and air).
Mantle	Silicate minerals, particularly olivine and pyroxene, (iron and magnesium silicates).	Solid
Core	Iron-nickel alloy.	Liquid (top) Solid (bottom)

Source: Mason, B., Principles of Geochemistry, 3rd Ed., J. Wiley and Sons, Inc., N.Y., 1966.

The solid earth is made up of a core, mantle and crust (Fig. 1.6). The inner core is probably solid and the outer core liquid, and composed mainly of iron and nickel. The mantle is principally composed of dense silicates of iron and magnesium. The crust, or the lithosphere, which is less than 0.4% of the mass of the earth, is the part we are most familiar with. Its composition is complex, being made up mainly of silicates (58.7% of the crust corresponds to the impirical formulation SiO_2). The other major component (15%) corresponds to the impirical formulation Al_2O_3.

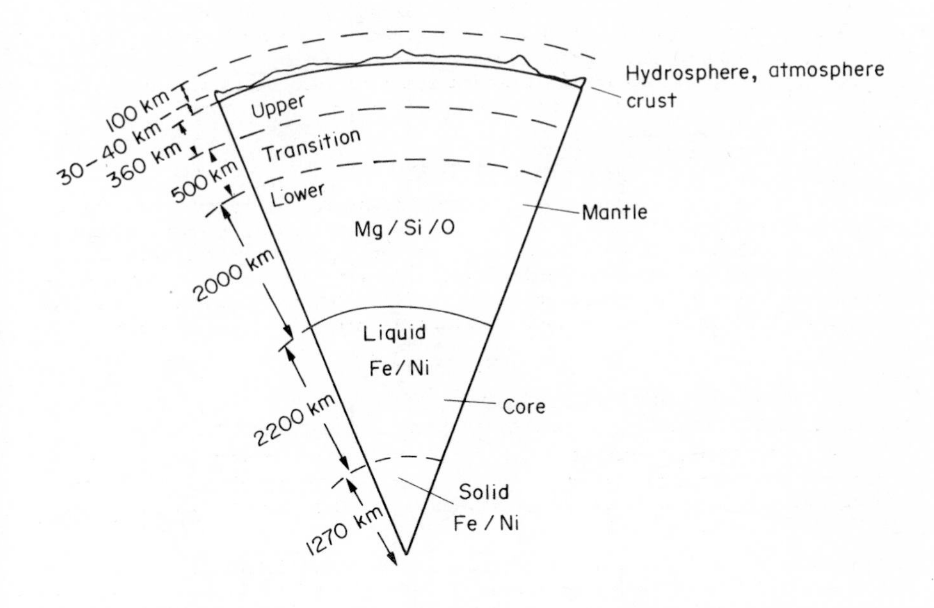

Fig. 1.6. The structure of the earth.

Comparison of the elemental abundances of the three major divisions of the earth, the biosphere and the crust (Fig. 1.7) and the data given in Table 1.2 indicates that in the formation of the earth some fractionation of the elements has occurred.

Formation of the Earth

We will now consider some of the broad chemical processes that have led to the formation of the earth as we know it today.

The solid earth. In one model it is proposed that during accretion of the earth, the silicates and iron condensed, and subsequent degassing and melting produced the internal structure and composition of the earth, the crust and also the atmosphere and the hydrosphere. In a second model the early stages of the earth's formation are likened to the phases occurring in a blast furnace, in which molten iron sinks to the bottom while the less dense, and also less volatile, silicates float on top.

When the disc of material, which gave rise to the earth started to condense the temperature was probably around 2000K, so that all the material, would be present in an atomic and gaseous state. On cooling the least volatile stable combinations of the elements in the disc would condense out. It is believed that the first materials were calcium aluminium silicates, followed by the more volatile iron-nickel system, and then followed by magnesium silicates (Table 1.6). Like a blast furnace, the materials would separate according to their densities producing the major components of the core, mantle and crust. On further cooling the more volatile materials, such as

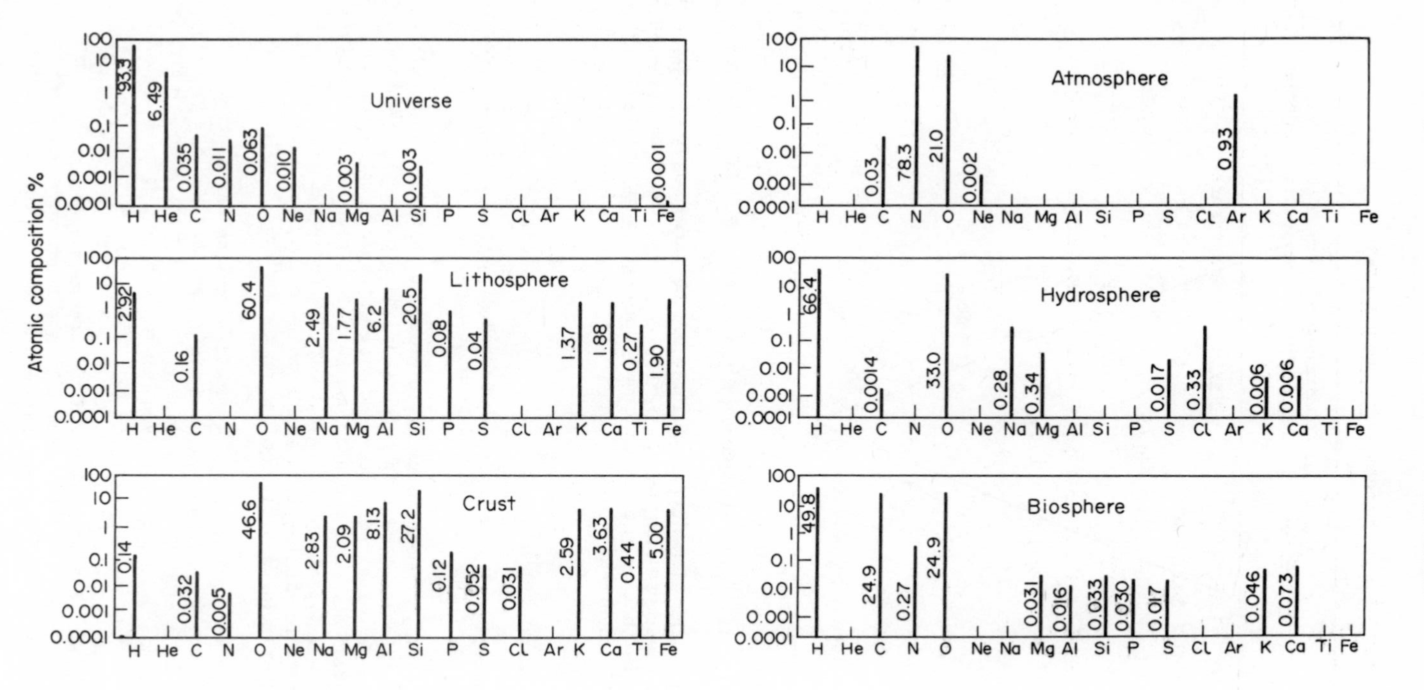

Fig. 1.7. Elemental abundances of the Universe, solid earth, atmosphere, hydrosphere and biosphere.

TABLE 1.6 Condensation Temperatures

Phase	Species	Condensation Temperature (K)	Phase	Species	Condensation Temperature (K)
Corundum	Al_2O_3	1758	Forsterite	Mg_2SiO_4	1444
Perovskite	$CaTiO_3$	1647	Anorthite	$CaAl_2Si_2O_8$	1362
Melilite	$Ca_2Al_2SiO_7$	1625	Enstatite	$MgSiO_3$	1349
Spinel	$MgAl_2O_4$	1515	Rutile	TiO_2	1125
Metallic iron	Fe+12.5 mol% Ni	1473	Magnetite	Fe_3O_4	405
Diopside	$CaMgSi_2O_6$	1450			

Source: Turekian,K.K.,Chemistry of the Earth,Holt,Rinehart,Winston, N.Y. 1972

lead, bismuth, thallium, hydrated silicates, iron compounds (sulphides and oxides) and water would condense, leaving gaseous materials in the atmosphere. These final stages of condensation would add only to the crust, but since the oldest rocks are around 3.5 billion years old, and since the earth was formed around 4.6 billion years ago it seems that during the first billion years a considerable reprocessing of the crust must have occurred. The reprocessing did not however, bring the core and crust materials together as the core materials remain in a reduced state, while crust materials are in an oxidised state.

The chemical and physical properties which influenced the chemical processes during the earth's formation were; reduction and oxidation, ionic size, bond strength, pressure, temperature, phase separation, elemental abundances and gravitational and magnetic fields. The earth's formation can be considered in three stages. The first, the primary differentiation of the elements, was influenced by the reduction potentials of the elements relative to iron. The secondary differentiation was influenced by the relative size of metal ions, the bonding tendencies of the elements, fractional crystallisation from the magma and the densities of compounds. The third process, still operating today, is the interaction of environmental factors with the earth's crust.

Primary differentiation of the elements. As insufficient oxygen and sulphur are available to oxidise all elements, and since iron is so abundant the primary distribution of the elements was influenced by the potential for the reduction $M^{n+} + ne \rightarrow M$ to proceed compared with the reaction $Fe^{2+} + 2e \rightarrow Fe$. Elements whose reduction potentials are more positive than iron are called *siderophiles* (Table 1.7).

As molten iron was pulled towards the centre of the earth it reduced siderophile metal ions to the metal which alloyed with the iron and were carried into the core. Nickel is a good example of this;

$$Fe - 2e \rightarrow Fe^{2+} \qquad -E^{o} = +0.429V, \qquad (1.42)$$

$$Ni^{2+} + 2e \rightarrow Ni \qquad E^{o} = -0.250V \qquad (1.43)$$

$$\text{i.e.} \quad Fe + Ni^{2+} \rightarrow Fe^{2+} + Ni \qquad E = +0.179V, \qquad (1.44)$$

TABLE 1.7 Reduction Potentials

Element	Reduction Potential	Element	Reduction Potential
Lithophiles		Chalcophiles	
Na^+/Na	-2.71	Zn^{2+}/Zn	-0.76
Ca^{2+}/Ca	-2.87	Pb^{2+}/Pb	-0.13
Al^{3+}/Al	-1.66	Cu^{2+}/Cu	+0.34
Cr^{3+}/Cr	-0.77	Ag^+/Ag	+0.80
Siderophiles			
Fe^{2+}/Fe	-0.43	Pd^{2+}/Pd	+0.99
Ni^{2+}/Ni	-0.25	Au^+/Au	+1.69
Sn^{2+}/Sn	-0.14		

As a consequence siderophiles tend to be of low abundance in the earth's crust.

The majority of the remaining elements, most of which have more negative reduction potentials than iron, are classified according to their bonding potential with oxygen or sulphur (lithophiles or chalcophiles respectively) (Table 1.8).

TABLE 1.8 Geochemical Classification of the Elements[a]

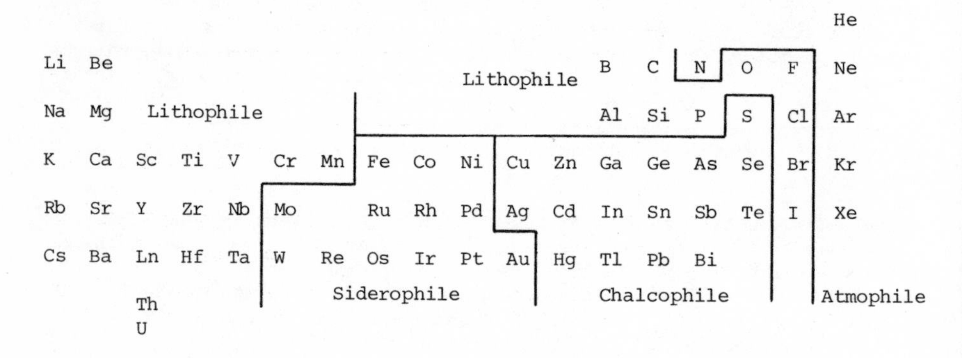

[a] Elements can display more than one type of behaviour

Lithophiles have more negative reduction potentials than chalcophiles (Table 1.7), their ions are relatively small in size and not readily polarised, whereas chalcophiles have larger ions more readily polarisable and therefore associate with the polarisable sulphide ion. This broad classification of the elements is recognizable in the dominant chemical form of the element that exists in nature (Table 1.1).

Secondary differentiation of the elements. As the magma cooled a variety of minerals crystallised depending on their melting point and abundance of the constituent elements. The number of phases produced may be predicted with use of the phase rule P + F = C + 2, provided the system is at equilibrium. For two degrees of freedom (pressure and temperature) P = C, and for six components (e.g. O, Si, Al, Fe, Mg and Na) there is the possibility of forming up to six phases, i.e. six different chemical species (minerals). More than 99% of igneous rocks are composed of seven principal minerals (the silica minerals, feldspars, feldspathoids, olivine, pyroxenes, amphiboles and micas).

Fractional separation of minerals and therefore the elements occurs as the magma cools producing in some cases ore bodies (see Chapter 3). The relative temperatures at which minerals crystallise will depend in part on their lattice energies which is related to the size of the ions and their charge, because of the form of the lattice energy equation;

$$\Delta H_L = -\frac{NMZ^+Z^-e^2}{4\,\varepsilon_0(r_+ + r_-)} . \qquad (1.45)$$

Minerals with small highly charged ions will have high lattice energies and may crystallise first. For example, of the two feldspars $CaAl_2Si_2O_8$ (anorthite) and $NaAlSi_3O_8$ (albite), anorthite has the higher melting point, and is more dense. This is because of the greater lattice energy of anorthite and because Ca^{2+} (At. wt. 40) is heavier than Na^+ (At. wt. 23). One finds that, of the two minerals, anorthite lies deeper in the crust.

The less abundant elements often occur as a constituent of an abundant minerals because of isomorphous replacement, that is, where one ion replaces another in a crystal lattice. For this to happen the radii of the atoms or ions must lie within 10 to 20% of each other (Table 1.9). Therefore, a

TABLE 1.9 Ionic Radii

Range (pm)	Ions (radii, pm)
10 - 29	B^{3+}(20).
30 - 49	Be^{2+}(31).
50 - 69	Al^{3+}(53), Ga^{3+}(62), Fe^{3+}(60), Ti^{4+}(61), Mn^{3+}(62), Cr^{3+}(64), Mg^{2+}(65), V^{3+}(66), Li^+(68), Sn^{4+}(69), Nb^{4+}(69).
70 - 89	Ni^{2+}(70), Co^{2+}(72), Cu^{2+}(73), Fe^{2+}(75), Zn^{2+}(75), Ti^{3+}(79), Mn^{2+}(80), Hf^{4+}(83), Zr^{4+}(84).
90 -109	Na^+(98), Eu^{2+}(98), Ca^{2+}(99), U^{4+}(100), La^{3+}(105), Th^{4+}(106).
110-139	Sr^{2+}(112), Pb^{2+}(118), K^+(133), Ba^{2+}(135).
140-170	Ra^{2+}(143), Rb^+(147), Tl^+(147), Cs^+(167).

range of feldspars with compositions lying between the extremes $NaAlSi_3O_8$ and $CaAl_2Si_2O_8$ exist, because the radius of the Na^+ and Ca^{2+} ions are similar (note also the change in the Al:Si ratio but that the number of Al + Si atoms remains constant). Those elements, whose ionic radius is similar to the ionic radius of one of the abundant elements (Mg, Fe, Ca, Al, Si, Na, K) occur as constituents in the common minerals. For example Ba^{2+} occurs in the potassium feldspars because Ba^{2+} and K^+ have similar ionic radii (135 and 133 pm respectively). Other common ion pairs that undergo isomorphous replacement are: K^+ (133 pm), Rb^+(147 pm); Ga^{3+}(62 pm), Al^{3+}(53 pm); Si^{4+}(42 pm), Ge^{4+} (47 pm). If the ionic radii of the less abundant elements are much greater or less than the common elements they are not involved in isomorphous replacement and tend to concentrate in the lowest melting fraction of the magma.

The formation of an ore body may be explained in terms of a solid-liquid freezing point or eutectic diagram (Fig. 1.8). If the magma containing two components A and B has a composition designated by point Y, then when the temperature reaches T, the solid A will crystallise out and will continue to do so until the composition reaches point Z.

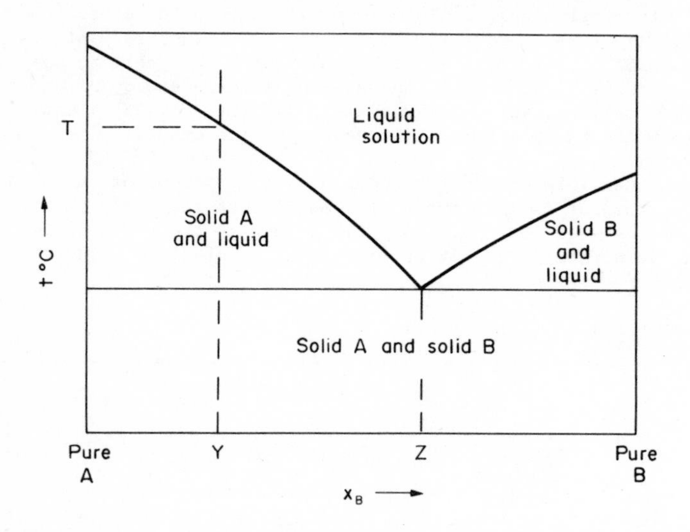

Fig. 1.8. Two component eutectic diagram.

The solid A that crystallised will settle out as a deposit at the bottom of the solution (if it is more dense) and when the remaining solution finally solidifies it will be rich in component B especially if the eutectic point lies on the right of the diagram. A large platinum deposit in Southern Africa was formed in this way.

Continuing changes. The earth's crust is undergoing continual change which, over the centuries, has produced much of the world's mineral deposits (see Chapter 3). The principal influence is weathering which is the interaction of the crust with water and air at different temperatures and pressures. Physical changes are produced by wide temperature variations, and by pressure and force. Chemical weathering is achieved with water at different pH's and temperatures and by the action of O_2 and CO_2.

Bauxite - Al_2O_3 rich deposits - have been produced by chemical weathering of aluminium rich rocks. The soluble ions, Na^+, K^+, Ca^{2+}, have been leached out by rainwater leaving kaolinite ($Al_4Si_4O_{10}(OH)_8$), and then the less soluble silica is slowly dissolved at a pH 5 to 9 leaving enriched Al_2O_3, which may be contaminated with insoluble Fe_2O_3. The need for large quantities of water is obvious, and explains why many bauxite deposits occur in the wet tropics, or areas that have been wet and tropical. A more detailed discussion of weathering is given in Chapter 2.

Atmosphere and Hydrosphere. Comparison of the elemental abundances in the earth with those of the universe is measured by a *deficiency ratio* (= log[cosmic abundance]/[earth's abundance]) relative to silicon. For the elements occurring predominantly in the solid earth the relative abundances are similar (Table 1.10). The lighter elements (H, C, N), and the noble gases (He, Ne, Ar, Xe, Kr) are less abundant on the earth.

TABLE 1.10 Deficiency Ratios[a]

Element	Atomic Number	Deficiency Ratio	Element	Atomic Number	Deficiency Ratio
H	1	6.6	Al	13	0
He	2	14	Si	14	0
C	6	4.0	P	15	0.1
N	7	5.9	S	16	0.5
O	8	0.8	Cl	17	0.7
F	9	1.0	Ar	18	6.3
Ne	10	10.6	Fe	26	-0.6
Na	11	0	Kr	36	7.2
Mg	12	0	Xe	54	6.5

a $\log_{10}$ [Cosmic abundance/Earth abundance]

relative to $\log_{10} [Si_{Cosmic}/Si_{Earth}] = 0$

Source: Mason, B., Principles of Geochemistry, 3rd Ed., J. Wiley and Sons, Inc., N.Y., 1966.

Two processes were probably a factor in the formation of the atmosphere. A degassing of the solid material, which still occurs in the activity of volcanoes, geysers and fumaroles. The primitive atmosphere of the earth (H_2, He, N_2, H_2O, CO_2, CO, NH_3, CH_4) altered as the vertical velocities of the materials were sufficient to allow them to escape from the gravitational pull of the earth. At the temperatures of the earth's atmosphere today, H_2 and He can still escape; whereas in the days of the earth's formation, when temperatures were higher, the other gases would also have escaped. The moon has no atmosphere because of its smaller mass, so all gaseous substances escape.

The predominance of N_2 and O_2 in our atmosphere, and H_2O in the hydrosphere arises from their chemical properties. As the earth cooled dinitrogen (N_2) could not escape and it would concentrate in the atmosphere because of its low reactivity. In the case of O_2, which is reactive, its presence in the atmosphere, where it did not occur initially, is probably a result of photochemical decomposition of water;

$$2H_2O \xrightarrow[\text{radiation}]{\text{UV}} 2H_2 + O_2 \qquad (1.46)$$

(note: H_2 will escape from the atmosphere), and a product of biochemical photosynthesis.

The high deficiency ratio of the heavy noble gases, Ar, Kr and Xe, is surprising. The elements are unreactive and gaseous, and would naturally be expected in the atmosphere. If the earth, initially, had no atmosphere then it would arise from degassing of the solid earth. Argon would also form from the decay of radioactive ^{40}K.

The hydrosphere formed as the earth cooled and the water condensed. In the early stages the water would quickly evaporate, and while hot would readily dissolve soluble ions producing the mineral content of the oceans.

Determining the Age of the Earth

It has been an interest of man to determine when events happened; fortunately when the earth was forming a natural clock, the decay of radioactive nuclei, produced as described earlier, was set into action.

The number of radioactive nuclei that decay within a time interval dt is proportional to the number of nuclei present as expressed by the decay law;

$$\frac{dN}{dt} = -\lambda N \qquad (1.47)$$

where N is the number of nuclei present during the time interval dt, dN is the number of nuclei that decayed in the time, and λ is the decay constant. The negative sign indicates that N is decreasing. Integration of the equation over the time interval t=o to t=t gives;

$$\ln\left(\frac{N}{N_o}\right) = -\lambda t, \qquad (1.48)$$

where N_o is the number of nuclei at t=o and N is the number at time t. The equation can be rewritten;

$$N = N_o e^{-\lambda t}. \qquad (1.49)$$

The half-life of radioactive nuclei, $t_{\frac{1}{2}}$, is the time taken for half of the nuclei to decay, i.e. when $N = \frac{1}{2}N_o$;

hence $$\ln\left(\frac{\frac{1}{2}N_o}{N_o}\right) = -\lambda t_{\frac{1}{2}}, \qquad (1.50)$$

which gives;

$$t_{\frac{1}{2}} = \frac{0.693}{\lambda}. \qquad (1.51)$$

All radioisotopes have their own distinctive decay constant λ and $t_{\frac{1}{2}}$ values, and it is these properties which allow radioisotopes to be used for dating purposes.

Equation (1.49) may be rearranged as follows;

$$N_o = Ne^{\lambda t}, \qquad (1.52)$$

and subtraction of N from each side gives;

$$N_o - N = N(e^{\lambda t} - 1), \quad (1.53)$$

i.e.
$$\frac{N_o - N}{N} = e^{\lambda t} - 1. \quad (1.54)$$

Replace (N_o-N) with d, where d is the number of daughter atoms produced during time t, and replace N with p, the number of parent atoms remaining at the time of measurement.

Hence;
$$\frac{d}{p} = e^{\lambda t} - 1 \quad (1.55)$$

i.e.
$$t = \frac{2.303}{\lambda} \log(\frac{d}{p} + 1). \quad (1.56)$$

The time t corresponds to the age that the material containing the radio-isotope was formed, and may be calculated from an accurate measurement of "d" and "p". As an example, consider an uranium-containing mineral in which the parent nucleus ^{238}U has decayed to give the inactive daughter ^{206}Pb. If the gram-atom ratio of ^{206}Pb to ^{238}U is measured (with a mass spectrometer) and found to be 0.048 we know that 0.048g of lead has been produced from the ^{238}U and that 1.000g of ^{238}U remains. The value of λ can be obtained using equation (1.51) since $t_{½} = 4.5 \times 10^9$ yrs, (this $t_{½}$ is the longest half-life of all the radioactive nuclei in the ^{238}U decay series and so controls the decay).

Hence,
$$\lambda = \frac{0.693}{4.5 \times 10^9} \quad (1.57)$$

and from equation (1.56) we can calculate t;

i.e.,
$$t = \frac{2.303}{0.693} \times 4.5 \times 10^9 \log[0.048 + 1] \quad (1.58)$$

$$t = 3.0 \times 10^8 \text{ years}, \quad (1.59)$$

which is an estimate of the age of the rock in which the ^{238}U occurs.

Radioactive nuclei that may be used for dating are divided into four categories (Table 1.11). The radioactive nucleus ^{40}K decays by two separate routes;

$$^{40}Ar \xleftarrow[\text{electron capture}]{12\%} {}^{40}K \xrightarrow{88\%} {}^{40}Ca + \beta^-. \quad (1.60)$$

The measurement of the $^{40}K/^{40}Ar$ ratio is a useful dating method because potassium is so widely distributed in nature. The amount of argon produced can be measured accurately, and since the half-life is reasonably short (1.3×10^9 years), young rocks can be dated (5000 to a few billion years). The one problem is that some of the argon may have escaped from the rock matrix over the years. The $^{87}Rb/^{87}Sr$ ratio does not suffer from this disadvantage and is also used for dating purposes.

The amounts of the radioactive isotope ^{14}C, and the stable isotope ^{12}C were in a constant ratio, and 15.3 β emissions occurred per minute per gram of carbon

$$^{14}C \rightarrow {}^{14}N + \beta^- \quad (1.61)$$

Since carbon is taken up by living material the same $^{14}C/^{12}C$ ratio will occur, but after death no further carbon is assimilated and the $^{14}C/^{12}C$ ratio will decrease as the ^{14}C decays, with a half life of 5760 years. Carbon-14 dating has been useful for biological and archaeological remains, but since the advent of nuclear explosions the $^{14}C/^{12}C$ ratio in the world has altered so no dating will be able to be carried out on material living during and since the nineteen forties. Carbon-14 dating allows estimates to be made of dates in the recent past. Other radioactive nuclei that can be used in this way are ^{3}H, ^{226}Ra, ^{32}Si, ^{10}Be and ^{284}U.

TABLE 1.11 Radioactive Nuclei Used in Dating

Category	Nuclei	Uses
Primary	^{235}U, ^{238}U, ^{232}Th, ^{40}K, ^{87}Rb, ^{187}Re	Rocks meteorites, shells.
Secondary	^{234}U, ^{230}Th, ^{226}Ra, ^{210}Pb, ^{231}Pa.	Deep sea sediments, ocean water, ice corals.
Cosmic-ray induced	^{14}C, ^{3}H, ^{32}Si, ^{10}Be, ^{7}Be.	Wood, bones, shells, sediments, ocean water.
Relict	^{129}I, ^{244}Pu.	Early events in the solar system, universe.

Source: Turekian, K.K., Chemistry of the Earth, Holt, Rinehart and Winston, Inc., N.Y., 1972.

Finally, the temperature at which events may have occurred can be estimated by measuring the $^{16}O/^{18}O$ ratio as some slight fractionation of isotopes occurs, depending on the temperature.

CHAPTER 2

The Earth Today

In this chapter we will discuss, in more detail, some features of the earth today, including the chemical species in the atmosphere, hydrosphere and lithosphere (crust), and their chemical interactions in the natural environment.

THE ATMOSPHERE

The earth's atmosphere extends approximately 2000 km above its surface, however, 50% of the material lies within 5 km, and 99.99% lies within 80 km of the surface. The atmosphere is divided into regions, each region having a distinctive temperature gradient (Fig. 2.1). The regions are the troposphere, stratosphere, mesosphere and thermosphere, and between each are the tropopause, stratopause and mesopause, regions of relatively constant temperature. The depth of each region varies daily, therefore the figures given in Fig. 2.1 are approximate.

Troposphere

The troposphere extends 10 to 16 km above the earth (polar latitudes ~10 km, tropical latitudes ~16 km) and the temperature falls from ambient at ground level to around 200 K. Air, warmed near the surface of the earth, rises and as the pressure falls it expands adiabatically (because of poor heat conduction) leading to a fall in temperature of the order of 6 K km^{-1}.

The weather patterns that we are familiar with are all contained within the troposphere, and mixing in the region is fast, of the order of weeks within a hemisphere, and of 1 to 2 years between hemispheres. Vertical mixing occurs in the troposphere but movement into the stratosphere is inhibited by the temperature inversion.

The major constituents of the dry atmosphere, near the earth's surface, are listed in Table 2.1. The two principal species, dinitrogen and dioxygen, account for 78.09% and 20.94% (by volume) of the air, respectively, and their concentrations fall off exponentially with height (Fig. 2.2).

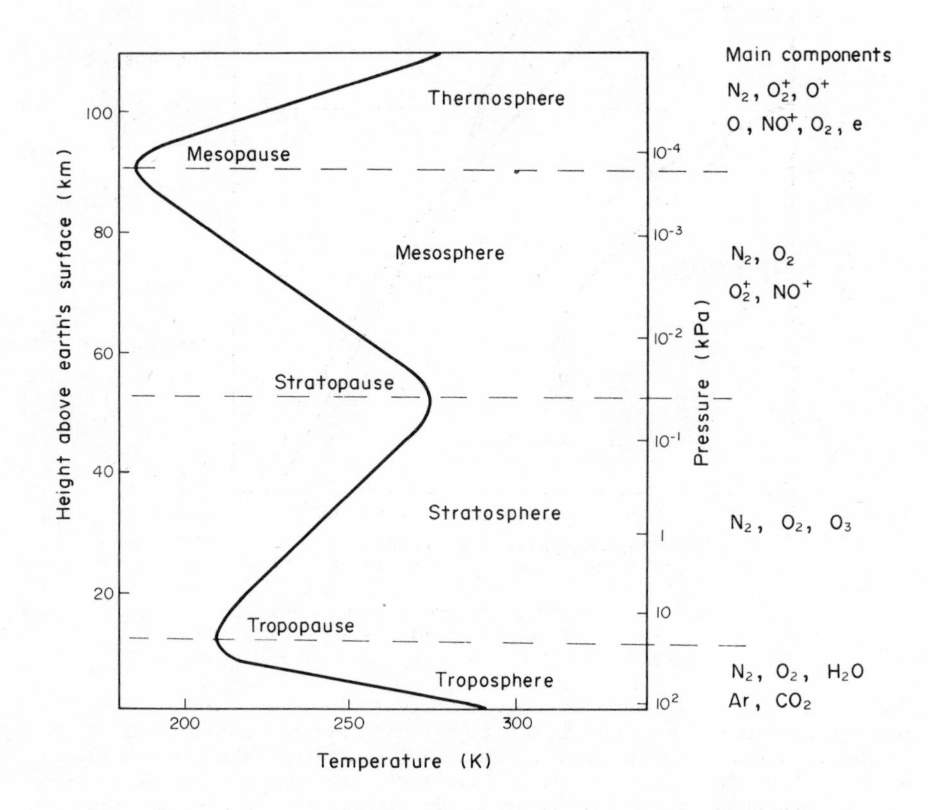

Fig. 2.1. The temperature variation and distribution of some chemical species in the atmosphere.

TABLE 2.1 Composition of Dry Air Near Sea Level

Component	Concentration(ppm)	Component	Concentration(ppm)
Major		Trace	
Dinitrogen, N_2	780,840	Neon, Ne	18.18
Dioxygen, O_2	209,460	Helium, He	5.24
		Methane, CH_4	1.4
		Krypton, Kr	1.14
		Dihydrogen, H_2	0.5
Minor		Nitrous oxide, N_2O	0.25
Argon, Ar	9340	Carbon monoxide, CO	0.08
Carbon dioxide, CO_2	325	Ozone, O_3	0.025
Water vapour, H_2O	variable	Ammonia, NH_3	6×10^{-3}
		Nitrogen dioxide, NO_2	4×10^{-3}
		Sulphur dioxide, SO_2	2×10^{-4}

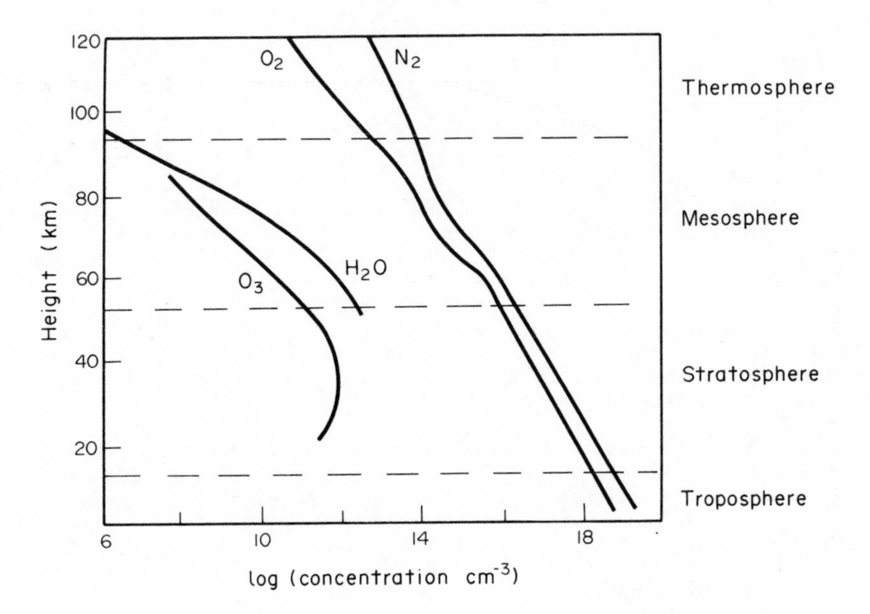

Fig. 2.2. The variation in the concentrations of N_2, O_2, H_2O and O_3 with height. Source: M.J. McEwan and L.F. Phillips, Chemistry of the Atmosphere, Edward Arnold, London (1975).

Though most of the energetic radiation from the sun has been filtered out at higher altitudes, some important chemical reactions take place in the troposphere, particularly at the atmosphere/biosphere and atmosphere/lithosphere interfaces. Two of these reactions are photosynthesis, the overall process given by the two equations;

$$2H_2O \rightarrow O_2 + 4H^+ + 4e^- \quad (2.1)$$

$$CO_2 + 4H^+ + 4e^- \rightarrow [CH_2O] + H_2O \quad (2.2)$$

and nitrogen fixation;

$$N_2 \underset{\text{hydrogenase}}{\overset{\text{nitrogenase}}{\rightarrow}} NH_3 \rightarrow \underset{\text{in the soil}}{NO_2^- \rightarrow NO_3^-}. \quad (2.3)$$

Chemical reactions, for example hydrolysis and oxidation, also occur at the surface of the earth (weathering, see later). The corrosion of iron is represented by the equations;

$$Fe - 2e \rightarrow Fe^{2+} \quad (2.4)$$

$$O_2 + 2H_2O + 4e \rightarrow 4OH^- \quad (2.5)$$

i.e. $$2Fe + O_2 + 2H_2O \rightarrow 2Fe^{2+} + 4OH^- \underset{\text{air}}{\overset{\text{in}}{\rightarrow}} \underset{\text{rust}}{Fe_2O_3xH_2O}. \quad (2.6)$$

Electrical discharges in the atmosphere initiate reaction between nitrogen and oxygen as follows;

$$N_2 + O_2 \xrightarrow{h\nu} 2NO \xrightarrow{O_2} 2NO_2 \qquad (2.7)$$

Stratosphere

The stratosphere is from 10 - 16 km to around 60 km above the earth's surface. It has a positive temperature gradient, reaching 280 - 290 K (Fig. 2.1). The principal chemical species in the stratosphere, N_2, O_2, O_3 and some H_2O, are chemically active because of their interaction with ultraviolet solar radiation.

The interaction of a species with radiation, producing chemical changes, is *photochemistry*. Radiation is absorbed forming an activated species and, provided the conditions are suitable, a new species is formed.

The far ultraviolet radiation (wavelength <190 nm) is removed before it reaches the stratosphere by atomic nitrogen and oxygen. A considerable portion of longer wavelength ultraviolet, (i.e. below 340 nm) is removed in the stratosphere, for example, radiation of wavelength <242 nm promotes the reaction;

$$O_2 + h\nu \rightarrow O + O^*. \qquad (2.8)$$

The atomic oxygen produced reacts with O_2 to form ozone;

$$O^* + O_2 + M \rightarrow O_3 + M^*, \qquad (2.9)$$

and the ozone absorbs ultraviolet radiation of wavelength <340 nm;

$$O_3 + h\nu \rightarrow O_2 + O, \qquad (2.10)$$

followed by the reaction;

$$O_3 + O \rightarrow 2O_2. \qquad (2.11)$$

Hence a cycle of ozone formation and destruction (Chapman mechanism) exits,

i.e. $$3O_2 \rightarrow 2O_3, \qquad (2.12)$$

and $$2O_3 \rightarrow 3O_2. \qquad (2.13)$$

Other reactions can also occur, the principal among which is the reaction of ozone with nitric oxide.

$$NO + O_3 \rightarrow NO_2 + O_2 \qquad (2.14)$$

$$NO_2 + O \rightarrow NO + O_2 \qquad (2.15)$$

$$NO_2 + h\nu \rightarrow NO + O \quad (\lambda < 400 \text{ nm}). \qquad (2.16)$$

The overall process is as before;

$$2O_3 \rightarrow 3O_2 \qquad (2.17)$$
(twice reaction (2.14) plus reactions (2.15) and (2.16)),

and accounts for removal of approximately 70% ozone. Species HO_x are involved in similar reactions.

The absorption of ultraviolet radiation in the stratosphere prevents virtually all radiation below 290 nm, and most between 290 and 320 nm, from reaching the earth (Fig. 2.3). Present day biological life has been protected from this radiation, a situation that may not continue as gaseous pollutants are added to the atmosphere.

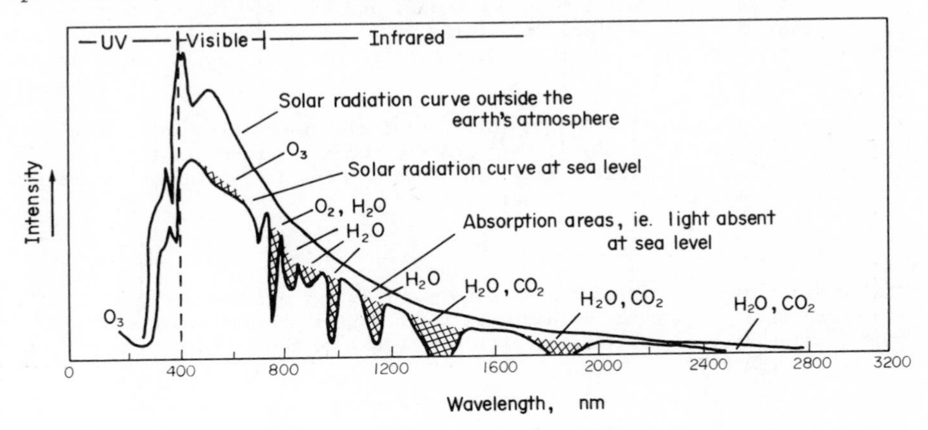

Fig. 2.3. Solar radiation outside the earth's atmosphere and at sea level. Areas of absorption and species contributing are indicated.

In reaction (2.9) an excited species, M*, is produced. The energy may be transferred as heat, which would explain the rise in temperature within the stratosphere. Infrared radiation also absorbed and re-emitted by ozone adds to the temperature rise.

Mesosphere and Thermosphere

The temperature falls in the next region of the atmosphere, the memosphere, to a minimum of 200 K, and rises again in the fourth region, the thermosphere, reaching 1300 K in the upper regions, 700 - 800 km above the earth. Photochemical reactions such as;

$$O_2 \xrightarrow{h\nu} O + O, \qquad (2.18)$$

which absorb radiation of wavelength 135 - 176 nm, produce the temperature rise.

The principal chemical species in the mesosphere are N_2, O_2, O_2^+ and NO^+, and in the thermosphere are N_2, O_2, O_2^+, O, O^+ and NO^+. The thermosphere is also called the ionosphere as ionization of molecular and atomic species takes place in the lower layers. These layers are responsible for the reflection of radio-waves.

THE HYDROSPHERE

The hydrosphere comprises the oceans, 98.3%; ice, 1.6% and fresh water, 0.1%, and covers over 70% of the earth's surface. There is also extensive quantities of water tied up in the earth's mantle and crust, in various hydrated compounds.

Water is of central importance in our world, and its ubiquitous presence ensures that the natural chemistry of the world is aqueous. All living materials are, to varying degrees, dependent on the availability of water, of a particular quality.

Sea water is approximately a 3.5% solution of dissolved salts, the ionic species; Cl^-, Na^+, Mg^{2+}, SO_4^{2-}, Ca^{2+} and K^+, account for 99.3% of all dissolved salts. Fresh water is a much less concentrated solution, averaging 0.012% (120 ppm), but varying widely with locality.

Water Cycle

The movement of water through the environment is summarised in Fig. 2.4. The residence time for a water molecule in the ocean is approximately 44,000 years, calculated as the time to fill the oceans to their present level from all known inputs.

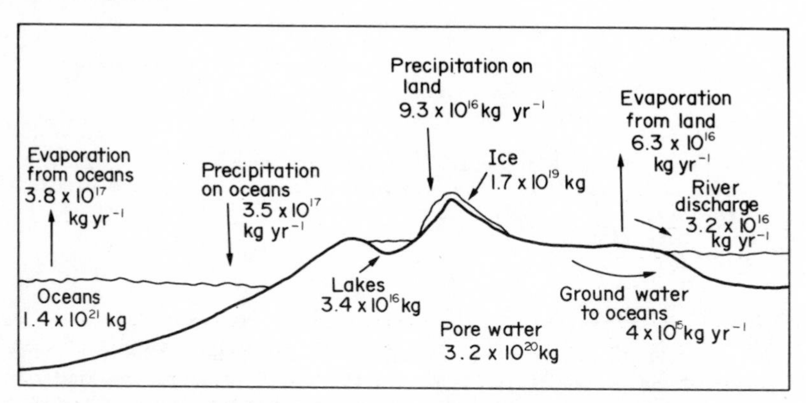

Fig. 2.4. Global water cycle; data approximate and in units kg and kg yr^{-1} (in terms of volume $1\ km^3 = 10^{12}$ kg).

More than water is involved, as water can dissolve and react with materials which it comes into contact with. Rain water contains CO_2 in equilibrium with atmospheric CO_2, and depending on the environment, may also contain small amounts of HNO_3 and H_2SO_4 from natural sources (and man-made sources). Rain water may have a pH of 5 to 6, and its acidity will increase as it percolates decaying organic debris. Such water will react with, and dissolve material in the earth's crust, becoming mildly alkaline in the process;

$$CaCO_3 + CO_2 + H_2O \rightarrow Ca^{2+} + 2HCO_3^- \quad (2.19)$$

$$Mg_2SiO_4 + 4H_2O \rightarrow 2Mg(OH)_2 + Si(OH)_{4(soln)} \quad (2.20)$$

$$\downarrow\uparrow$$

$$2Mg^{2+} + 4OH^-$$

The dissolved ions are finally carried into the oceans.

Dissolved Materials in Natural Water

The principal components of sea water and river water (greater than 1 ppm) are listed in Table 2.2. The sea water values vary somewhat with locality, though the relative amounts do not differ greatly. A wide range of concentrations occur for river water, depending on the environment of the river.

TABLE 2.2 Dissolved Materials in Natural Water

Sea Water		River Water[a]	
Species	Concentration ($g\ kg^{-1}$)[b]	Species	Concentration (ppm)
Cl^-	18.98	HCO_3^-	58.4
Na^+	10.54	Ca^{2+}	15.0
SO_4^{2-}	2.46	SiO_2	13.1
Mg^{2+}	1.27	SO_4^{2-}	11.2
Ca^{2+}	0.40	Cl^-	7.8
K^+	0.38	Na^+	6.3
HCO_3^-	0.14	Mg^{2+}	4.1
Br^-	0.06	K^+	2.3
H_3BO_3	0.02	Fe^{2+}	0.67

a Concentrations can vary widely. *b* $1g\ kg^{-1} = 1000$ ppm

The relative amounts of the major constituents in sea water follow the order; $Na^+ > Mg^{2+} > Ca^{2+} > K^+$ and $Cl^- > SO_4^{2-} > HCO_3^-$; while in river water the order is almost the reverse; $Ca^{2+} > Na^+ > Mg^{2+} > K^+$ and $HCO_3^- > SO_4^{2-} > Cl^-$. Since approximately 90% of the chlorine in the world is found in the oceans as chloride ions, one would not expect its concentration to be high in river water. The amounts of Ca^{2+} and HCO_3^- ions are influenced by biological and chemical processes in the sea, and the concentration of each is less than expected (on the basis of additions from rivers).

More complete data on dissolved species in the sea are given in Table 2.3, including suggested chemical forms, and residence times. The sea is assumed to be in a "steady state", i.e. the rate of addition and removal of an element are the same. In general the abundant elements in the oceans have the longest residence time. Factors that control the removal of a chemical species from the sea are, pH, redox properties, solubility and the biochemistry of the chemical species.

TABLE 2.3 Composition of Sea Water

Element	Concentration (ppm)	Principle Species	Residence Time (years)
Li	0.17	Li^+	2.3×10^6
B	20	$B(OH)_3$, $[B(OH)_4]^-$	1.3×10^7
C	28	HCO_3^-, $CO_{2(aq)}$, CO_3^{2-}, organic	
N	0.5	NO_3^-, NO_2^-, NH_4^+, N_2, organic	
F	1.3	F^-, MgF^+	5.2×10^5
Na	10540	Na^+	6.8×10^7
Mg	1270	Mg^{2+}, $Mg^{2+}SO_4^{2-}$	1.2×10^7
Al	0.01	$[Al(OH)_4]^-$	1.0×10^2
Si	3.0	$Si(OH)_4$, $[Si(OH)_3O]^-$	1.8×10^4
P	0.07	HPO_4^{2-}, $H_2PO_4^-$, $Mg^{2+}PO_4^{3-}$, PO_4^{3-}, H_3PO_4	1.8×10^5
S	2460	SO_4^{2-}, $Na^+SO_4^{2-}$, $Mg^{2+}SO_4^{2-}$, $Ca^{2+}SO_4^{2-}$	
Cl	18980	Cl^-	1×10^8
K	380	K^+	7×10^6
Ca	400	Ca^{2+}, $Ca^{2+}SO_4^{2-}$	1×10^6
Mn	0.002	Mn^{2+}, $Mn^{2+}SO_4^{2-}$, $MnCl^+$	1×10^4
Fe	0.01	$[Fe(OH)_2]^+$, $[Fe(OH)_4]^-$	2×10^2
Cu	0.003	Cu^{2+}, $Cu^{2+}SO_4^{2-}$, $CuOH^+$, $CuCO_3$	2×10^4
Zn	0.01	Zn^{2+}, $Zn^{2+}SO_4^{2-}$, $ZnOH^+$	2×10^4
Br	60	Br^-	1×10^8
Rb	0.12	Rb^+	4×10^6
Mo	0.01	MoO_4^{2-}	2×10^5
Cd	0.00011	Cd^{2+}, $CdCl^+$	5×10^5
I	0.06	IO_3^-, I^-	4×10^5
Cs	0.0005	Cs^+	6×10^5
Hg	0.00003	$HgCl_3^-$, $HgCl_4^{2-}$	8×10^4
Pb	0.00003	Pb^{2+}, $Pb^{2+}SO_4^{2-}$, $[Pb(CO_3)_2]^{2-}$	4×10^2

Sources: B. Mason, Principles of Geochemistry, 3rd Ed., John Wiley & Sons, N.Y. (1966); W. Stumm, and P.A. Brauner, Chemical Speciation, Chemical Oceanography, 2nd Ed., Vol. 1, Eds. J.P. Riley, and G. Skirrow, Academic Press (1975); and P.G. Brewer, Minor Elements in the Sea, ibid.

The chemical forms of an element in the sea depend on the nature of other species present, the pH, the nature of the metal ions and bonded groups and the relative stability of the species formed. Small cations and cations with a high charge (e.g. Co^{3+}, Na^{+}, Al^{3+}, Fe^{3+}) which are not very polarizable ("hard acids"), tend to associate with small ligands, which are also difficult to polarise (e.g. F^{-}, OH^{-}, H_2O). On the other hand, large cations (e.g. Hg^{2+}, Bi^{3+}, Cu^{+}) which are polarizable ("soft acids"), tend to bond to large, polarizable ligands such as I^{-}, S^{2-} and CN^{-}. This preference is reflected in the stability constant data listed in Table 2.4. A stability constant which is underlined indicates that the corresponding species exists

TABLE 2.4 Stability Constants of Species in Sea Water

(log K at 25°C)

Ligand / Metal ion	F^{-}	OH^{-}	CO_3^{2-}	SO_4^{2-}	Cl^{-}
Mg^{2+}	1.3	1.5	2.2	1.0[a]	0.2
Ca^{2+}	0.6	0.8	1.9	1.0	-
Fe^{3+}	5.0	11[b]	-	2.3	0.5
Cu^{2+}	0.7	~7	5.5	1.2	~0.7
Pb^{2+}	0.3	~6.8	6.2	-	~0.8
Zn^{2+}	0.7	~4.7	~4	1.2	~0
Cd^{2+}	0.5	~4	~4	1.2	1.5
Hg^{2+}	1.0	1.0	-	1.3	6.7

a --- indicates some ion-pair formation occurs in sea water

b ___ indicates some complex formation occurs in sea water

Source: W. Stumm, and P.A. Brauner, Chemical Speciation, Chemical Oceanography, 2nd Ed., Vol. 1, Eds. J.P. Riley and G. Skirrow Academic Press (1975).

in sea water. Ion-pair formation is also important, e.g. the $Mg^{2+}SO_4^{2-}$ ion-pair accounts for approximately 11% of Mg^{2+} in sea water (Table 2.5).

Chlorine. Chlorine occurs almost entirely as the Cl^{-} ion, as it does not form stable complexes with metal ions in water.

Sodium. Sodium as the Na^{+} ion has one of the longest residence times of any element in the sea, implying its lack of reactivity in the marine environment. It is likely that a Na^{+} balance is achieved through ion-exchange with sedimentary clays.

$$\underset{\text{soln.}}{Na^{+}} + \underset{\text{solid}}{\text{M-clay}} \rightleftharpoons \underset{\text{solid}}{\text{Na-clay}} + \underset{\text{soln.}}{M^{+}} \qquad (2.21)$$

Potassium, on the other hand, is less readily leached from rocks and more readily removed from solution, by a similar ion-exchange process.

TABLE 2.5 Ion Pairs in Sea Water[a]

% ion / with	Na^+	K^+	Mg^{2+}	Ca^{2+}	SO_4^{2-}	HCO_3^-	CO_3^{2-}
Na^+	99				21	8	17
K^+		99			0.5	-	-
Mg^{2+}			87		22	19	67
Ca^{2+}				91	3	4	7
SO_4^{2-}	1	1	11	8	54		
HCO_3^-	-	-	1	1		69	
CO_3^{2-}	-	-	0.3	0.2			9

[a] Figures on the diagonal correspond to the percentage of free ion.

Source: W. Stumm, and J.J. Morgan, Aquatic Chemistry, Wiley-Interscience, (1970); R.M. Garrels, and M.E. Thompson, *Amer. J. Sci.*, 260, 57 (1962)

Magnesium. Magnesium comes from the weathering of silicate and carbonate rocks;

i.e.
$$MgCO_3 + CO_2 + H_2O \rightarrow Mg^{2+} + 2HCO_3^-, \qquad (2.22)$$

$$MgSiO_3 + 2CO_2 + 3H_2O \rightarrow Mg^{2+} + 2HCO_3^- + Si(OH)_4. \qquad (2.23)$$

The Mg^{2+} balance is maintained by its removal as carbonate minerals, and by ion-exchange processes. For example, in the presence of bacteria that reduce sulphate, the following reactions take place;

$$SO_4^{2-} + 2\{CH_2O\} + 2H^+ \xrightarrow{\textit{Desulphovibrio}} H_2S + 2CO_2 + 2H_2O \qquad (2.24)$$
(carbohydrate)

$$\text{Fe-clay} + Mg^{2+} \rightleftharpoons \text{Mg-clay} + Fe^{2+} \qquad (2.25)$$

$$Fe^{2+} + 2H_2S \rightarrow FeS_2\downarrow + 4H^+ \qquad (2.26)$$

Calcium. The calcium ions which are added to the sea are removed principally through biological processes, such as the formation of $CaCO_3$ (both calcite and aragonite) in sea shells and corals. The surface waters of the sea are close to saturation with $CaCO_3$, that is the degree of saturation;

$$D = \frac{[Ca^{2+}][CO_3^{2-}]}{K_{SP(CaCO_3)}} \qquad (2.27)$$

is approximately 1. Removal of Ca^{2+} ions by $CaCO_3$ precipitation will not maintain $D \approx 1$. However, biological removal achieves this. From the solubility product relationship for $CaCO_3$ it follows that if the $[CO_3^{2-}]$ is increased the concentration of Ca^{2+} will decrease. Therefore the properties, and amount of Ca^{2+} in sea water, are tied to the amount of carbonate-bicarbonate ions available. At depths where the pressure is greater than one atmosphere, $CaCO_3$ becomes more soluble (Fig. 2.5) and sinking fragments of

$CaCO_3$ tend to dissolve.

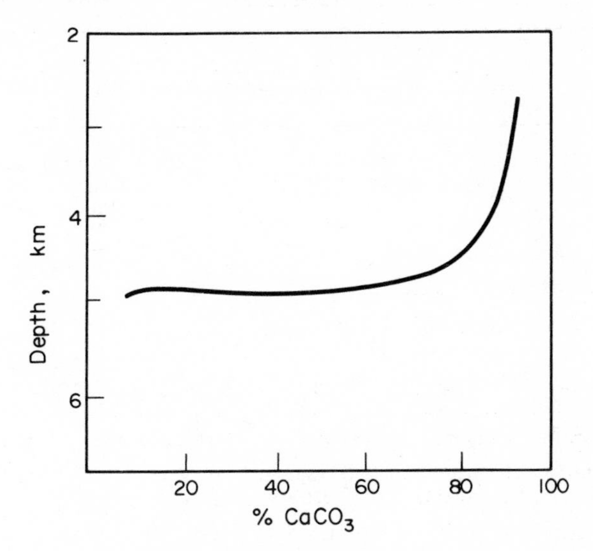

Fig. 2.5. The approximate percent saturation of $CaCO_3$ with depth in sea water.

Sulphur. The principal form of sulphur in sea water is SO_4^{2-}, approximately 60% of which, is tied up in ion-pair formation with Na^+, Mg^{2+} and Ca^{2+} (Table 2.5). In anaerobic conditions (e.g. in ocean floor sediments) bacteria reduce sulphate to sulphide (equation 2.24 above).

Carbon. The principl forms of dissolved carbon in river and sea water are CO_3^{2-}, HCO_3^- and $CO_{2(aq)}$. The major sources are the air, decaying organic material and the weathering of carbonate rocks. Any one of the three carbon species will give rise to the other two, depending on the pH of the water. We will investigate this feature, because the $CO_2/HCO_3^-/CO_3^{2-}$ system influences a number of other aspects of natural water chemistry. We will use $CO_{2(aq)}$ rather than H_2CO_3 in the discussion, because the equilibrium constant for the reaction;

$$CO_{2(aq)} + H_2O \rightleftharpoons H_2CO_3 \tag{2.28}$$

is approximately 2×10^{-3}. The equilibria between $CO_{2(aq)}$, HCO_3^- and CO_3^{2-} in water are;

$$CO_{2(aq)} + 2H_2O \rightleftharpoons H_3O^+ + HCO_3^-, \tag{2.29}$$

$$K_1 = \frac{[H_3O^+][HCO_3^-]}{[CO_{2(aq)}]} = 4.2 \times 10^{-7},\ pK_1 = 6.38; \tag{2.30}$$

and

$$HCO_3^- + H_2O \rightleftharpoons H_3O^+ + CO_3^{2-}, \tag{2.31}$$

$$K_2 = \frac{[H_3O^+][CO_3^{2-}]}{[HCO_3^-]} = 4.8 \times 10^{-11},\ pK_2 = 10.32. \tag{2.32}$$

The total concentration of inorganic carbon, $[CO_2]_T$, is given by;

$$[CO_2]_T = [CO_{2(aq)}] + [HCO_3^-] + [CO_3^{2-}], \tag{2.33}$$

and the fraction present of any one of the species is;

$$[X] = \alpha_X[CO_2]_T. \tag{2.34}$$

Therefore for $[CO_{2(aq)}]$ we have;

$$\alpha_{CO_{2(aq)}} = \frac{[CO_{2(aq)}]}{[CO_2]_T}. \tag{2.35}$$

The sum of all the fractions $\Sigma\alpha_X = 1$, i.e.

$$\alpha_{CO_{2(aq)}} + \alpha_{HCO_3^-} + \alpha_{CO_3^{2-}} = 1. \tag{2.36}$$

In equation (2.35);

i.e. $$\alpha_{CO_{2(aq)}} = \frac{[CO_{2(aq)}]}{[CO_{2(aq)}] + [HCO_3^-] + [CO_3^{2-}]}, \tag{2.37}$$

$[HCO_3^-]$ and $[CO_3^{2-}]$ can be replaced by terms in K_1, K_2, $[CO_{2(aq)}]$ and $[H_3O^+]$. From equations (2.30) and (2.32) we have;

$$[HCO_3^-] = \frac{K_1[CO_{2(aq)}]}{[H_3O^+]}, \tag{2.38}$$

and $$[CO_3^{2-}] = \frac{K_1K_2[CO_{2(aq)}]}{[H_3O^+]^2}, \tag{2.39}$$

hence $$\alpha_{CO_{2(aq)}} = \frac{[CO_{2(aq)}]}{[CO_{2(aq)}] + \frac{K_1[CO_{2(aq)}]}{[H_3O^+]} + \frac{K_1K_2[CO_{2(aq)}]}{[H_3O^+]^2}}, \tag{2.40}$$

which simplifies to;

$$\alpha_{CO_{2(aq)}} = \frac{[H_3O^+]^2}{[H_3O^+]^2 + K_1[H_3O^+] + K_1K_2}. \tag{2.41}$$

Similar expressions can be obtained for $\alpha_{HCO_3^-}$ and $\alpha_{CO_3^{2-}}$ i.e.;

$$\alpha_{HCO_3^-} = \frac{K_1[H_3O^+]}{[H_3O^+]^2 + K_1[H_3O^+] + K_1K_2}, \tag{2.42}$$

and

$$\alpha_{CO_3^{2-}} = \frac{K_1K_2}{[H_3O^+]^2 + K_1[H_3O^+] + K_1K_2}. \tag{2.43}$$

If values of $[H_3O]^+$ are placed in the three equations, the corresponding α_x values are obtained, and the results presented as a distribution diagram (Fig. 2.6) show which species predominates at a particular pH. In the pH range found for most natural water systems the principal species is the bicarbonate ion, HCO_3^-.

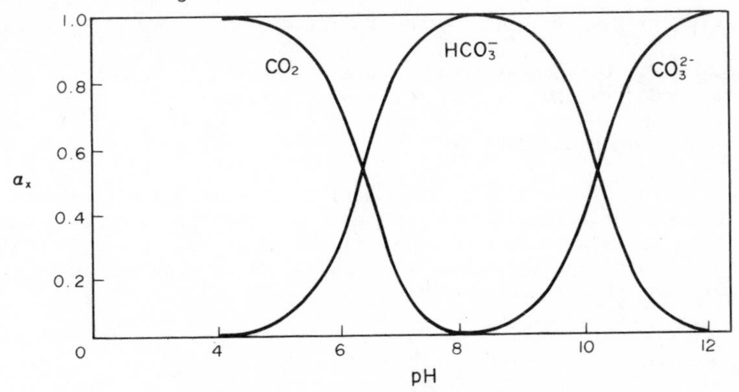

Fig. 2.6. Distribution diagram for the species $CO_{2(aq)}$ - HCO_3^- - CO_3^{2-} in water.

Although CO_2 from the air is only one (minor) source of inorganic carbon, it is instructive to calculate how much dissolves in sea water, assuming equilibrium has been reached. At one atmosphere pressure (101.3 kPa), and a temperature of 298 K, the pressure due to the dry atmosphere is;

$$P_{dry\ gas} = P_{total} - P_{water\ vapour} \quad (2.44)$$

$$= 101.3 - 3.2 \text{ kPa} = 98.1 \text{ kPa}.$$

The CO_2 in the air is 0.0325%, and has a partial pressure of;

$$P_{CO_2} = P_{dry\ gas} \times \frac{0.0325}{100} = 0.032 \text{ kPa}. \quad (2.45)$$

We can now make use of Henry's Law;

$$[CO_2] = kP_{CO_2} \quad (2.46)$$

to find the concentration of dissolved CO_2, using $k = 3.34 \times 10^{-4}$ mol l^{-1} kPa^{-1}.

Hence; $$[CO_2] = 3.34 \times 10^{-4} \text{ mole } l^{-1} \text{ kPa}^{-1} \times 0.032 \text{ kPa} \quad (2.47)$$

$$= 1.07 \times 10^{-5} \text{ mole } l^{-1} (= 0.47 \text{ mg } l^{-1}). \quad (2.48)$$

The total concentration of inorganic carbon in sea water is much greater, around 2.3×10^{-3} mole l^{-1}.

Alkalinity of Water

The alkalinity (or acidity) of natural water depends on the amounts of dissolved CO_2, HCO_3^-, CO_3^{2-}, OH^- and minor species, such as phosphate and silicate. The bicarbonate ion is often the dominant acid-base species in water, particularly in the pH range 6 - 8. (Fig. 2.6). The bicarbonate ion is involved in two equilibria;

$$HCO_3^- + H_2O \rightleftharpoons H_3O^+ + CO_3^{2-}, \ K_a = 4.8 \times 10^{-11}, \tag{2.49}$$

$$HCO_3^- + H_2O \rightleftharpoons H_2CO_3 + OH^-, \ K_b = 2.4 \times 10^{-7}, \tag{2.50}$$

and since $K_b > K_a$, bicarbonate solutions are observed to be alkaline due to the OH^- produced (reaction (2.50)). The equilibrium;

$$H_2CO_3 \rightleftharpoons H_2O + CO_{2(aq)} \tag{2.51}$$

lies mainly on the right, but if the position of equilibrium is altered it will influence the alkalinity of the water. For example, dissolved carbon dioxide ($CO_{2(aq)}$) is used in photosynthesis by algae, which drives reaction (2.51) to the right, and therefore reaction (2.50) is also pushed to the right, increasing the alkalinity of water.

Ice-Water System

Liquid water has its maximum density at 4°C, which is more dense than ice. An important consequence of this is that ice floats on water, a significant factor for the existence of marine life when temperatures drop below freezing. The floating ice insulates the water below, and prevents complete freezing of rivers and lakes.

Ice has a relatively open structure (Fig. 2.7) owing to extensive hydrogen-bonding. On melting some hydrogen bonds break allowing water molecules to collapse into the cavities increasing the density until 4°C.

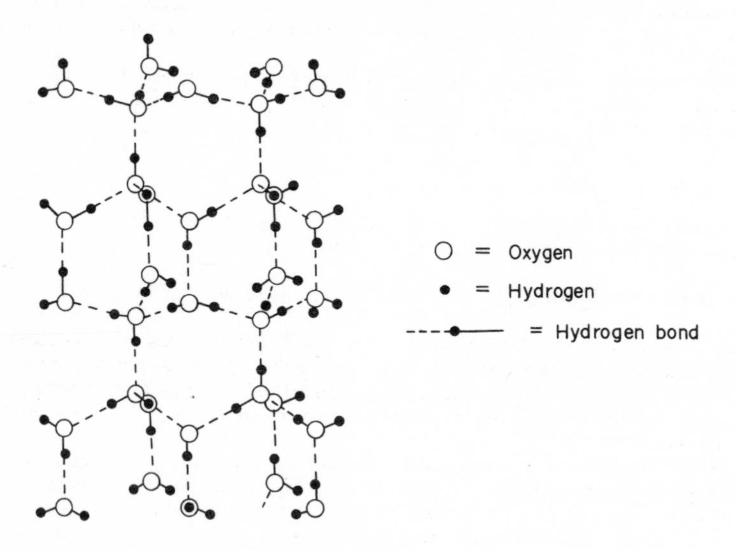

Fig. 2.7. Structure of ice (I), note the tetrahedral arrangement around each oxygen and the open structure with hexagonal shaped channels.

Air-Water Interface

Exchange of material occurs at the interface between the atmosphere and the hydrosphere, gases can dissolve, while water and dissolved materials can escape into the atmosphere. Under ideal conditions the solubility of gases can be estimated using Henry's Law, i.e. the pressure of a volatile solute (P_B) above a solution is proportional to the mole fraction of the solute (x_B) in solution.

$$P_B = k_B x_B, \tag{2.52}$$

where k_B is Henry's Law constant. The vapour pressure of the solvent (P_A) for a dilute solution and the vapour of the pure liquid (P_A^o) are connected by;

$$P_A = x_A P_A^o \text{, (Raoult's Law),} \tag{2.53}$$

where x_A is the mole fraction of the solvent in the solution. However, the air-water system is not ideal because the conditions of equilibrium are infrequently realised, due to constantly changing temperature and pressure, and movement of air and water.

Vapour pressure and water. Water vapour consists of individual water molecules, and the amount in the atmosphere is measured as the relative humidity, i.e. the ratio of the quantity of water in the air to the maximum quantity the air can contain, at the same temperature. The amount of water vapour, in equilibrium with liquid water, increases with temperature, and a plot of log P against 1/T(K) is linear with a slope of $\Delta H_V/R$. Fortunately, for our comfort, the liquid/vapour system in the world is not often at equilibrium owing to continual movement of the atmosphere. A sudden, or a large, temperature drop can lead to excess water vapour in the air producing condensation (dew or precipitation). Many wet tropical regions approach a state close to equilibrium, during certain seasons, and high humidities are experienced.

Sea water, being a solution, has a lower vapour pressure, at a particular temperature, than pure water (equation 2.53). For a system with just one substance dissolved in water;

$$x_A + x_B = 1 \tag{2.54}$$

where x_A and x_B are the mole fractions of the water and solute respectively. Combining equations (2.53) and (2.54) gives;

$$P_A = P_A^o(1 - x_B), \tag{2.55}$$

i.e. the vapour pressure of A(water) is lowered by an amount dependent on the mole fraction of the solute. For sea water this is approximately a 2% reduction in the vapour pressure. The dissolved solids also reduce the freezing point of the solution, and a temperature around -2°C is required to freeze sea water.

Solubility of gases in water. Aquatic life depends on the availability of gases such as O_2 and CO_2 in water. These gases (especially O_2) are not freely available in water, and the amount dissolved can vary considerably with the environmental conditions. For example, decaying organic material uses oxygen to produce CO_2;

$$\underset{\text{(carbohydrate)}}{\{CH_2O\}} + O_2 \rightarrow CO_2 + H_2O. \quad (2.56)$$

Loss of oxygen in this way can have a serious effect on fish life. The stoichiometry of the reaction (2.56) indicates that 1 mole of oxygen reacts with 1 mole of $\{CH_2O\}$, which means that approximately 7.5 mg of organic material will remove dissolved oxygen (8 mg) from 1 litre of water.

Assuming equilibrium for a gas between air and water has been achieved,

i.e. $$A_{(g)} \rightleftharpoons A_{(aq)} \quad (2.57)$$

then the amount of dissolved gas can be calculated using Henry's Law. For moist air at a pressure of 101.3 kPa the partial pressure of oxygen at 298 K is;

$$P_{O_2} = (101.3 - 3.2)0.209 \text{ kPa} \quad (2.58)$$
$$= 20.5 \text{ kPa}.$$

The partial pressure of water vapour at 298 K is 3.2 kPa, and the figure of 0.209 is the fraction of O_2 in the dry air. Hence for a Henry's Law constant of 1.26×10^{-5} mole l^{-1} kPa^{-1}, we have;

$$O_{2(aq)} = 1.26 \times 10^{-5} \times 20.5 \text{ mole } l^{-1} \quad (2.59)$$
$$= 2.58 \times 10^{-4} \text{ mole } l^{-1}$$
$$= 8.26 \text{ mg } l^{-1}$$

The solubility of some gases in sea water at 273 and 298 K are given in Table 2.6. The gases are actually more soluble than the figures suggest,

TABLE 2.6 Solubilities of Some Gases in Sea Water (surface)

Gas	Partial Pressure in Dry Air kPa	Concentration of Gas (at equilibrium) mg l^{-1}	
		273 K	298 K
N_2	79.1	17.5	11.3
O_2	21.2	12.6	8.3
Ar	0.95	0.64	0.39
CO_2	3.25×10^{-2}	0.92	0.47
Ne	1.8×10^{-3}	1.5×10^{-4}	1.3×10^{-4}
He	5.24×10^{-4}	7.3×10^{-6}	6.1×10^{-6}
Kr	1.11×10^{-4}	3.0×10^{-4}	1.8×10^{-4}

because the values listed are the equilibrium solubilities based on the partial pressure of the particular gas in the atmosphere.

The transfer of a gas across the air-sea water interface depends on turbulence, size of the gas bubbles, the temperature and the concentration gradient. The transfer is diffusion controlled, but will proceed from the air to water, i.e. from high to low concentration. The rate of the gas flow, or flux, is given by the expression:

$$D([gas]_{air} - [gas]_{aq})/t \qquad (2.60)$$

where D is the diffusion coefficient and t the boundary layer thickness which is of the order of 1.7×10^{-5} m (17μm).

THE LITHOSPHERE

Modification of the earth's crust, as a result of its interaction with the atmosphere and hydrosphere is called weathering. The visible changes are the breakdown of materials, and their dispersal around the world. Weathering falls into three categories - physical, chemical and biological.

Physical Weathering

The outcome of physical weathering can be quite dramatic, e.g. when large rocks split open, but this is often the result of many years of slow activity. When rocks submerged in the lithosphere, are exposed to the atmosphere the release of pressure can produce small cracks allowing water to enter. If the climate is sufficiently cold the ice produced expands exerting a pressure of 150 kg cm^{-2}, greater than the average tensile strength of rocks (120 kg cm^{-2}).

Alternating hot and cold periods can break up rocks, especially if the temperature variation is large and the change occurs rapidly. Rocks, with some surfaces exposed to high temperatures while others are cool, are under pressure from different degrees of expansion. Abrasion produced by wind, glaciers and erosion, also produces physical weathering. Finally roots of plants can enter cracks in rocks, and the pressure exerted can increase the crack size, and produce new ones.

Chemical Weathering

Chemical weathering of the earth's crust involves relatively simple chemical reactions, such as dissolving, hydration, carbonation, hydrolysis and oxidation and reduction. However, because the materials are complex; e.g. silicate rocks, mixtures of minerals, and water of variable mineral composition and pH; the chemistry can be complicated.

Weathering agents. The two principal chemical agents are water and air (in particular O_2 and CO_2). The mineral composition of natural water depends on the local environment, for example coastal fresh water will contain more Na^+ and Cl^- than inland water. Also the mineral composition of water changes as it percolates the earth's crust. The "best" average for the composition of rain water is, Na^+ 2.0, K^+ 0.3, Mg^{2+} 0.3, Ca^{2+} 0.1, Cl^- 3.8, SO_4^{2-} 0.6 and HCO_3^- 0.12 ppm, and pH ~5.7. The pH of rain water generally increases with time as acid gases (e.g. SO_2, NO_2) are washed out of the atmosphere.

The amount of CO_2 in rain water will be in equilibrium with atmospheric CO_2, but in contact with the land the CO_2 content can become quite high in the presence of decaying vegetation. The quantity of dissolved oxygen also varies and, in particular, will decrease if significant oxidation takes place.

Solubility. Ionic materials in the earth's crust that are relatively soluble in water are; NaCl (350 g l^{-1}) from halite deposits and from rocks, and $CaSO_4$ (gypsum and anhydrite) (2 g l^{-1}). To a lesser extent silica is also soluble (6.5 mg l^{-1});

$$SiO_{2(s)} + 2H_2O \rightleftharpoons \underset{\text{silicic acid}}{H_4SiO_4}, \tag{2.61}$$

but the amount of H_4SiO_4 in solution is greater, as it forms from hydrolysis of silicate minerals. Silicic acid is a weak acid with pK = 9.8 for the reaction;

$$H_4SiO_4 + H_2O \rightleftharpoons H_3O^+ + H_3SiO_4^- \tag{2.62}$$

The solubility of solids is a function of their lattice energies and the hydration energy of the ions. The two energies work against each other as the ionic radius-ionic charge diagram (Fig. 2.8) indicates. A low ionic

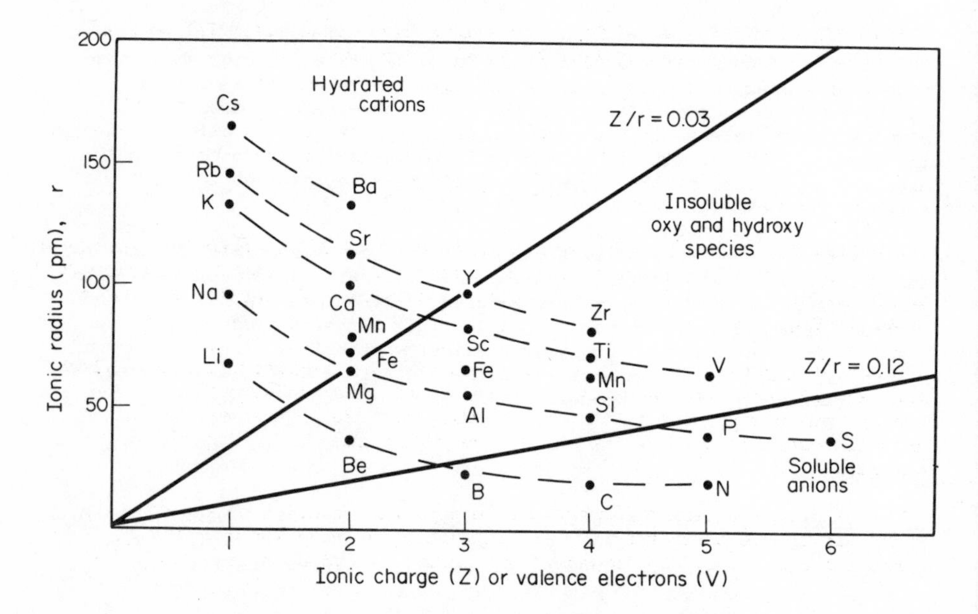

Fig. 2.8. Relationship between ionic charge (z) or valence electrons (v) and ionic radius (pm), and also solubility in water.

charge and a large ionic radius is associated with the most soluble species. Ions with low ionic potentials (i.e. charge/radius) are the more soluble (Table 2.7). Ions possessing a high charge and small ionic radius, (i.e. high ionic potential) may hydrolyse in water giving insoluble hydroxides.

TABLE 2.7 Ionic Potentials[a]

Ion	Ionic Potential	Ion	Ionic Potential	Ion	Ionic Potential
Cs^{+}	0.6	Fe^{2+}	2.6	Mn^{4+}	7.4
Rb^{+}	0.7	Co^{2+}	2.7	Si^{4+}	9.8
K^{+}	0.75	Zn^{2+}	2.1	Mo(VI)	9.7
Na^{+}	1.0	Mg^{2+}	3.1	B(III)	15.0
Li^{+}	1.5	Fe^{3+}	5.6	P(V)	14.7
Ba^{2+}	1.55	Al^{3+}	6.6	S(VI)	20.7
Sr^{2+}	1.8	Be^{2+}	6.6	C(IV)	26.7
Ca^{2+}	2.1	Ti^{4+}	5.9	N(V)	45.5
Mn^{2+}	1.6				

[a] ionic charge (z)/ionic radius ($pm^{-1} \times 10^{2}$)

The interaction of water with an ion with a high ionic potential weakens a O - H bond of water, and produces a metal hydroxide;

$$Al^{3+}\text{-----}O\begin{smallmatrix}H\\ H\end{smallmatrix} \rightleftharpoons [Al-O-H]^{2+} + H^{+} \quad \downarrow \quad Al(OH)_{3(s)} + 3H^{+} \tag{2.63}$$

The solubility of many minerals is also pH dependent, for example an increase in the pH of water increases the solubility of SiO_2, by pushing reaction 2.62 to the right. The change in solubility of a number of species with pH is shown in Fig. 2.9.

The actual crystalline form of a compound may also influence its solubility, for example aragonite $CaCO_3$ is 16% more soluble than calcite $CaCO_3$. The reason is, in part, due to the overcrowding and steric pressure in the aragonite structure reducing its lattice energy compared with that of calcite.

Carbonation. The term *carbonation* refers to the reaction of H_2O/CO_2 with materials in the earth's crust, and in particular reaction with $CaCO_3$. As established earlier, CO_2 in water is a source of hydrogen ions;

$$CO_{2(g)} + H_2O \rightleftharpoons H^{+}_{(soln)} + HCO^{-}_{3(soln)}, \tag{2.64}$$

which react with $CaCO_3$;

$$CaCO_{3(s)} + H^{+}_{(soln)} \rightleftharpoons Ca^{2+}_{(soln)} + HCO^{-}_{3(soln)}. \tag{2.65}$$

The overall reaction is,

$$CaCO_{3(s)} + CO_{2(g)} + H_2O \rightleftharpoons Ca^{2+}_{(soln)} + 2HCO^{-}_{3(soln)}. \tag{2.66}$$

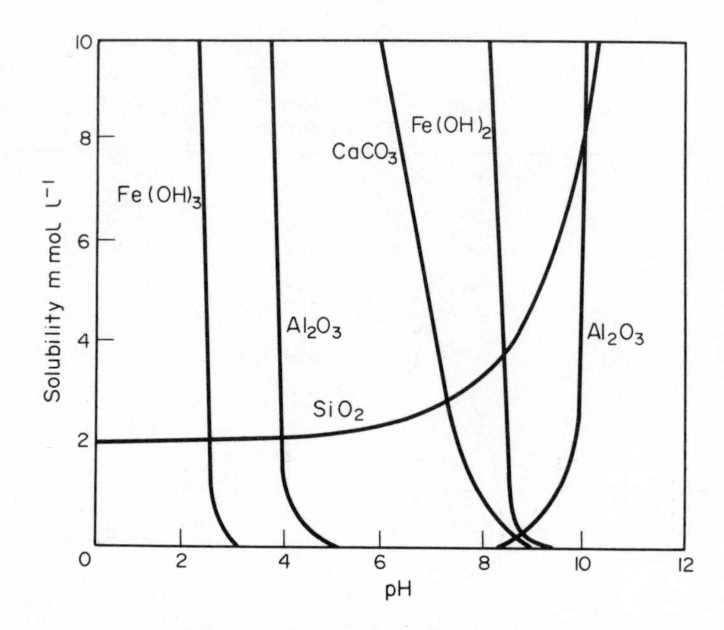

Fig. 2.9. The change in solubility with pH. Note the amphoteric nature of Al_2O_3, and the change in solubility of SiO_2.

This is a common and rapid chemical weathering process. The formation of limestone caves is represented by the forward direction of reaction (2.66), and limestone growths (stalactites and stalagmites) by the reverse reaction.

The equilibrium constant for reaction (2.66) is;

$$K = \frac{[Ca^{2+}][HCO_3^-]^2}{p_{CO_2}} = 1.56 \times 10^{-8}. \tag{2.67}$$

Two moles of HCO_3^- are produced for each mole of Ca^{2+}, i.e.;

$$K = \frac{(x)(2x)^2}{p_{CO_2}}$$

where $x = [Ca^{2+}]$.

For p_{CO_2} = 0.032 kPa (the partial pressure of CO_2 in the atmosphere at 298 K) we have;

$$1.56 \times 10^{-8} = \frac{4x^3}{0.032},$$

i.e. $$x = \sqrt[3]{\frac{(1.56 \times 10^{-8})(0.032)}{4}}.$$

Hence, $$[Ca^{2+}] = 10^{-3.3} = 5.0 \times 10^{-4} \text{ mol l}^{-1}, \tag{2.69}$$

i.e. a concentration of Ca^{2+} of 20 ppm (20 mg l^{-1}). A CO_2 pressure of 1.01 kPa, typical for soil water, would produce a concentration of Ca^{2+} ions of 1.58×10^{-3} mole l^{-1}, i.e. 63 ppm. For one atmosphere of pure CO_2, the concentration would be around 320 ppm.

Hydrolysis. The reaction between water and other species which leads to breaking of a O - H bond, and the formation of either alkaline or acid solutions, is called *hydrolysis*. Two examples are;

$$CaO_{(s)} + H_2O \rightleftharpoons Ca(OH)_{2(soln\ or\ s)} \quad \text{(alkaline)}, \qquad (2.70)$$

and

$$SO_{2(g)} + H_2O \rightleftharpoons 2H^+ + SO_3^{2-}{}_{(soln)} \quad \text{(acid)}. \qquad (2.71)$$

The mechanism is;

$$CaO + 6H_2O \underset{\text{hydration}}{\rightleftharpoons} [Ca(H_2O)_6]^{2+}O^{2-} \rightarrow \left[Ca \text{----} O \begin{matrix} H \\ H \end{matrix}\right]^{2+} + O^{2-}$$

bond formation, bond breaking

$$Ca^{2+} + 2OH^-_{(soln)} \rightleftharpoons Ca(OH)_{2(s)} \xleftarrow{H_2O} [Ca\text{-}OH]^+ + H^+ + O^{2-} \rightarrow OH^- \qquad (2.72)$$

(alkaline)

Hydrolytic weathering of minerals is similar, and may also include the breaking of M - O bonds, i.e.

$$\underset{\text{mineral}}{\overset{+\delta\ \ -\delta}{O - M}} + O\begin{matrix} H \\ H \end{matrix} + \delta \rightarrow O\text{-----}M\text{-----}O\begin{matrix} H \\ H \end{matrix}$$

bond breaking

$$\underset{\text{mineral}}{O - H} + MOH \leftarrow \underset{\text{mineral}}{[>O]^-} + M - OH + H^+ \qquad (2.73)$$

Species with the weakest M - O bonds are expected to weather first. The strengths of a number of M - O bonds are given in Table 2.8, and, as we will see below, the metal ions which are most readily removed from a rock have the weakest M - O bonds.

TABLE 2.8 M-O Bond Strengths

Bond	Strength ($kJ\ mol^{-1}$)	Bond	Strength ($kJ\ mol^{-1}$)
Ti-O	674	Mn-O[a]	389
Al-O	582	Fe-O[a]	389
Si-O	464	Mg-O	377
Ca-O	423		

a Divalent oxidation state

The change in the composition of rocks with time (Table 2.9) indicates different rates of weathering of the components. In columns 'A' and 'B' the analytical data for a weathered rock are listed, while column 'I' contains the composition of weathered rock B assuming there is no loss or gain of Al_2O_3. In the last column the percentage mass gain or loss of each

TABLE 2.9 Changes in Rock Composition with Weathering

Compound	Fresh Rock[a]	Increasing Weathering → A	B	I	Loss or Gain %
SiO_2	71.54[b]	68.09[b]	55.07[b]	30.8[b]	-56.9
Al_2O_3	14.62	17.31	26.14	14.62	0.0
Fe_2O_3	0.69	3.86	3.72	2.08	+201
FeO	1.64	0.36	2.53	1.41	-14.0
CaO	2.08	0.06	0.16	0.09	-95.7
MgO	0.77	0.46	0.33	0.18	-76.6
Na_2O	3.84	0.12	0.05	0.03	-99.2
K_2O	3.92	3.48	0.14	0.08	-98.0
H_2O	0.32	5.61	10.39	5.8	+1712.5
other	0.65	0.56	0.58	0.32	-50.8
Rock	100.07			55.41	-44.6

a Morton Granite Gneiss

b Columns 1 to 4 from S.S. Goldrich, *J. Geol.*, 1938, *46*, 17-58.

constituent;

$$\text{i.e.} \quad \left[\frac{\text{Fresh Rock} - \text{I}}{\text{Fresh Rock}} \times \frac{100}{1}\right], \tag{2.74}$$

is given and the data are plotted in Fig. 2.10. The results reveal that Al_2O_3 and Fe_2O_3 do not weather, in fact there is a gain in Fe_2O_3, probably from the weathering and release of FeO followed by oxidation to Fe_2O_3. Comparison of the data for the fresh rock and that given in column A indicates that MgO and K_2O weather more slowly than CaO and Na_2O.

Hydrolytic weathering can be summarised by the following equations:

$$\underset{\text{orthoclase}}{4KAlSi_3O_{8(s)}} + 22H_2O \rightarrow \underset{\text{kaolinite}}{Al_4Si_4O_{10}(OH)_{8(s)}} + 8H_4SiO_{4(aq)} + 4K^+_{(aq)} + 4OH^-_{(aq)}. \tag{2.75}$$

Note that the weathering solution which contains H_4SiO_4, K^+ and OH^- becomes more alkaline. If H^+ ions are involved (from organic acids or H_2SO_4 and HNO_3 from aerial sources of SO_2 and NO_2) the reaction will approximate to;

$$4KAlSi_3O_{8(s)} + 4H^+ + 18H_2O \rightarrow Al_4Si_4O_{10}(OH)_{8(s)} + 8H_4SiO_{4(aq)} + 4K^+_{(aq)}. \tag{2.76}$$

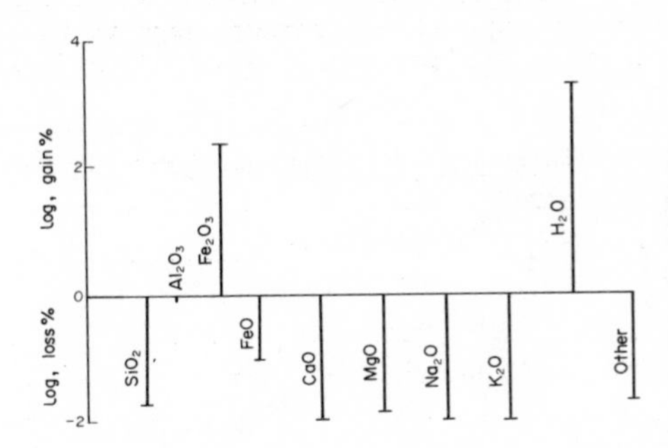

Fig. 2.10. Change in rock composition with weathering

The first step is probably hydration, followed by hydrolysis releasing the group I and II cations. The process involved in the structural change, from the 3-D aluminosilicate lattice of the rock, to the layer lattice of the clay (also an aluminosilicate) is not clear. One suggestion is that the Si and Al oxyanions dissolve and reprecipitate as the clay, which is, in general, more thermodynamically stable (see below). Also, the Al^{3+} ion undergoes a favourable coordination number change from 4 to 6. Dissolved CO_2 is also a source of hydrogen ions, as is clear from the reaction;

$$4KAlSi_3O_{8(s)} + \underbrace{4CO_2 + 22H_2O}_{4H^+ + 4HCO_3^- + 18H_2O} \rightarrow Al_4Si_4O_{10}(OH)_{8(s)} + 8H_4SiO_{4(aq)} + 4K^+_{(aq)} + 4HCO^-_{3(aq)} \quad (2.77)$$

The rocks which crystallised first from the magma weather most readily, i.e. the Ca-feldspars and olivines, while quartz and the K-feldspars, which crystallise at a lower temperature, weather more slowly. This is unexpected, as the species that crystallise first are those with the higher lattice energies. However, the conditions of magmatic crystallisation (high temperatures and non-aqueous) are quite different from the conditions of weathering (ambient temperatures and an aqueous environment).

The clays formed are to some extent dependent on the climatic conditions, for example, in temperate climates, with a moderate to heavy rainfall, the formation of kaolinite dominates, while montmorillonite and illite (high in Mg^{2+}, K^+ and Fe^{2+}) form in dry arid soils, and the free oxides Al_2O_3, Fe_2O_3 are produced in the humid tropics and subtropics. These generalizations are summarised in Fig. 2.11.

Thermodynamic data can be used to determine which minerals are the more stable for a particular weathering reaction. For example, the weathering of potassium feldspar to mica represented by the equation;

$$3KAlSi_3O_{8(s)} + 12H_2O + 2H^+_{(aq)} \rightleftarrows KAl_3Si_3O_{10}(OH)_{2(s)} + 6H_4SiO_{4(aq)} + 2K^+_{(aq)} \quad (2.78)$$

has a free energy of 43 kJ obtained from the following data:

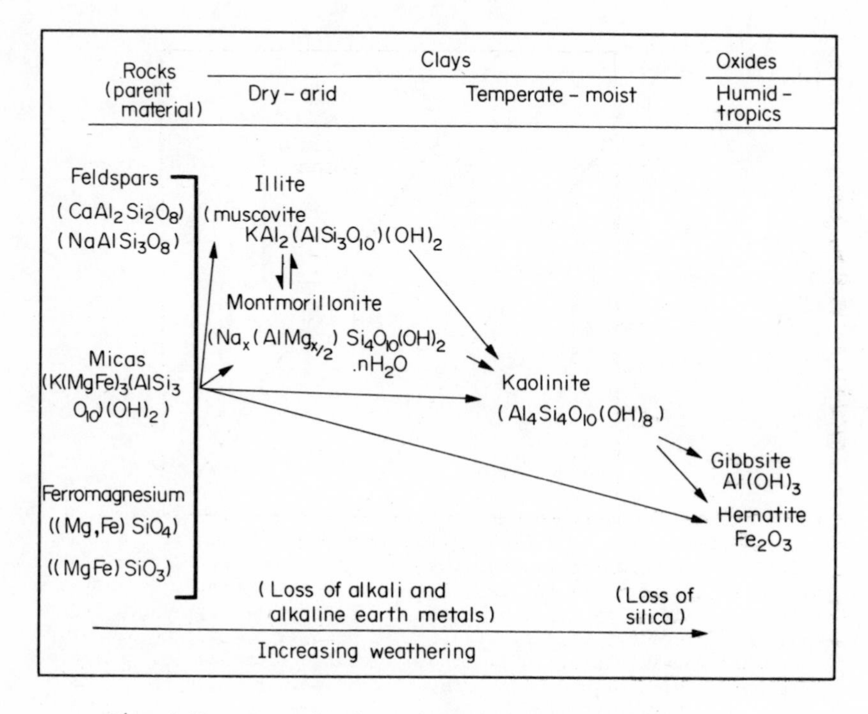

Fig. 2.11. A weathering scheme for rocks and clays

Substance	ΔG_f^o(kJ mol^{-1})	Substance	ΔG_f^o(kJ mol^{-1})
$KAlSi_3O_8$	-3582	$KAl_3Si_3O_{10}(OH)_2$	-5439
H_2O	-238	H_4SiO_4	-1255
		K^+	-280

Since $\log k = -\dfrac{\Delta G}{2.303RT}$, at 298 K we have $\log k = -\dfrac{43000}{2.303 \text{x} 298 \text{x} 8.284} = -7.6$

where $k = \dfrac{[K^+]^2[H_4SiO_4]^6}{[H^+]^2}$,

hence $$-7.6 = 2\log\frac{[K^+]}{[H^+]} + 6\log[H_4SiO_4]. \tag{2.79}$$

The plot of $\log([K^+]/[H^+])$ against $\log[H_4SiO_4]$ (Fig. 2.12) gives the boundary line between K-feldspar and mica. Similar equations can be derived for the other equilibria, for example the conversion of mica to gibbsite;

$$2KAl_3Si_3O_{10}(OH)_{2(s)} + 18H_2O + 2H^+_{(aq)} \rightleftarrows 3Al_2O_33H_2O_{(s)} + 6H_4SiO_{4(aq)} + 2K^+_{(aq)}. \tag{2.80}$$

$$\Delta G^o = -[2(-5439)+18(-238)] + [3(-2322)+6(-1255)+2(-280)] \tag{2.81}$$

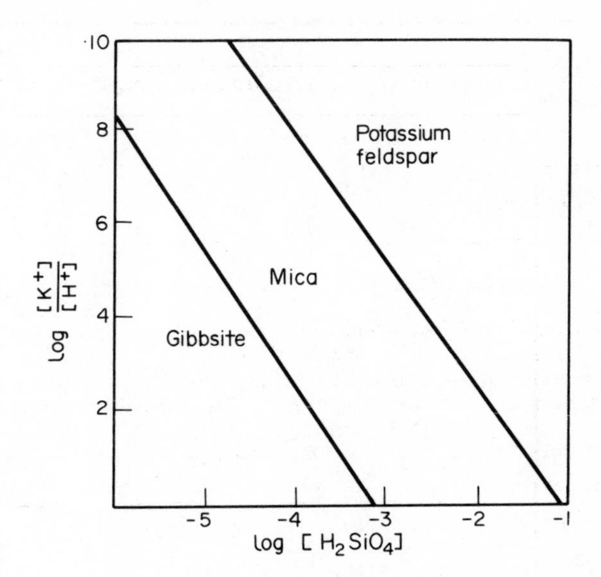

Fig. 2.12. Diagram of stability fields of the minerals potassium feldspar, mica and gibbsite, as a function of the activities of K^+, H^+ and H_4SiO_4.

i.e. ΔG^o = 106 kJ, and log k = -18.6.

$$\text{Therefore } -18.6 = 2\log\frac{[K^+]}{[H^+]} + 6\log[H_4SiO_4] \qquad (2.82)$$

If the composition of the weathering water, i.e. its $[K^+]/[H^+]$ ratio, and $[H_4SiO_4]$ are known, it is possible from a diagram such as Fig 2.12 to determine the stable mineral product.

Oxidation and Reduction. Another important chemical process in weathering is *oxidation*, and consequential *reduction*. This occurs where cations and anions have more than one accessible oxidation state,

e.g. Fe(II),(III); Mn(II),(IV); Cu(I),(II); S(-II),(IV),(VI).

Oxidation is the removal of electrons from a species, e.g.;

$$Fe^{2+} - e \rightarrow Fe^{3+}, \quad -E^o = -0.77 \text{ V}, \qquad (2.83)$$

$$SO_3^{2-} + H_2O - 2e \rightarrow SO_4^{2-} + 2H^+, \quad -E^o = -0.17 \text{ V}, \qquad (2.84)$$

and reduction is addition of electrons, i.e. the reverse reactions.

The magnitude and sign of the standard potentials, and often the pH of the reaction medium, are important in determining the course of a reaction. It is clear from the E^o values given above that the reaction;

$$2Fe^{3+} + SO_3^{2-} + H_2O \rightarrow 2Fe^{2+} + SO_4^{2-} + 2H^+, \quad E = +0.60 \text{ V} \qquad (2.85)$$

will proceed, particularly in a medium which removes the protons.

The weathering of a material such as fayalite, Fe_2SiO_4, may be represented by the following equations. The reaction:

$$Fe_2SiO_4 + 4H_2O + 4CO_2 \rightarrow 2Fe^{2+}_{(aq)} + 4HCO^{-}_{3(aq)} + H_4SiO_{4(aq)}, \quad (2.86)$$

releases the iron, and then oxidation occurs;

$$Fe^{2+}_{(aq)} - e \rightarrow Fe^{3+}_{(aq)}, \quad -E^{o} = -0.77\ V, \quad (2.87)$$

$$O_2 + 4H^{+} + 4e \rightarrow 2H_2O, \quad E^{o} = +1.23\ V, \quad (2.88)$$

i.e. $$4Fe^{2+}_{(aq)} + O_2 + 4H^{+} \rightarrow 4Fe^{3+}_{(aq)} + 2H_2O, \quad E = +0.46\ V \quad (2.89)$$

The ferric ion is readily hydrolysed to form insoluble $Fe(OH)_3$ and Fe_2O_3, the formation of which, assists in driving the reaction to the right.

Both iron and sulphur are oxidised when FeS_2 (pyrite and marcasite) is weathered.

$$2FeS_2 + 7O_2 + 2H_2O \rightarrow 2Fe^{2+}_{(aq)} + 4SO_4^{2-} + 4H^{+} \quad (2.90)$$

$$2Fe^{2+}_{(aq)} + \tfrac{1}{2}O_2 + 2H^{+} \rightarrow 2Fe^{3+}_{(aq)} + H_2O \quad (2.91)$$

$$2Fe^{3+}_{(aq)} + 6H_2O \rightarrow 2Fe(OH)_3 + 6H^{+} \quad (2.92)$$

i.e. $$4FeS_2 + 15O_2 + 14H_2O \rightarrow 4Fe(OH)_3 + 8SO_4^{2-} + 16H^{+}. \quad (2.93)$$

This system produces a strongly acid medium which will increase further by;

$$FeS_2 + 14Fe^{3+}_{(aq)} + 8H_2O \rightarrow 15Fe^{2+}_{(aq)} + 2SO_4^{2-} + 16H^{+}. \quad (2.94)$$

(i.e. $$Fe^{3+}_{(aq)} + e \rightarrow Fe^{2+}_{(aq)} \quad (2.95)$$

and $$S_2^{2-} + 8H_2O - 14e \rightarrow 2SO_4^{2-} + 16H^{+}). \quad (2.96)$$

The Fe^{2+} (equation 2.94) can then recycle through the above sequence (equation 2.91). Strongly acid mine waters are formed by the above reactions, and it has been estimated that in the USA 8 million tonnes of H_2SO_4 are produced annually in coal mines. Removal of oxygen, addition of bacteria that oxidise Fe^{2+} to Fe^{3+}, and neutralization of the acid (say with lime), have all been used to reduce the acidity.

Examples of other oxidative weathering processes are given below;

$$\underset{\text{(rhodonite)}}{MnSiO_3} + \tfrac{1}{2}O_2 + 2H_2O \rightarrow MnO_2 + H_4SiO_4, \quad (2.97)$$

and $$\underset{\text{(rhodochrosite)}}{MnCO_3} + \tfrac{1}{2}O_2 + H_2O \rightarrow MnO_2 + H_2CO_3. \quad (2.98)$$

The Mn^{2+} ion is first released, followed by its oxidation.

$$Mn^{2+} + 2H_2O - 2e \rightarrow MnO_2 + 4H^+, \quad -E^o = -1.23 \text{ V}. \tag{2.99}$$

$$O_2 + 4H^+ + 4e \rightarrow 2H_2O, \quad E^o = +1.23 \text{ V}. \tag{2.100}$$

$$2Mn^{2+} + 2H_2O + O_2 \rightarrow 2MnO_2 + 4H^+, \quad E = 0\text{V}. \tag{2.101}$$

Sulphide undergoes oxidation, for example from PbS;

$$\underset{\text{(galena)}}{PbS} + 2H^+ + 2HCO_3^- \rightarrow Pb^{2+} + 2HCO_3^- + H_2S. \tag{2.102}$$

This reaction releases Pb^{2+} from an ore body close to the earth's surface, and the hydrogen sulphide is then oxidised.

$$H_2S + 4H_2O - 8e \rightarrow SO_4^{2-} + 10H^+. \tag{2.103}$$

$$2O_2 + 8H^+ + 8e \rightarrow 4H_2O. \tag{2.104}$$

$$Pb^{2+} + SO_4^{2-} \rightarrow PbSO_{4(s)}. \tag{2.105}$$

Addition of equations (2.102) to (2.105) gives;

$$PbS + 2O_2 \rightarrow PbSO_4, \tag{2.106}$$

which has a favourable free energy change of -719 kJ ($\Delta G_f^o(PbSO_4) - \Delta G_f^o(PbS)$ = - 811 - (-92) = 719 kJ.

E_h - pH diagrams. Data on redox reactions and pH dependent reactions can be summarised by plotting the potential of a redox reaction (E_h) against the pH of the solution. The reduction potential of the reaction;

$$\text{oxidised state} + ne \rightarrow \text{reduced state}, \tag{2.107}$$

is given by the Nernst equation;

$$E_h = E^o - \frac{RT}{nF} \ln \frac{[\text{reduced state}]}{[\text{oxidised state}]}. \tag{2.108}$$

This expression relates E_h and pH, as often the $[H^+]$ occurs in the $\ln\frac{[\text{red. state}]}{[\text{ox. state}]}$ expression.

The principal areas of the diagram are illustrated in Fig. 2.13a, and the three types of E_h - pH lines are given in Fig. 2.13b. Line (a) represents a situation where no oxidation/reduction occurs, i.e. an E_h independent function. Line (b) is a pH independent function, corresponding to a redox reaction which does not involve protons and line (c) represents a situation which is both E_h and pH dependent.

In the natural environment the limiting factor is the stability of water towards oxidation and reduction. Water is oxidised according to the reaction;

$$2H_2O \rightleftharpoons O_{2(g)} + 4H^+ + 4e \tag{2.109}$$

Since, in this treatment, all reactions will be written as reductions, we have;

$$O_{2(g)} + 4H^+ + 4e \rightleftharpoons 2H_2O, \quad E^o = 1.23 \text{ V}. \tag{2.110}$$

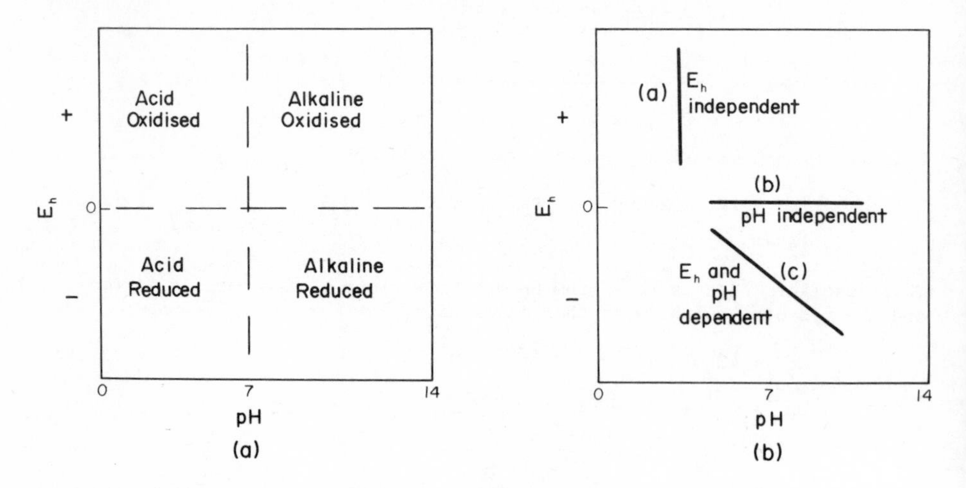

Fig. 2.13. Principal features of an E_h - pH diagram.

Hence, $$E_h = E^o - \frac{RT}{4F} \ln \frac{[H_2O]^2}{p_{O_2}[H^+]^4} , \qquad (2.111)$$

and for p_{O_2} = 1 atms. in an aqueous medium;

$$E_h = 1.23 - \frac{0.059}{4} \log \frac{1}{[H^+]^4} \qquad (2.112)$$

i.e. $$E_h = 1.23 - 0.059 \text{ pH}. \qquad (2.113)$$

The reduction of water is given by the equation;

$$2H^+ + 2e \rightleftharpoons H_{2(g)}, \; E^o = 0, \qquad (2.114)$$

(the protons coming from $H_2O \rightleftharpoons H^+ + OH^-$).

Hence, $$E_h = E^o - \frac{0.059}{2} \log \frac{p_{H_2}}{[H^+]^2} . \qquad (2.115)$$

For a hydrogen pressure of 1 atmosphere we have;

$$E_h = -0.059 \text{ pH}. \qquad (2.116)$$

Both these lines (equations (2.113) and (2.116)) are the dashed lines on Fig. 2.14. The area bounded by the lines and pH 4 to 10 represents the natural aqueous environment.

We will consider as an example the Fe/H_2O system, containing the species Fe, Fe^{2+}, Fe^{3+}, $Fe(OH)_2$ and $Fe(OH)_3$. There are seven equilibria inter-relating these species, which are listed in Table 2.10. For each equation a relationship can be derived in terms of E_h and pH, which may then be plotted on an E_h - pH diagram. We will consider three of the equations (1,2 and 7 in Table 2.10).

TABLE 2.10 Fe/H_2O System

	Chemical Equation	E^o or log K	E_h - pH equation	Dependency
1.	$Fe^{2+}_{(aq)} + 2e \rightleftharpoons Fe_{(s)}$	-0.44 V	$E_h = -0.44 + 0.0295 \log a_{Fe^{2+}}$	E_h
2.	$Fe(OH)_{2(s)} + 2H^+ \rightleftharpoons Fe^{2+}_{(aq)} + 2H_2O$	12.9	$\log a_{Fe^{2+}} = 12.9 - 2\ pH$	pH
3.	$Fe^{3+}_{(aq)} + e \rightleftharpoons Fe^{2+}_{(aq)}$	+0.77 V	$E_h = 0.77 - 0.059 \log \frac{a_{Fe^{2+}}}{a_{Fe^{3+}}}$	E_h
4.	$Fe(OH)_{3(s)} + 3H^+ \rightleftharpoons Fe^{3+}_{(aq)} + 3H_2O$	3.9	$\log a_{Fe^{3+}} = 3.9 - 3\ pH$	pH
5.	$Fe(OH)_{2(s)} + 2H^+ + 2e \rightleftharpoons Fe_{(s)} + 2H_2O$	-0.047 V	$E_h = -0.047 - 0.059\ pH$	E_h, pH
6.	$Fe(OH)_{3(s)} + H^+ + e \rightleftharpoons Fe(OH)_{2(s)} + H_2O$	+0.27 V	$E_h = 0.27 - 0.059\ pH$	E_h, pH
7.	$Fe(OH)_{3(s)} + 3H^+ + e \rightleftharpoons Fe^{2+}_{(aq)} + 3H_2O$	+1.06 V	$E_h = 1.06 - 0.177\ pH - 0.059 \log a_{Fe^{2+}}$	E_h, pH

$$Fe^{2+}_{(aq)} + 2e \rightleftharpoons Fe_{(s)}, \quad E^o = -0.44 \text{ V. (Equation 1)} \tag{2.117}$$

The Nernst equation for this equilibrium is;

$$E_h = E^o - \frac{RT}{nF} \ln \frac{1}{a_{Fe^{2+}}}, \tag{2.118}$$

i.e.
$$E_h = E^o - \frac{0.059}{2} \log \frac{1}{a_{Fe^{2+}}} \tag{2.119}$$

i.e.
$$E_h = -0.44 + 0.0295 \log a_{Fe^{2+}}. \tag{2.120}$$

This equation, for a particular value of $a_{Fe^{2+}}$, is a straight line parallel to the pH axis, i.e. independent of pH. For $a_{Fe^{2+}} = 1 \times 10^{-5}$ mole l^{-1} we get;

$$E_h = -0.59. \quad \text{(line 1, Fig. 2.14).} \tag{2.121}$$

$$Fe(OH)_{2(s)} + 2H^+ \rightleftharpoons Fe^{2+}_{(aq)} + 2H_2O. \text{ (Equation 2)} \tag{2.122}$$

$$K = \frac{a_{Fe^{2+}}}{(a_{H^+})^2} = 7.9 \times 10^{12}, \tag{2.123}$$

hence
$$\log K = \log a_{Fe^{2+}} - 2 \log a_{H^+} \tag{2.124}$$

i.e.
$$12.9 = \log a_{Fe^{2+}} + 2pH \tag{2.125}$$

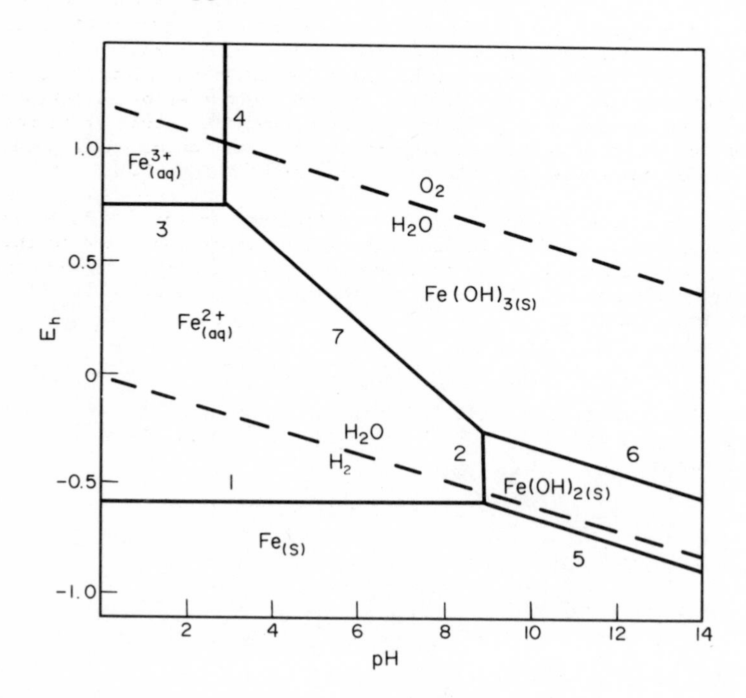

Fig. 2.14. Fe - H_2O, E_h - pH diagram.

This equation is a line parallel to the E_h axis, and for $a_{Fe^{2+}} = 1 \times 10^{-5}$ mol l^{-1};

$$pH = \frac{12.9 + 5}{2} = 8.9. \quad \text{(line 2, Fig. 2.14)}. \tag{2.126}$$

$$Fe(OH)_{3(s)} + 3H^+ + e \rightleftharpoons Fe^{2+} + 3H_2O, \quad E^o = +1.06 \text{ V. (Equation 7)}. \tag{2.127}$$

The Nernst equation is;

$$E_h = E^o - \frac{RT}{nF} \ln \frac{a_{Fe^{2+}}}{(a_{H^+})^3}, \tag{2.128}$$

i.e. $$E_h = 1.06 - 0.059 \log \frac{a_{Fe^{2+}}}{(a_{H^+})^3}, \tag{2.129}$$

i.e. $$E_h = 1.06 - 0.059 \log a_{Fe^{2+}} - 0.177 \text{ pH} \tag{2.130}$$

For $a_{Fe^{2+}} = 1 \times 10^{-5}$ mole l^{-1}, we have $E_h = 1.355 - 0.177$ pH, (2.131)

both E_h and pH dependent (line 7, Fig. 2.14).

When the seven lines are plotted in Fig. 2.14 the areas in which each species exists becomes defined. If the concentration of iron is different from 1×10^{-5} mole l^{-1} the lines dependent on this concentration will be in a different but parallel position. The diagram reveals that metallic iron does not occur naturally in the environment, as its region of stability lies below the water reduction line, i.e. water will reduce before either Fe^{2+} or $Fe(OH)_2$ are reduced to iron. Putting it another way, metallic iron will oxidise in the natural environment, which is called the corrosion (rusting) of iron. The occurrence of Fe^{3+} depends on having a low pH, i.e. $\leqslant 3$, which rules out its existence in natural waters. The E_h - pH properties of different natural water systems have been investigated, and from these it is possible to determine which iron species will exist in a particular environment. Some data for environmental waters are presented in Fig. 2.15.

The iron system described in Fig. 2.14 is limited, in that it does not include iron minerals found in rocks and soils; it is more akin to the situation found in the hydrosphere. However, iron minerals can be considered in exactly the same way. For example, the magnetite-hematite equilibrium is given by the equation;

$$3Fe_2O_{3(s)} + 2H^+_{(aq)} + 2e \rightleftharpoons 2Fe_3O_{4(s)} + H_2O \tag{2.132}$$

The reduction potential for the reaction can be determined from free energy data, viz. Fe_2O_3, $G^o_f = -741$ kJ mol^{-1}; Fe_3O_4, $G^o_f = -1014$ kJ mol^{-1} and H_2O, $G^o_f = -237$ kJ mol^{-1}.

Hence, $$\Delta G^o = 2 \times (-1014) - 237 - (3 \times (-741)) = 42 \text{ kJ}. \tag{2.133}$$

Since, $$E^o = -\frac{\Delta G}{nF}, \tag{2.134}$$

we have, $$E^o = -\frac{-42 \times 10^3}{2 \times 96.5 \times 10^3} \text{ volts} = 0.22 \text{ volts}. \tag{2.135}$$

Now, $$E_h = E^o - \frac{0.059}{2} \log \frac{1}{(a_{H^+})^2} \tag{2.136}$$

hence, $$E_h = 0.22 - 0.059 \text{ pH}. \tag{2.137}$$

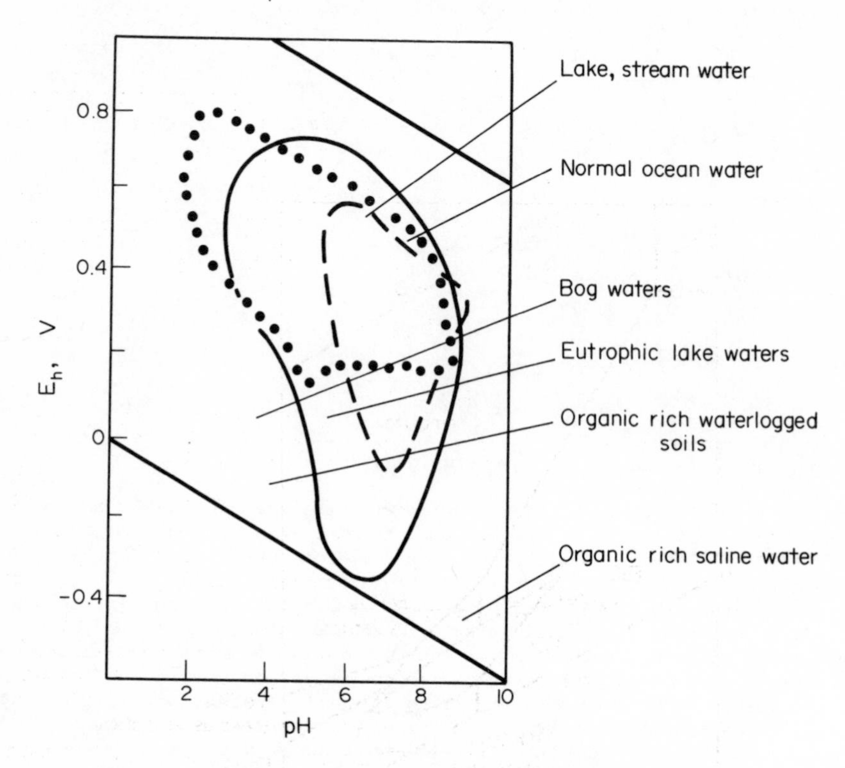

Fig. 2.15. The approximate E_h - pH values for some environmental waters,— soil, water, --- subsurface waters, ···· oxidised mine waters.

An E_h - pH diagram showing the stability relationship of iron oxides, sulphides and carbonate in water is given in Fig. 2.16, while systems for manganese (Fig. 2.17) and lead (Fig. 2.18) are also presented.

We have touched briefly on some aspects of the chemistry of the natural world in this chapter, without attempting to be exhaustive in coverage. In later chapters we will investigate some aspects further, particularly in relation to the influence of chemicals added to the environment through the activity of man.

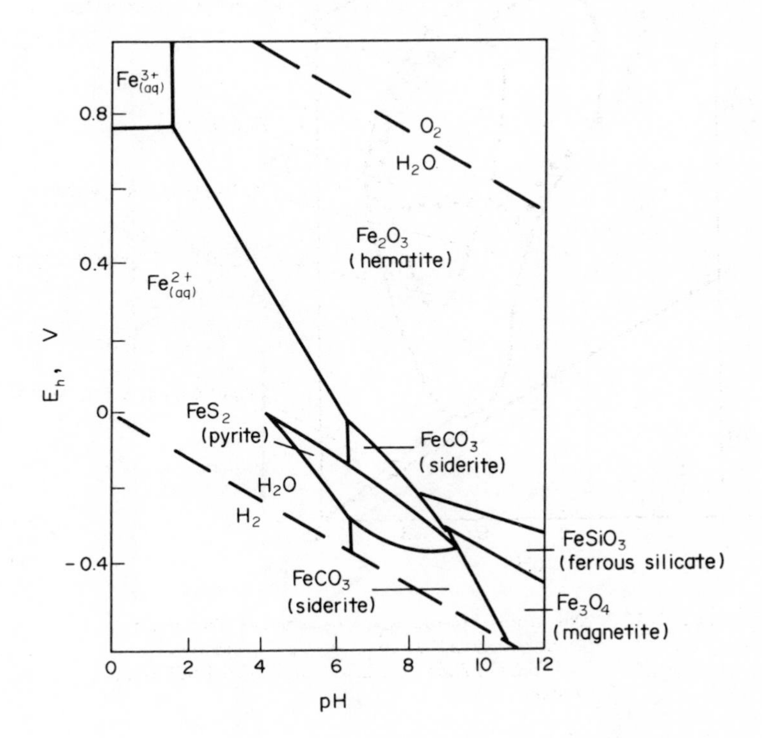

Fig. 2.16. E_h-pH diagram for FeS_2, $FeCO_3$ and Fe^{n+}(aq). Source: Garrels, R.M. (1960) Mineral Equilibria, Harper and Brothers, New York. Garrels, R.M. and Christ, C.L. (1965) Solutions, Minerals and Equilibria, Harper, and Row, New York.

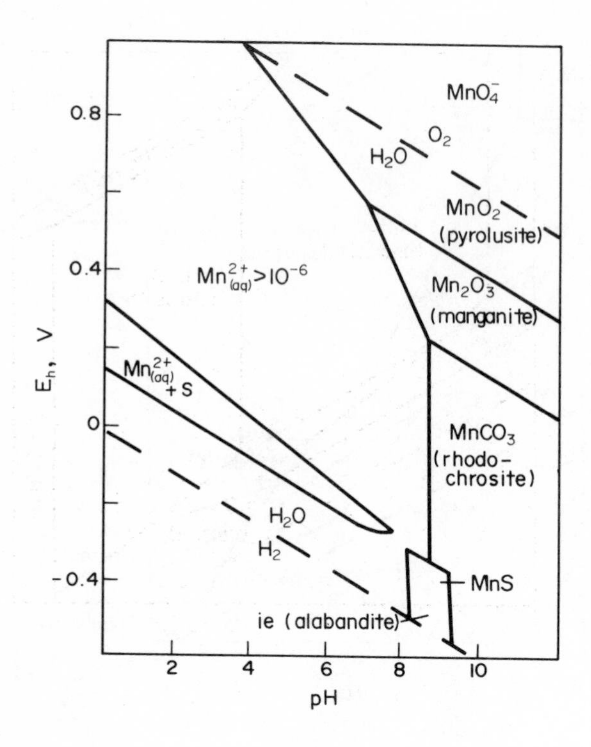

Fig. 2.17. E_h-pH diagram for some manganese species. Source: Garrels, R.M. (1960) Mineral Equilibria, Harper and Brothers, New York. Garrels, R.M. and Christ, C.L. (1965) Solutions, Minerals and Equilibria, Harper and Row, New York.

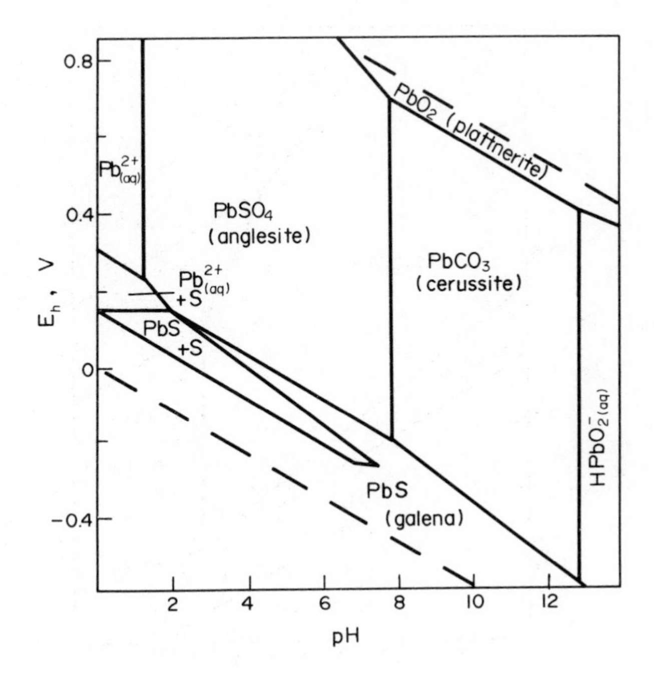

Fig. 2.18. E_h-pH diagram for some lead species. Source: Garrels, R.M. (1960) Mineral Equilibria, Harper and Brothers, New York. Garrels, R.M. and Christ, C.L. (1965) Solutions, Minerals and Equilibria, Harper, and Row, New York.

PART II

Resources

CHAPTER 3

Mineral Resources

In this chapter we will consider the mineral resources of the earth. We will refer to the formation of minerals and the principal methods of extraction of the elements from their ores.

A relatively common article, a telephone hand-set, is made up of components containing up to 40 chemical elements (Table 3.1). This highlights the diverse uses that chemical elements can be put to. If we are to continue to use the earth's resources this will depend on our knowledge of the amounts available, the technology of extraction, and of how to best make use of the material without wastage and with maximum recycling.

RESOURCES

It is of interest that the nine most abundant metals also have the greatest production figures (Table 3.2). More effort is being given to the investigation of substitutes for scarce elements with technological uses. For example a list of some substitutes for mercury, a scarce metal, is given in Table 3.3. Numerous attempts have been made to estimate the world's mineral reserves, and their expected life-times based on some model of consumption. The estimates range from pessimistic to optimistic. Estimated reserves and projected life-times for some elements are listed in Table 3.4. The data are subject to large errors and variability. The estimates of known reserves (column 2) vary with the method of estimation, and often with time as new resources are discovered, or technical developments enable a metal to be extracted economically from low concentration ores. The projected life-times are also conjecture and are only a guide. If, for example, a new alloy is developed which reduces the use of certain metals then the life-time of the metals may be significantly increased.

The geographic distribution of the minerals is well established but the figures of the percentage of the total reserves suffer the same limitations as for the total reserves. The situation changes, for example during the last century 40% of the world's copper, and significant quantities of tin and lead was produced in the United Kingdom, but since then other sources have been discovered and exploited and the prominence of the United kingdom has declined. The geographical situation will continue to change as reserves are worked out in certain areas.

TABLE 3.1 Elements in the telephone handset

Element	How used
Aluminium	Metal alloy in dial mechanism, transmitter, and receiver
Antimony	Alloy in dial mechanism
Arsenic	Alloy in dial mechanism
Beryllium	Alloy in dial mechanism
Bismuth	Alloy in dial mechanism
Boron	Touch-Tone dial mechanism
Cadmium	Colour in yellow plastic housing
Calcium	In lubricant for moving parts
Carbon	Plastic housing, transmitter steel parts
Chlorine	Wire insulation
Chromium	Colour in green plastic housing, metal plating, stainless steel parts
Cobalt	Magnetic material in receiver
Copper	Wire, plating, brass piece parts
Fluorine	Plastic piece parts
Germanium	Transistors in Touch-Tone dial mechanism
Gold	Electrical contacts
Hydrogen	Plastic housing, wire insulation
Indium	Touch-Tone dial mechanism
Iron	Steel, magnetic materials
Krypton	Ringer in Touch-Tone set
Lead	Solder in connections
Lithium	In lubricant for moving parts
Magnesium	Die castings in transmitter, ringer
Manganese	Steel in piece parts
Mercury	Colour in read plastic housing
Molybdenum	Magnet in receiver
Nickel	Magnet in receiver, stainless steel parts
Nitrogen	Hardened heat-treated piece parts
Oxygen	Plastic housing, wire insulation
Palladium	Electrical contacts
Phosphorus	Steel in piece parts
Platinum	Electrical contacts
Silicon	Touch-Tone dial mechanism
Silver	Plating
Sodium	In lubricant for moving parts
Sulphur	Steel in piece parts
Tantalum	Integrated circuit in Trimline set
Tin	Solder in connections, plating
Titanium	Colour in white plastic housing
Tungsten	Lights in Princess and key sets
Vanadium	Receiver
Zinc	Brass, die casting in transmitter, ringer

Source, A.C. Chynoweth, *Science*, 191, 726 (1976)

The geographical distribution has geopolitical implications (Table 3.5). For example 51% of the world's aluminium, 40% of the copper and 46% of the nickel are located in so-called "Third World" countries, while 56% of the world's zinc, 87% of the uranium and 61% of the lead are located in the "Western" nations. The unequal distribution of the world's resources, raises questions concerning the availability of resources, irrespective of the reserves.

TABLE 3.2 Production and Reserves of the Nine Most Abundant Metals

Metal	Reserves x 10^6 tonnes	Rank	Production x 10^6 tonnes (1973)	Rank
Fe	109,000	1	475	1
Al	2,950	2	10	2
Mg	2,090	3	4	6=
Mn	800	4	9	3
Cr	780	5	2	9
Ti	440	6	4	6=
Cu	310	7	6	4
Zn	120	8	5	5
Pb	85	9	4	6=

TABLE 3.3 Substitutes for Mercury

Uses	Substitute	Consumption of Hg in USA 1964-73 %
$NaOH/Cl_2$ production	Diaphragm cell	34.0
Electrical (esp. batteries)	$Zn-MnO_2$-graphite dry cell	26.5
Biocidal paints	Plastic paints, copper oxide paint	13.2
Dental amalgams	Metal powders, porcelain, plastic	4.4
Agriculture[a]	Organic fungicides	3.9
Catalysts	Ethylene process for PVC manufacture	2.4

[a] The use of methyl mercury on seeds has been banned. Source: H.E. Goeller and A.M. Weinberg, *Science*, 191, 683 (1976).

The consumption of metals has steadily increased, as portrayed by the graphical data in Fig. 3.1 for the United States of America. Consumption has been more rapid than population growth, reflecting a growth in industrialisation and consumerism. The annual consumption per capita is 3 to 6 times greater in industrialised nations, such as the United Kingdom (Table 3.6), compared with the world as a whole. The same scene occurs, but with a different set of figures, when comparing mine production and consumption of metals such as copper, zinc and lead (Table 3.7). In addition to an uneven distribution of the world's mineral resources there is an unequal geographic distribution of consumption.

TABLE 3.4 Reserves, Lifetimes and Location of Principal Mineral Resources

Mineral Resources	Known Reserves[1] x10[6] tonnes	Projected Lifetime[2] year A.D. Current Trends	Static	Countries with Major Reserves (% total)	Principal Chemical Source
1. Structural Metal					
Iron[3]	109,000	2060	2210	USSR(33), Sth America(18) Canada(14)	hematite(Fe_2O_3), goethite($HFeO_2$) magnetite(Fe_3O_4), siderite($FeCO_3$)
2. Base Metals					
Copper	310[4]	1990	2010	USA(24), Chile(17), Canada(9) USSR(10), Zambia(8)	chalcopyrite($CuFeS_2$), chalcocite(Cu_2S)
Zinc	120	1990	1995	Canada(27), USA(23), USSR(11)	sphalerite(ZnS)
Lead	85	1990	1995	USA(38), Canada(13), Australia(11)	galena(PbS)
Tin	4.5	soon		Thailand(33) Malaysia(14)	cassiterite(SnO_2)
3. Precious Metals					
Silver	0.007	soon		Cent. Planned Econ.(36), USA(24)	silver sulphide(Ag_2S)
Platinum metals	0.004	2015	2100	Sth Africa(47), USSR(47)	platinum ores
Gold	0.0004	soon		Sth Africa(40), USSR(14)	gold(Au), gold telluride (AuTe)
4. Light Metals					
Magnesium[3]	2090		2500		seawater(Mg^{2+}), magnesite($MgCO_3$)

Aluminium[3]	2950	2000	2070	Australia(34),Guinea(34) Surinam(11),Jamaica(5)	bauxite(Al_2O_3)
Titanium[3]	440			Canada	ilmenite($FeTiO_3$),rutile(TiO_2)
5. Ferroalloys					
Manganese[3]	800[4]	2015	2065	Sth Africa(38),USSR(25)	pyrolusite(MnO_2), psilomelane($Mn_2O_32H_2O$)
Chromium[3]	780	2065	2390	Sth Africa(75)	chromite($(Mg,Fe)_2CrO_4$)
Nickel	70	2025	2120	Cuba(24),New Caledonia(22) USSR(13),Canada(13)	pentlandite($(Ni,Fe)_9S_8$), garnierite($H_4Ni_3S_{12}O_9$)
Molybdenum	5	2005	2050	USA(58),USSR(20)	molybdenite(MoS_2)
Cobalt	2.5	2030	2080	Congo(31),Zambia(16)	cobalt sulphides and arsenides
Tungsten	1.5	2000	2010	China(73)	scheelite($CaWO_4$), wolframite($FeWO_4$)
6. Nuclear					
Uranium	1[5]		1995	USA(30),Australia(23), Sth Africa(17),Canada(13)	uranite(UO_2,
7. Electronics					
Mercury	1.1		soon	Spain(30),Italy(21)	cinnabar(HgS)
Indium			2000	Canada,USSR,Peru,Japan, India,Brazil,Malagasy	

[1] These figures vary with different estimates, and change almost yearly; [2] Lifetimes approximate; [3] Abundant elements; [4] Large reserves of manganese and copper occur in manganese nodules on the ocean sea-bed; [5] For non-communist countries. Sources: Numerous sources including U.N. Statistical Yearbooks; Warren, K. (1973) Mineral Resources, Penguin Books, Harmondsworth; Meadows, D.H., Meadows, D.K., Randers, J. and Behrens III, W.W. (1972) The Limits of Growth, Universe Books, New York; Robinson, F.A., Chem. Soc. Revs.,(London) 317 (1976); Sugden, T.M. and Briffa, F.E., Chem. and Ind., 669 (1975).

TABLE 3.5 World Production of Al, Cu, Ni, Zn, Pb, U. (1971)

Aluminium (2950 x 10^6 tonnes)

Country	Amount[a]	%
Australia	1,000	34
Guinea	1,000	34
Surinam	323	11
Jamaica	153	5.2
USSR	76	2.6
Guyana	20	0.7
USA	12	0.4

Nickel (68 x 10^6 tonnes)

Country	Amount[a]	%
Cuba	16	23.5
N. Caledonia	15	22.0
Canada	9	13.2
USSR	9	13.2
Australia	1	1.5

Uranium (1.080 x 10^6 tonnes) (non-communist)

Country	Amount[a]	%
USA	.320	29.6
Australia	.243	22.5
Sth Africa	.186	17.2
Canada	.144	13.3
Niger	.040	3.7
France	.037	3.4
Algeria	.028	2.6
Gabon	.020	1.9
Spain	.010	0.9

Copper (310 x 10^6 tonnes)

Country	Amount[a]	%
USA	74	23.9
Chile	51	16.5
USSR	32	10.4
Canada	27	8.7
Zambia	25	8.1
Peru	20	6.5
Zaire	18	5.8
Phillipines	9	2.9
Rest (W. Germany Japan, U.K.)	47	15.2

Zinc (120 x 10^6 tonnes)

Country	Amount[a]	%
Canada	31	25.8
USA	27	22.5
Australia	9	7.5
Peru	7	5.8
Mexico	4	3.3
Communist	18	15.0
Other Non-Communist	21	17.5

Lead (85 x 10^6 tonnes)

Country	Amount[a]	%
USA	32	37.6
Canada	10.8	12.7
Australia	9	10.6
USSR	4.5	5.3
Peru	4.5	5.3

[a] Amount x 10^6 tonnes. Source: Various, especially U.N. statistical Yearbooks

TABLE 3.6 Annual Consumption of Metals - tonnes per capita (1970)

Metal	World	U.K.	Factor[a]
Steel	0.17	0.5-1	3-6
Aluminium	0.0025	0.007-0.01	3-4
Copper	0.002	0.01	5
Lead and Zinc	0.0025	0.01	4

[a] Ratio, U.K. consumption/World consumption. Source: D.S. Davies, *Chemtech*, 1974, 135.

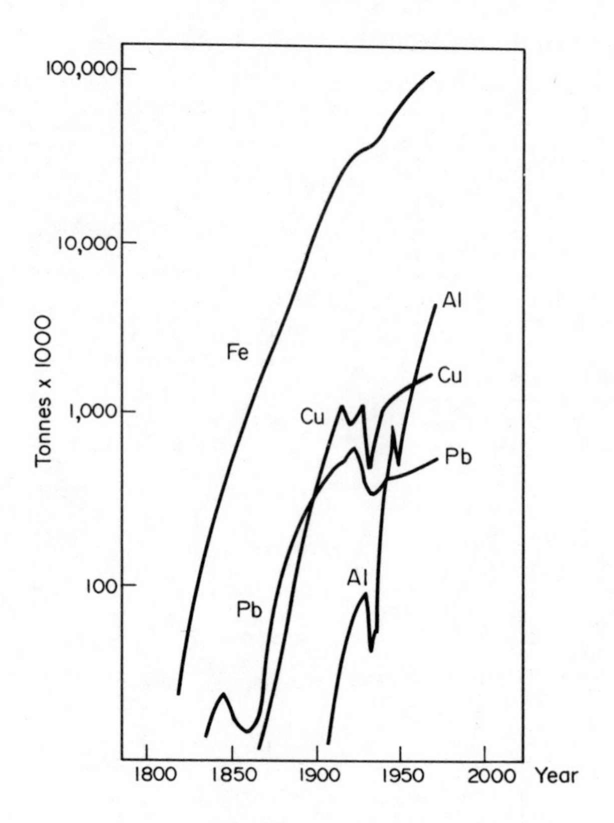

Fig. 3.1. Growth in consumption of some metals in the United States of America. Source: S.V. Radcliffe, *Science*, 191, 700 (1976).

TABLE 3.7 Production and Consumption of Cu, Pb and Zn (% of total)

Country	Cu		Pb		Zn	
	Production	Consumption	Production	Consumption	Production	Consumption
USA	25	29	22	27	11	29
USSR	15	13	30[a]	26[a]	32[a]	31[a]
W. Europe	3	38	15	36	17	33
Canada	11	4	11	2	29	4
Japan	3	9	2	6		
Australia	3	1	14	2	11	3
Chile	14	2				
Zambia	13	-				
Zaire	6	-				

[a] Also includes E. Europe. Sources: A.K. Barbour, *Chemistry in Britain*, 13, 213 (1977); S.E. Kesler, Our Finite Mineral Resources, McGraw-Hill 1976.

Limitations on Mineral Resources

In addition to limitation of reserves and geopolitical factors, the quality of the ore and the energy required to extract the metal are of import. The amount of an element in a rock type can be depicted on a percent frequency versus concentration curve, which usually is log normal shape and positively skewed. In Fig. 3.2 this effect is shown for the distribution of magnesium in a series of granites. The curve shape is the result of a few samples with high Mg content. Material from which a metal can be obtained economically lie to the far right of the curve. The cut-off grades for the extraction of a number of elements listed in Table 3.8 indicate that minimum levels vary with metal. The ratio depends on the difficulty of extracting an element from its ore, and the abundance of high grade ores.

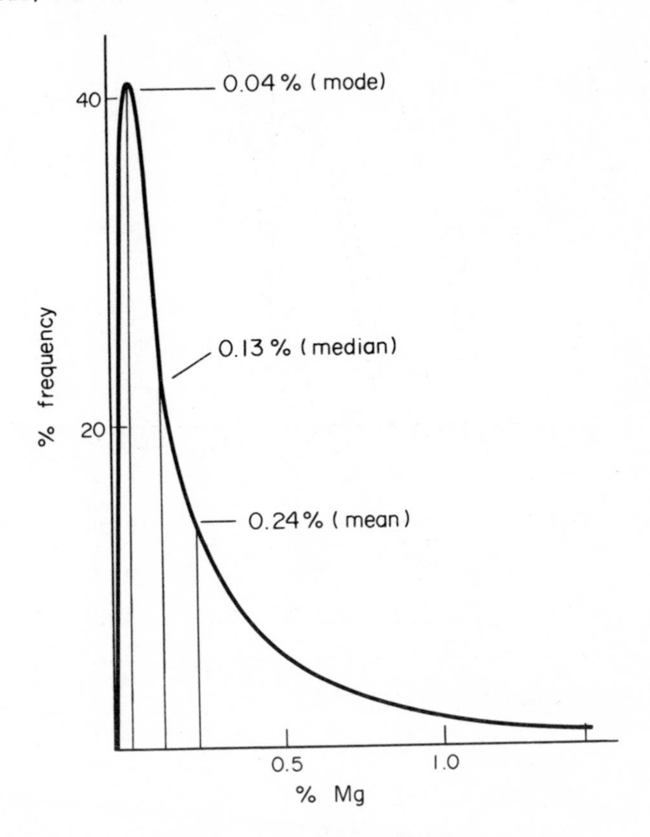

Fig. 3.2. Percent frequency/concentration plot for magnesium in granites. Source: L.H. Ahrens, Distribution of the Elements in our Planet, McGraw-Hill, N.Y., 1965.

TABLE 3.8 Crustal Abundance and Extraction Cut-Off Grade for Selected Elements

Element	Crustal Abundance ppm	Cut-Off Grade ppm	Ratio Cut-Off Grade / Crustal Abundance
Mercury	0.089	1,000	11,200
Lead	12	40,000	3,300
Chromium	110	230,000	2,100
Tin	1.7	3,500	2,000
Zinc	94	35,000	370
Uranium	1.7	700	350
Manganese	1,300	250,000	190
Nickel	89	9,000	100
Cobalt	25	2,000	80
Copper	63	3,500	56
Titanium	6,400	100,000	16
Iron	58,000	200,000	3.4
Aluminium	83,000	185,000	2.2

Source: E. Cook, *Science*, 191, 677 (1976).

An exponential relationship exists between the percent of metal in an ore and the amount of ore required to obtain the metal (Fig. 3.3). A major limiting factor in extracting from low grade ores is the amount of material to be handled. Figures for copper, available for the Cuajone Mine in Peru (Table 3.9) demonstrates that a state of diminishing returns is quickly reached as

TABLE 3.9 Copper Ore Grade, Total Ore and Copper Content and Energy Required to Mine and Mill at Cuajore Mine, Peru

Average Composition (% Cu)	Ore		Copper Content		Energy to mine ore / Energy to mine 1% ore
	$Mg^a \times 10^3$	%	$Mg^a \times 10^3$	%	
1.32	18,140	1	239	5	0.76
0.99	390,090	27	3862	87	1.01
0.32	92,530	6	296	7	3.13
<0.20	958,900	66	60	1	16,000

[a] $Mg = 10^6$ grams. Source: E. Cook, *Science*, 191, 677 (1976).

the grade of ore drops. Around 87% of the copper is accounted for by the ore containing 0.99% Cu or more, and only a further 7% Cu is obtained using ores down to 0.32% Cu. The last 1% of copper, obtainable from ores < 0.2% Cu, requires the handling of around 66% of the total ore body. The poorer the

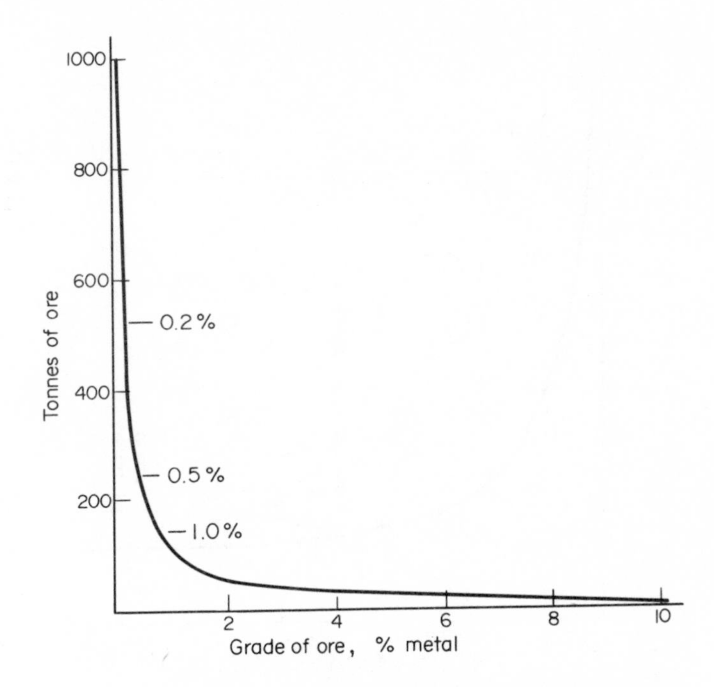

Fig. 3.3. Tonnes of ore required to produce one tonne of metal at different grades of ore.

grade of ore the greater the proportion of energy used in the mining and concentration stages (Table 3.10). The same information is given in Fig. 3.4, which shows the equivalent coal energy required to extract Cu from a number of different grades of copper sulphide ores. Attempts to work anything less rich than 0.35% Cu, lies on the steeply rising part of the energy/concentration curve.

TABLE 3.10 Energy Inputs in Mining and Concentrating Selected Ores

Metal	Ore Grade (% metal)	Mg[a] ore per Mg metal	Energy per Mg metal (GJ)[b]		% of total energy used
			Mining	Concentration	
Iron	30	3.3	0.6	2.1	11
Zinc	10	10	5.6	5.5	7
Lead	10	10	4.5	4.7	30
Copper	0.7	143	22.8	44.6	57
Uranium	0.2	500	351	260	58

[a] Mg = 10^6 grams, [b] GJ = 10^9 joules. Source: E.T. Hayes, *Science*, 191, 661 (1976).

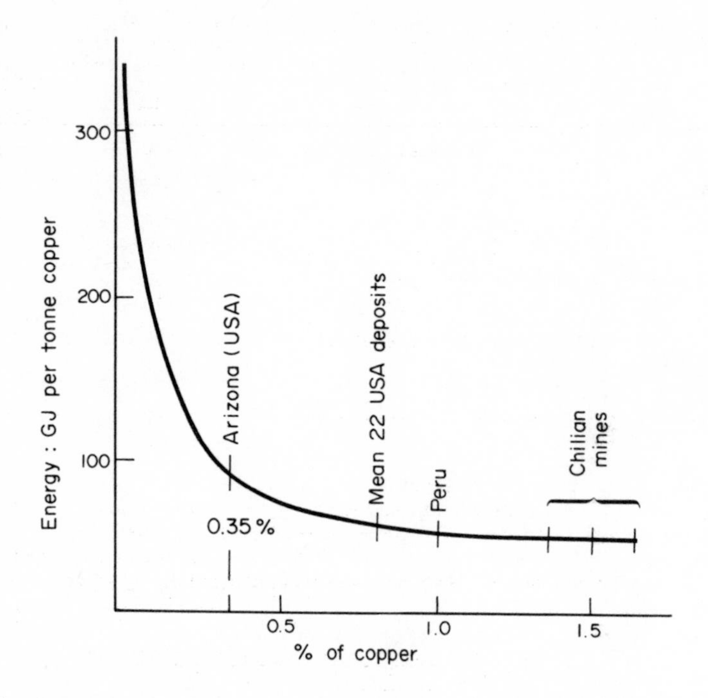

Fig. 3.4. Energy required per tonne of copper produced from different grades of copper sulphide ores. Source: E. Cook, *Science*, 191, 677 (1976).

Technological developments correlate with a fall in the quality of an ore that can be successfully used. For example, in Roman times copper ores had to contain around 10% Cu in order to extract the metal. Because of this the Romans were concerned at the declining supply of copper. By 1900 the copper ore grade which could be handled was 3-4% Cu, and this has steadily diminished until ores with around 0.35% Cu are being mined today.

Numerous attempts at providing comparative information on the energy requirements for the production of different metals have been made. The data given in Table 3.11 are examples, but must be considered only as a guide to the relative magnitude of the energy requirements. The energy depends on the grade of ore, the final product, as well as local mining conditions. The data in Table 3.12 lists the different energy requirements depending on the type of ore, and it becomes clear why bauxite is preferred over clay for the production of aluminium. The data in Tables 3.11 and 3.12 allowing for their limitations, are however, a guide for the future, as regards the best areas to concentrate research endeavour in order to achieve the best materials for the minimum energy.

One further limiting factor on the exploitation of mineral resources is the effect on the environment. The problems associated with mining large quantities of low grade ores, and the disposal of corresponding large quantities of waste are but two of many environmental problems that may arise in the future. These problems will only be resolved by balancing the material needs of mankind and environmental quality.

TABLE 3.11 Energy Requirements for Selected Metals

Metal	Product	Energy Required GJ Mg^{-1}	Energy Relative to Steel Slabs
Titanium	metal	430	17.2
Magnesium	metal	378	15.1
Aluminium	ingot	257	10.3
	rolled	219	8.8
Copper	rolled	67	2.7
	refined	118	4.7
Chromium	low-carbon ferroalloy	136	5.4
Zinc	ingot	69	2.8
Manganese	electric furnace	55	2.2
Lead	ingot	28	1.1
Steel	slabs	25	1
	carbon steel castings	44	1.8

Sources: E.T. Hayes, *Science*, 191, 661 (1976); Energy Resources and the Environment, Eds. J. Lenihan and W.W. Fletcher, Blackie, Glasgow.

TABLE 3.12 Energy Requirements for Selected Metals Related to Source

Metal	Source	Energy required GJ Mg^{-1}
Titanium ingot	rutile	500
	ilmenite	593
	Ti-rich soils	817
Magnesium ingot	seawater	360
Aluminium ingot	bauxite	202
	clay	261
Copper (refined)	porphyry ore (1% Cu)	50
	porphyry ore (0.3% Cu)	98
Steel (raw)	magnetic taconites	36
	iron laterites	43

Source: H.E. Goeller & A.M. Weinberg, *Science*, 191, 683 (1976).

FORMATION OF ORES

We will consider in more detail some of the ways ore bodies have formed in the earth's crust.

Occurrence of the Elements in Nature

A broad classification of the principal ore types that occur in nature, for each element, is given in Table 1.1. It must be noted that the boundaries are not sharp divisions, and the classification given is restricted to principal ore types. The difference between the ores for Group I and II elements relates to their ionic charges M^+ and M^{2+} respectively. The divalent alkaline earth metal ions form salts with high lattice energies and low solubility, while salts of the monovalent alkali metal ions have lower lattice energies and are more soluble. The metals that mainly occur as oxides tend to have compact ions, with few outer d-electrons and are not readily polarizable. They associate with the compact O^{2-} ion. On the other hand metals that mainly occur as sulphides are more polarizable, especially the heavier members, and unite with the polarizable S^{2-} ion. The oxide ion is described as a hard base and the sulphide ion as a soft base, each associating with hard and soft acids (metals) respectively. The platinum group metals that may occur free in nature, have oxidation states with favourable reduction potentials. A list of the principal ores for some abundant and rare elements are listed in Table 3.13.

Ore Formation

The geo-chemical processes involved in ore formation are diverse; different chemical and physical processes have lead to an enrichment of elements in certain areas. The processes fall into four principal catagories; cooling, heating, weathering and transport, and oxidation and reduction.

Cooling processes: *Magmatic concentration*. Materials crystallizing from the magma at different temperatures may settle out in bands, for example chromite deposits. At a later stage in the cooling process residual molten magma, may inject into rock crevices. The magnetite deposits at Kiruna, Sweden, were formed in this way.

Sublimation. Sublimation is the phase change from solid to gas, without passing through a liquid phase. The reverse process (gas to solid) is the way that some sulphur deposits were formed, for example in Sicily.

Hydrothermal processes. Hydrothermal fluid is heated, or superheated, aqueous solutions under pressure, which derive from the magma and which ascend through the crust. The many materials that the fluid contains may either re-deposit in rock cavities or substitute for materials already in rocks. The crystallization or precipitation of materials from solution occurs because of reduced solubility as the temperature and pressure drop during the ascent. Examples of cavity filling deposits are gold quartz at Bendigo, Australia, zinc and lead deposits in various areas and copper at Lake Superior. Reactions may also occur between the aqueous solutions and rocks, with substitution or replacement of a rock, such as limestone, with hydrothermal materials. Porphyrycopper deposits in a number of areas, sulphide ores in Spain, gold quartz in the South Island, New Zealand, are examples of such deposits. Such deposits can be quite massive.

TABLE 3.13 Principal Ores of the Technologically Important Metals

The Abundant Metals

Metal	Principal Ores	Metal	Principal Ores
Iron	Magnetite, Fe_3O_4; hematite, Fe_2O_3; goethite, $HFeO_2$ siderite, $FeCO_3$	Titanium	Ilmenite, $FeTiO_3$; rutile, TiO_2
		Manganese	Pyrolusite, MnO_2; psilomelane, $BaMn_9O_{18}2H_2O$ cryptomelane, KMn_8O_{16}; rhodocrosite, $MnCO_3$
Aluminium	Gibbsite, H_3AlO_3; diaspore, $HAlO_2$; boehmite, $HAlO_2$; kaolinite, $Al_4Si_4O_{10}(OH)_8$	Magnesium	Magnesite, $MgCO_3$; dolomite, $CaMg(CO_3)_2$

The Scarce Metals

Commonly forming sulfide minerals (Chalcopile)

Metal	Principal Ores	Metal	Principle Ores
Copper	Covellite, CuS, chalcocite, Cu_2S; digenite, Cu_9S_5 chalcopyrite, $CuFeS_2$; bornite, Cu_5FeS_4; tetrahedrite, $Cu_{12}Sb_4S_{13}$	Cadmium	Substitution for Zn in sphalerite
		Cobalt	Linnaeite, Co_3S_4; substitution for Fe in pyrite, FeS_2;
Zinc	Sphalerite, ZnS	Mercury	Cinnabar, HgS; metacinnabar, HgS
Lead	Galena, PbS	Silver	Acanthite, Ag_2S; substitution for Cu and Pb in their common ore minerals
Nickel	Pentlandite $(NiFe)_9S_8$; garnierite, $H_4Ni_3Si_2O_9$		
Antimony	Stibnite, Sb_2S_3	Bismuth	Bismuthinite, Bi_2S_3
Molybdenum	Molybdenite, MoS_2		
Arsenic	Arsenopyrite, FeAsS; orpiment, As_2S_3; realgar, AsS		

Commonly found in the native form (Sideropile)

Metal	Principle Ores	Metal	Principle Ores
Gold	Calaverite, $AuTe_2$; krennerite, $(Au,Ag)Te_2$; sylvanite, $AuAgTe_4$; petzite, $AuAg_3Te_2$	Palladium	Arsenopalladinite, Pd_3As; michenerite, $PdBi_2$; froodite, $PdBi_2$
		Rhodium	-
Platinum	Sperrylite, $PtAs_2$; braggite, PtS_2; cooperite, PtS	Iridium	-
		Ruthenium	Laurite, RuS_2
		Osmium	-

TABLE 3.13 Principal Ores of the Technologically Important Metals

The Scarce Metals

Commonly associated with oxygen (Lithophile)

Metal	Principal Ores	Metal	Principal Ores
Chromium	Chromite, Fe_2CrO_4	Niobium and Tantalum	Columbite, $FeNb_2O_6$; pyrochlore, $NaCaNb_2O_6F$; tantalite, $FeTa_2O_6$
Tin	Cassiterite, SnO_2		
Tungsten	Wolframite, $FeWO_4$; scheelite, $CaWO_4$	Thorium	Monazite, a rare-earth phosphate containing thorium by atomic substitution
Uranium	Uraninite(pitchblende), UO_2; carnotite, $K_2(UO_2)_2(VO_4)_2.3H_2O$; substituting for Fe in magnetite, Fe_3O_4	Beryllium	Beryl, $Be_3Al_2(SiO_3)_6$

Source: Skinner, B.J., *Earth Resources* (2nd Ed.) Prentice-Hall,New Jersey(1976)

Heating processes: *Evaporation*. Ores formed by evaporation of seawater are called evaporites. They consist of the more soluble minerals such as sodium chloride, gypsum, $CaSO_4H_2O$, and Chili nitrates.

Metamorphism. Deposits of materials such as asbestos, talc and graphite were formed as metamorphic deposits through the action of heat and pressure on rock materials.

Weathering and transport. *Sedimentation*. Weathered material in solution or in suspension can be deposited in sufficient concentration to become an ore body. Iron deposits have occurred in this way, such as the detrital magnetite iron-sands of New Zealand.

Mechanical concentration. Minerals resistant to weathering can be moved around, principally by water, and concentrate in specific regions - to give placer deposits. Alluvial gold is an example of such an ore, as are deposits of ilmenite, $FeTiO_3$, and cassiterite, SnO_2.

Residual concentration. Material left behind after weathering, and resistant to further changes can become an ore. Examples are bauxite deposits, and iron deposits such as those at Lake Superior.

Oxidation and Reduction processes: The importance of oxidation and reduction in environmental chemistry has been mentioned in Chapter 2. We will now discuss two examples leading to ore formation.

Supergene enrichment. Surface weathering followed by the leaching of materials into the crust, and subsequent ore enrichment is a feature of the formation of certain copper ore bodies, for example in Chile and Arizona (U.S.A.).

Copper(II) ions released from copper minerals near the earth's surface by hydrolytic weathering (especially in acid conditions), can move into the crust through leaching. If the Cu^{2+} ions come into contact with sulphide minerals of other metals the sulphides Cu_2S and CuS will tend to form due to their low solubility ($K_{sp} = 1.6 \times 10^{-48}$ and 8.7×10^{-36} respectively).

$$Cu^{2+}_{(soln)} + ZnS_{(s)} \rightarrow CuS_{(s)} + Zn^{2+}_{(soln)}. \quad (3.1)$$

Reduction may also oocur;

i.e. $$Cu^{2+} + e \rightarrow Cu^{+}. \quad (3.2)$$

$$S^{2-} + 4H_2O - 8e \rightarrow SO_4^{\ 2-} + 8H^{+}, \quad (3.3)$$

$$2Cu^{+}_{(soln)} + S^{2-}_{(soln)} \rightleftharpoons Cu_2S_{(s)} \quad (3.4)$$

$$ZnS_{(s)} \rightleftharpoons Zn^{2+}_{(soln)} + S^{2-}_{(soln)}, \quad (3.5)$$

the overall reaction being;

$$8Cu^{2+}_{(soln)} + 5ZnS_{(s)} + 4H_2O \rightarrow 4Cu_2S_{(s)} + 5Zn^{2+}_{(soln)} + SO_4^{\ 2-} + 8H^{+}. \quad (3.6)$$

The zinc ions will be transported away.

Microbiological processes. Micro-organisms can contribute to ore body formation through their involvement in accumulating particular elements, modifying environmental conditions, causing oxidations or reductions by metabolic processes and producing organic material. Bacteria are either autotrophic, that is not dependent on organic material for their carbon, or heterotrophic, that is dependent on organic material. Autotrophic, bacteria are involved in mineralization reactions.

Sulphate reducing bacteria, such as *desulphovibrio desulphuricans*, are instrumental in producing sulphide minerals of metals such as Cu, Pb, Zn and Ag. The reduction of sulphate can be represented by the equation;

$$SO_4^{\ 2-} + \underset{\text{carbohydrate}}{2(CH_2O)} + 2H^{+} \xrightarrow{\text{bacteria}} H_2S + 2CO_2 + 2H_2O, \quad (3.7)$$

and this is followed by precipitation of an insoluble sulphide, e.g.,

$$Pb^{2+} + H_2S \rightarrow PbS + 2H^{+}. \quad (3.8)$$

Deposits of elemental sulphur (in Sicily), may have been formed by bacterial reduction of SO_4^{2-} to sulphide, similar to above, followed by oxidation to sulphur by bacteria such as *thiobacillus thiooxidans*;

$$2H_2S + O_2 \xrightarrow{\text{bacteria}} 2S + 2H_2O. \quad (3.9)$$

Bacteria that metabolise, Fe and Mn, are important in mineralization of these metals. Bacteria, such as *Ferrobacillus* and *Gallionella*, catalyse the oxidation of Fe(II) to Fe(III), the energy released is used by the bacteria in their metabolism. The reaction;

$$4FeCO_3 + O_2 + 6H_2O \xrightarrow{\text{bacteria}} 4Fe(OH)_3 + 4CO_2 \quad (3.10)$$

is an example. Production of 1 gram of cell carbon (from the CO_2 at 2% efficiency) requires 220 g of Fe(II). Therefore areas where iron oxidising bacteria abound, may be associated with large deposits of $Fe(OH)_3$. It is likely that bacteria are also involved in the formation of manganese nodules on the sea bed.

Trace metal concentration. The distribution of trace metals in the solid magma depend on a number of factors, including the relative size of ions, and stabilization in the solid compared with the melt. Trace metal fractionation has been studied on the Skaergaard intrusion in Greenland, and from the data a quantity;

$$\log R = \frac{[M^{n+}] \text{ after X\% solidification}}{[M^{n+}] \text{ before solidification}} \qquad (3.11)$$

may be determined. A negative value of log R indicates concentration of the metal ion in the solid, while a positive value represents concentration in the residual melt. For the transition metals the distribution order in the solid (-log R) is;

$$Ni^{2+} > Co^{2+} > Mn^{2+} > Fe^{2+} > Cu^{2+} \qquad (M^{2+} \text{ ions}) \qquad (3.12)$$

$$Cr^{3+} > V^{3+} > Sc^{3+} > Fe^{3+} \qquad (M^{3+} \text{ ions}) \qquad (3.13)$$

The trends correlate with the ligand field stabilization energies for the metal ions in a weak octahedral field. The ligand field stabilization energy (LFSE) is given by the equation;

$$LFSE = x(-\tfrac{2}{5}\Delta) + y(+\tfrac{3}{5}\Delta) \qquad (3.14)$$

where x and y are the number of electrons in the t_{2g} and e_g orbitals respectively, and Δ is the ligand field splitting energy.

The correlation suggests that LFSE has some influence on the distribution. It is, however, surprising that this energy, which is only 10% of the total energy of an ion in a crystal lattice, has such a pronounced effect. It may be because the LFSE is stereospecific rather than spherical in its influence. A similar correlation occurs between the distribution of the transition metals and the ligand field stabilization achieved on crystallization.

$$M^{n+}_{(melt)} \rightarrow M^{n+}_{(crystal)} \qquad (3.15)$$

Stabilization is greater in the solid because the average coordination number of a metal ion is higher in the solid than in the melt because of reduced kinetic energy.

Ore Formation of Some Important Metals

We will now discuss briefly the formation of some ore deposits of the more important metals.

Iron. The geochemistry of iron is dominated by oxidation and reduction and the pH of the surrounding medium. Iron is immobilised in oxidising conditions and in alkaline reducing conditions, but is mobilized in acidic reducing conditions. Only in strong acid is Fe(III) mobilized.

The process which has produced most iron ore is the leaching of SiO_2 from primary Precambrian banded-iron-formations, which were formed from marine chemical precipitation. The Lake Superior deposits contain the four minerals FeS_2, $FeCO_3$, $FeSiO_3$ and Fe_2O_3, which all exist within a narrow band of E_h and pH values (Fig. 2.16). In wet tropical areas iron laterites (mainly goethite, $Fe_2O_3H_2O$) are found, their quality depends on the amount of aluminium in the original rocks and the amount of organic material which would preferentially stabilize Fe(II) and mobilise it. Hydrothermal iron ores, magnetite and hematite were formed by replacement of rocks such as limestone, e.g. Marmora, Canada. The Kiruna deposits in Sweden are probably magmatic in origin.

Aluminium. The principal, and widespread, source of aluminium is bauxite which is a general name for different mixtures of hydrated alumina: gibbsite, γ-$Al_2O_3.2H_2O$; boehmite, γ-$Al_2O_3.H_2O$ and diaspore, α-$Al_2O_3.H_2O$. Bauxite is readily purified by the Bayer process except when the ore is mainly diaspore. As discussed before bauxite is the residual product of weathering in warm ($>25^oC$ most of the time) humid climates.

Titanium. Titanium ores occur by two processes. Magmatic ores of ilmenite, $FeTiO_3$, appear as layers low down in the magma because of the mineral's high melting point and density (e.g. Allard Lake, Canada). The three principal minerals, rutile, TiO_2, ilmenite and sphene (or titanite), $CaTiSiO_5$, are all resistant to weathering and occur as placer deposits, again the high densities help in forming the deposits.

Copper. Copper is mainly associated with sulphur in its ore minerals, examples are chalcocite, Cu_2S, chalcopyrite, $CuFeS_2$ and bornite, Cu_5FeS_4. The majority of the world's copper ores are low grade hydrothermal deposits, mainly as $CuFeS_2$ and are called porphyry copper (cf. Chilian and Peruvian deposits). Sedimentary material is layered and may be associated with organic rich sediments called *Kupferschiefer*, that is red-bed copper deposits. Coordination of organic ligands to copper(II) may have been involved in formation of the ores. Central European deposits, and possibly the Zambian copper belt are examples of sedimentary ores. Supergene enrichment of copper ores has already been discussed. Finally native copper occurs in basic igneous rocks, as found in the Upper Michigan lava.

Nickel. Being a siderophile, nickel is associated with metallic iron, e.g. in meteorites. In ores nickel is associated mainly with sulphur, arsenic or in silicates. Early magmatic segregation produced ores containing the mineral pentlandite $(Fe,Ni)_9S_8$ (e.g. Sudbury, Canada). The other main ore source, found in the humid tropics, is due to the weathering of ultrabasic rocks (peridotites or serpentinites) which frees the Ni^{2+} ion. Because of its stability in aqueous solutions (in contrast to Fe^{2+} and Mn^{2+}) it can be transported long distances. Eventually as the pH rises sufficiently nickel is precipitated with hydrous silicates, e.g. garnierite $(Ni,Mg)_3Si_2O_5(OH)_4$ (in Cuba, New Caledonia and the Philippines).

Chromium. The principal chromium ore is chromite, $FeO.Cr_2O_3$, a member of the spinel group of structures $M^{2+}M^{3+}O_4$, i.e. $(Mg,Fe)O.(Al,Fe,Cr)_2O_3$, where the oxide ions are cubic close packed and half the octahedral cavities are filled with the trivalent cations. A range of compositions occur due to isomorphous replacement of cations, as the radii of the M^{3+} ions, Al^{3+} 53 pm, Cr^{3+} 64pm and Fe^{3+} 60 pm all lie within 20% of each other. Chromite ores occur as early magmatic crystallizations and as placer deposits, due to resistance to weathering of $FeO.Cr_2O_3$ and its high density.

Uranium. Pitchblende or uranite, UO_2, is the principal uranium ore. The oxide is insoluble and exists in granitic magmas. Weathering can lead to oxidation of the U(IV) to the stable U(VI) uranyl ion, $UO_2{}^{2+}$. This forms soluble salts and allows uranium to be transported. Reduction of UO_2^{2+} in the presence of organic material produces deposits of UO_2 in black carbonaceous rich shales as found in Sweden.

Zinc and Lead. The two metals have a similar geochemistry and very often their ores occur together. The two principal minerals are sphalerite, ZnS, and galena, PbS. Zinc has the same ionic radius size as Fe^{2+}(75 pm) and the two ions often occur together. The larger Pb^{2+} ion (ionic radius 118 pm) can replace K^+ ions (ionic radius 133pm) in materials such as potassium feldspar. The richest zinc and lead ores are of hydrothermal origin, where limestone has been replaced by the fluid which eventually produces crystalline minerals. A good example of this is the ore body in the Mississippi Valley (U.S.A.).

EXTRACTION OF METALS

The first metals, or alloys, used by man were those which were, either available as rich ores and could be readily obtained from the ore by heating at a relatively low temperature, or metals which occurred free in nature. Developments in extraction technology continues and new approaches are sought, especially as richer finds are being worked out. In this section we will consider some aspects of extraction of metals from their ores, including some new techniques.

Extraction Processes for Metals

The primary reaction is reduction, as all metals in their ores are in an oxidised state. Reduction can be considered as follows;

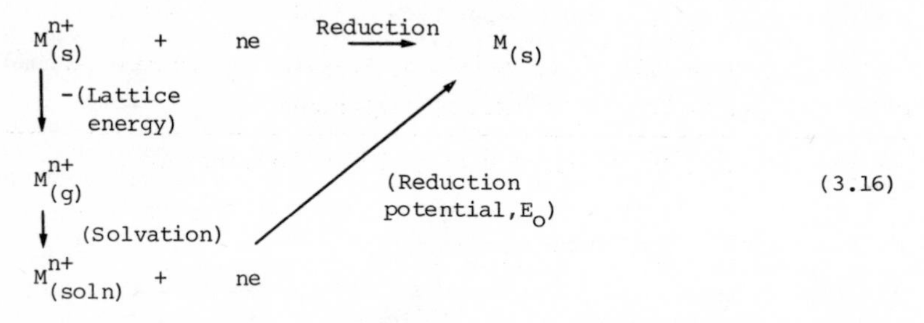

Since lattice and solvation energies are often of the same order of magnitude, and opposite in sign, the value of E^o, the reduction potential, may be used as a guide to the difficulty or ease of achieving the reduction. The more negative the potential the harder it is to reduce the metal ion. Reduction potentials for common metals are listed in Table 3.14, for metal ions with $E^o < -1.66$ V, electrolytic reduction is necessary, while for less negative potentials chemical reducing agents, such as Al, C and H_2, can be used. Metal ions having positive reduction potentials may occur free in nature (e.g. Cu) or they can be obtained by heating the ore without the need to add a separate reducing agent:

$$2HgS_{(s)} + 3O_{2(g)} \rightarrow 2HgO_{(s)} + 2SO_{2(g)}, \qquad (3.17)$$

$$2HgO_{(s)} \xrightarrow{500^oC} 2Hg_{(l)} + O_{2(g)}, \qquad (3.18)$$

TABLE 3.14 Reduction Potential of Some Metal Ions

Element	E^o(n)*	Main Source	Extraction
K	-2.92 (+1)	KCl, $KCl.MgCl_2$	Electrolysis of fused salts
Na	-2.71 (+1)	NaCl	
Ca	-2.87 (+2)	$CaSO_4$, $CaCO_3$	
		$Ca_3(PO_4)_2$	
		$CaCl_2$ (Solvay process)	
Mg	-2.36 (+2)	Mg salts	
Al	-1.66 (+3)	Al_2O_3	Electrolysis of Al_2O_3 in Na_3AlF_6
Mn	-1.18 (+2)	MnO_2	Reduction with Al (Thermite process)
Cr	-0.74 (+3)	$FeO.Cr_2O_3$	
Zn	-0.76 (+2)	ZnS	Chemical reduction with carbon (or hydrogen). First convert sulphides to oxides by roasting.
Fe	-0.44 (+2)	Fe_2O_3, Fe_3O_4	
Co	-0.28 (+2)	CoAsS, Co_3S_4	
Ni	-0.25 (+2)	Sulphides	
Sn	-0.14 (+2)	SnO_2	
Pb	-0.13 (+2)	PbS	
Cu	+0.34 (+2)	Metal, sulphide	
Hg	+0.85 (+2)	HgS	
Ag	+0.80 (+1)	Metal, Ag_2S, AgCl	Cyanide extraction
Au	+1.7 (+1)	Metal, tellurides	

* E^o for reaction, $M^{n+}_{(aq)} + ne \rightarrow M$ (n given in parenthesis).

(i.e. $2Hg^{2+}_{(s)} + 4e \rightarrow 2Hg_{(l)}$ reduction

$2O^{2-} - 4e \rightarrow O_{2(g)}$ oxidation.)

The information can also be represented diagramatically in terms of the periodic table (Fig. 3.5). Comparison of Figs 3.5 and Table 1.1 indicates correlations between the principal ore types for the elements and methods of extraction.

Chemical Reducing Agents

Most metals are obtained from their oxides or oxy-anions directly, or after conversion of the ore to an oxide. The reduction of an oxide, represented by the equation;

$$yMO_{x(s)} + xA_{(s)} \rightarrow yM_{(s)} + xAO_{y(g,l,\text{ or } s)}, \quad (3.19)$$

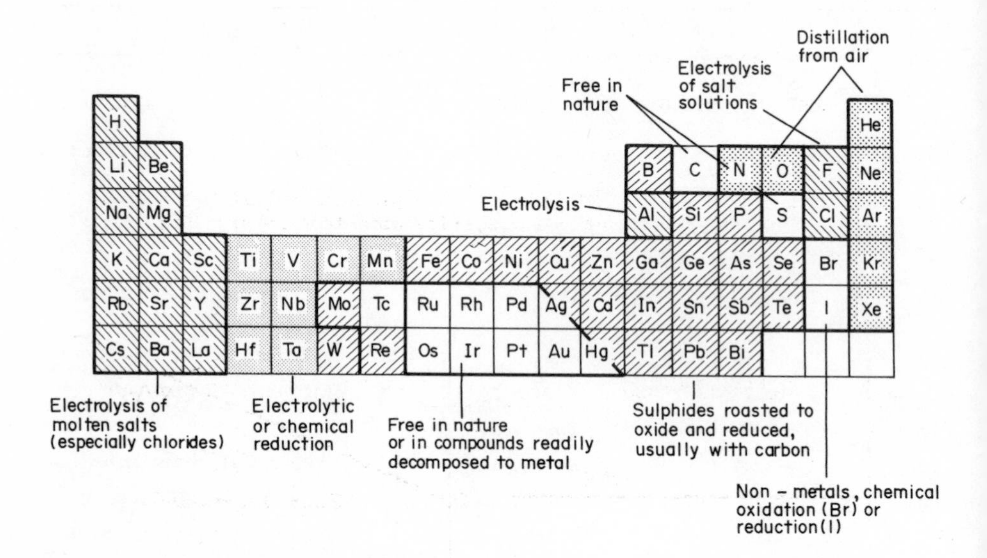

Fig. 3.5. A portrayal of the extraction of the elements in terms of the periodic table.

proceeds if the overall free energy change is negative. The standard state free energies of elements M and A are zero, hence the reaction depends on the relative standard free energies of formation of the two oxides.

i.e. $$\Delta G^o = x\Delta G^o_f(AO_y) - y\Delta G^o_f(MO_x) \qquad (3.20)$$

For the reaction;

$$2CaO_{(s)} + C_{(s)} \rightarrow 2Ca_{(s)} + CO_{2(g)}, \qquad (3.21)$$

$$\Delta G^o = (-394.5) - 2(-604.2) \text{ kJ}, \qquad (3.22)$$

i.e. $\Delta G^o = 814.1$ kJ, an unfavourable reaction.

Similarly for the reaction;

$$Fe_2O_3 + 3C \rightarrow 2Fe + 3CO, \qquad (3.23)$$

$$\Delta G^o = 3(-137)-(-741) \text{ kJ}, \qquad (3.24)$$

i.e. $\Delta G^o = 330$ kJ, also an unfavourable reaction. However if the reductions are carried out at high temperatures, the circumstances become more favourable, because of the influence of the $T\Delta S^o$ term on ΔG^o.

The reaction;

$$C_{(s)} + O_{2(g)} \rightarrow CO_{2(g)}, \quad (3.25)$$

has a standard entropy change of;

$$\Delta S^o = (213.6 - 5.7 - 205)J, \quad (3.26)$$

i.e. $\Delta S^o = 2.9J$.

If the quantities ΔH^o(-393.5 kJ) and $T\Delta S^o$ are plotted against T (K) then $\Delta G^o = \Delta H^o - T\Delta S^o$ shows that ΔG^o hardly varies with temperature (Fig. 3.6a). For the reaction;

$$2C_{(s)} + O_{2(g)} \rightarrow 2CO_{(g)}, \quad (3.27)$$

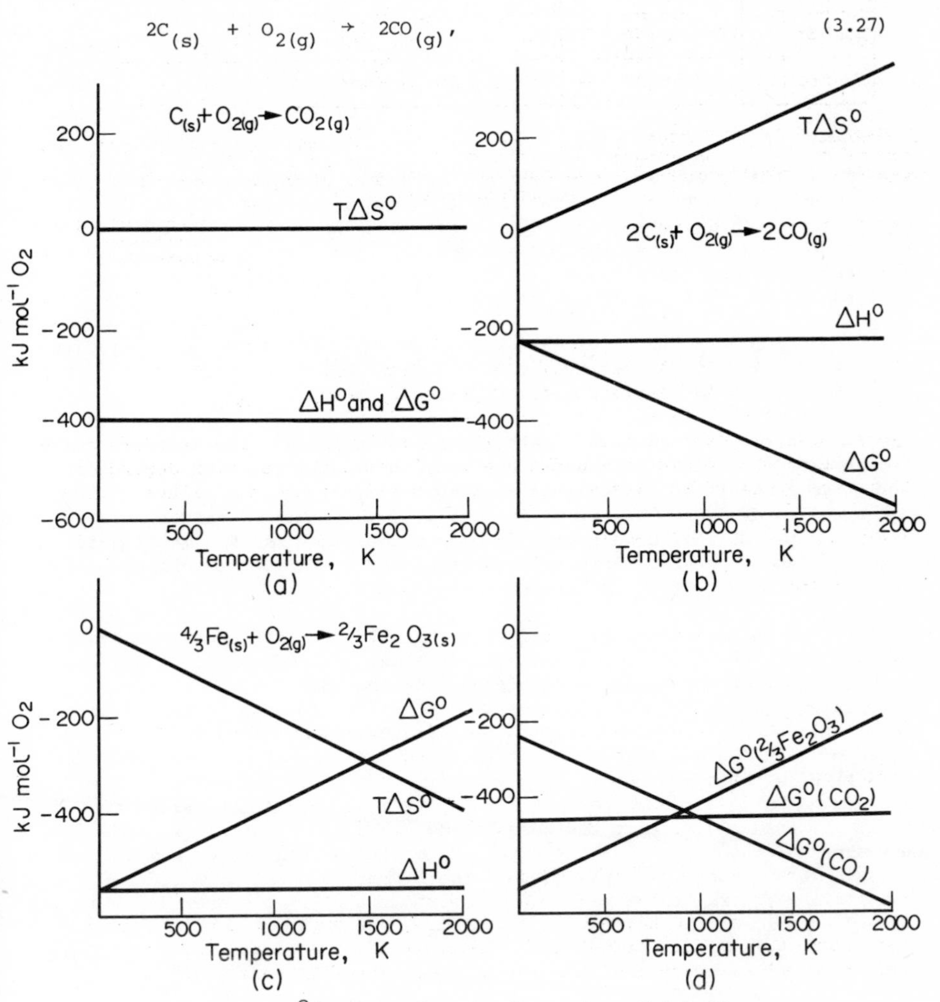

Fig. 3.6. $\Delta G^o/T$ plots for the formation of (a) CO_2, (b) 2CO, (c) $\frac{2}{3}Fe_2O_3$ and (d) combined.

the standard entropy change is;

$$\Delta S^o = (2 \times 198.5 - 2 \times 5.7 - 205)J \quad (3.28)$$

i.e. $\Delta S^o = 178.6J$

The ΔH^o, ΔG^o and $T\Delta S^o$ terms plotted against T(K) are given in Fig. 3.6b, and in this case $\Delta G/T$ line has a steep negative slope.

For the formation of Fe_2O_3, (writing the equation so one mole of O_2 is consumed) we have;

$$\frac{4}{3}Fe + O_2 \rightarrow \frac{2}{3}Fe_2O_3. \quad (3.29)$$

The standard entropy change is;

$$\Delta S^o = (\frac{2}{3} \times 90 - \frac{4}{3} \times 27.2 - 205) \quad (3.30)$$

i.e. $\Delta S^o = -181.3$ J.

The plots of ΔH^o, ΔG^o and $T\Delta S^o$ versus T (per mole of O_2 consumed) are given in Figure 3.6c; in this case the $\Delta G^o/T$ line has a positive slope. Comparison of the three $\Delta G^o/T$ lines (Fig. 3.6d) shows that, because of the opposing slopes of the $\Delta G^o/T$ lines for 2CO and $\frac{2}{3}Fe_2O_3$, there is a temperature at which the free energy of formation of 2CO is more negative than that of $\frac{2}{3}Fe_2O_3$ (around 900 K). This means the reaction,

$$\frac{2}{3}Fe_2O_{3(s)} + 2C_{(s)} \rightarrow \frac{4}{3}Fe_{(s)} + 2CO_{(g)}, \quad (3.31)$$

has a favourable ΔG^o above 900 K.

The free energy of formation of all metal oxides become less negative as the temperature increases, because of the unfavourable large entropy change ($-\Delta S^o$), as a result of the loss of a mole of oxygen. The $\Delta G^o/T$ lines for a number of metal oxides are given in Fig. 3.7 where the ΔG^o values are expressed per mole of O_2 consumed. Some of the lines for the metal oxides have more than one slope, since discontinuities occur at the melting points and boiling points of the metals and/or oxides.

Useful information can be obtained from the Ellingham diagram. Because the carbon lines (negative slope) cut the metal oxide lines (positive slopes), it becomes clear why carbon is a useful reducing agent; since at some temperature (above the point of cross-over) carbon is able to reduce many metal oxides. But in some cases the temperature required is so high, greater than 2000 K for Al_2O_3, that it is not economical to carry out the reduction, without some other energy input. The lines for metal oxides that are around $\Delta G^o = 0$, indicate that with sufficient heat these oxides become thermally unstable without the need to add a reducing agent (see above for mercury; equation (3.18)). Any metal oxide will be reduced by a metal whose oxide's $\Delta G^o/T$ line lies below that of the former, i.e. Al will reduce Cr_2O_3, but a high temperature is required to overcome the kinetic energy barrier. At 1000K ΔG for the reaction;

$$2Al + Cr_2O_3 \rightarrow 2Cr + Al_2O_3, \quad (3.32)$$

is -530 kJ. This method of extraction is called the thermite process.

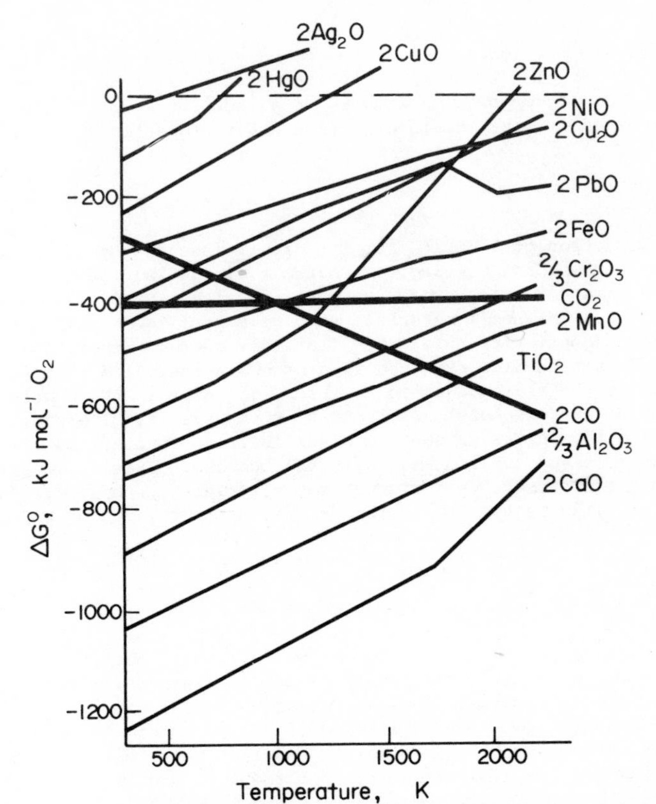

Fig. 3.7. Ellingham diagram, $\Delta G^{o}/T$, for oxide formation (discontinuities are due to phase changes of either metal or oxide). Source: D.J.G. Ives The Chemical Society, London, Monographs for Teachers, No. 3, 1969.

The two $\Delta G^{o}/T$ lines for the oxidation of carbon cross at approximately 925 K. Below this temperature CO is unstable relative to C and CO_2;

i.e. $$2CO \rightarrow C + CO_2, \tag{3.33}$$

while above 925 K the reverse reaction occurs. Hence below 925 K the principal reducing agent is CO, while above 925 K it is carbon.

Many metals occur as sulphides, but because there is no CS analogue to CO there is no $\Delta G^{o}/T$ line with a steep negative slope. Therefore carbon is not a good reducing agent for sulphides. Roasting a sulphide ore to obtain the oxide is a necessary first step before reduction with carbon;

i.e. $$2MS + 3O_2 \rightarrow 2MO + 2SO_2. \tag{3.34}$$

The same applies to H_2 as a reducing agent, as the $\Delta G^{o}/T$ line for

$$2H_2 + O_2 \rightarrow 2H_2O \tag{3.35}$$

has a positive slope and runs parallel to many metal oxide lines. Only oxides whose formation lines lie above the H_2 line will be reduced by dihydrogen. The other problem with H_2 as a reducing agent is that it can remain dissolved in many metals affecting their properties.

Ore Processing

Prior to extraction of a metal from its ore it is often necessary to process the ore by crushing and sorting to a correct size, also the ore may need to be enriched. Many sorting processes make use of the physical properties of the crushed ore. Magnetic material, such as magnetite, Fe_3O_4, can be concentrated magnetically. In froth flotation certain minerals attach themselves to air bubbles produced in a bath of water, and then float off in the froth. Frothers are added to the bath, to maintain the bubbles (e.g. pine oil or higher alcohols). Collectors, such as salts of organic acids or bases, attach themselves to certain minerals in the pulverised ore making them hydrophobic and carried away with the bubbles. Gravity separation depends on the different densities of the materials in the ore. Chemical processes may also be used to enrich an ore, such as leaching with acid or alkali.

Extraction of Iron

The blast furnace is the main method by which iron is obtained from its ores. The furnace (20 - 30m high and around 6 - 7m in diameter) lined with refractory bricks is loaded from the top with ore (Fe_2O_3, Fe_3O_4), coke and limestone. Hot air ($>800^oC$) is blasted in at the bottom through "tuyères". Molten iron is tapped off at the bottom and less dense molten slag is also removed. The furnace gases are vented at the top, and if any CO remains it is burnt to preheat the air blast. The chemical reactions that take place at different temperatures and at different positions in the furnace are given in Fig. 3.8.

Below 700—800^oC the principal reducing agent is CO, and the reactions are called "indirect reductions". The reactions around 1000^oC are called direct reductions because the overall stoichiometry is;

$$FeO_{(s)} + C_{(s)} \rightarrow Fe_{(s)} + CO_{(g)}. \quad (3.36)$$

However, this does not represent the true situation because at 1000^oC the two reactions;

$$FeO_{(s)} + CO_{(g)} \rightarrow Fe_{(s)} + CO_{2(g)}, \quad (3.37)$$

and

$$CO_{2(g)} + C_{(s)} \rightarrow 2CO_{(g)}, \quad (3.38)$$

make up equation (3.36), i.e. the CO acts as an intermediate.

Since the reductions are of the type;

$$\text{Metal oxide(A)}_{(s)} + CO \rightarrow \text{metal oxide(B)}_{(s)} \text{ or metal} + CO_2 \quad (3.39)$$

the position of equilibrium is given by the ratio P_{CO_2}/P_{CO}. At 600 - 700^oC the ratio is around 1 (the temperature of cross-over of the CO_2 and CO $\Delta G^o/T$

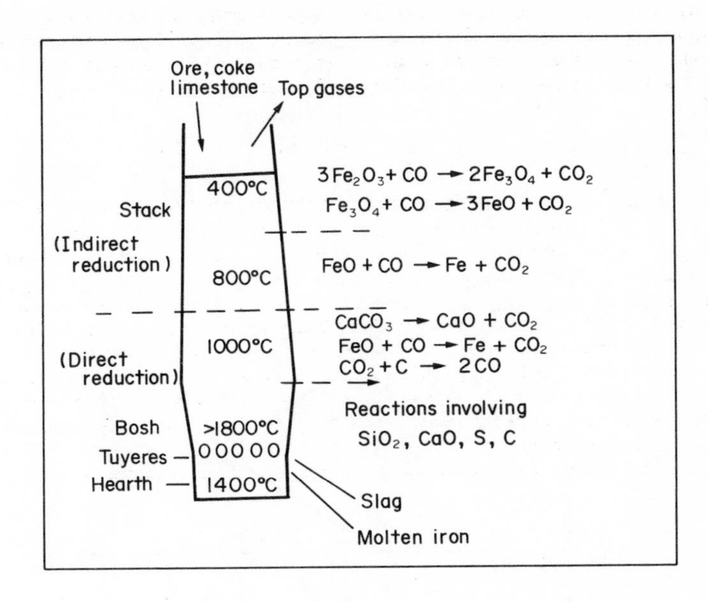

Fig. 3.8. A schematic diagram of a blast furnace, showing temperatures and chemical reactions. Source: T. Rosenqvist, Principals of Extractive Metallurgy. McGraw-Hill, N.Y., 1974.

lines in the Ellingham diagram, Fig. 3.7. The ratio is temperature dependent and increases up the furnace as the gases become less reducing. For an efficient furnace the exit gases will have a large ratio.

As the reaction $2CO \rightarrow C + CO_2$ occurs below 500-600°C, soot may be deposited on the ore, which at a higher temperature is useful for reduction. However, the soot tends to deteriorate the refractory bricks and is considered harmful.

Limestone is added to remove acidic impurities, such as SiO_2, as slag;

$$CaO + SiO_2 \rightarrow CaSiO_3. \tag{3.40}$$

Other material in the ore undergo various reactions in the blast furnace, for example above 1000°C, Fe_3C forms which dissolves in the molten iron, SiO_2 is reduced to Si above 1500°C which also dissolves in the iron. Reduction of phosphate produces finally Fe_3P, and manganese oxides reduce to the metal, which alloys with the iron. Any sulphur, mainly from the coke, reacts with the iron to give FeS, but this is removed by the reaction;

$$FeS + CaO + C \rightarrow Fe + CO + CaS, \tag{3.41}$$

the CaS appearing in the slag. The final product, brittle "pig iron", may contain up to 4% C, 2.5% Mn, 2.5% Si, 2% P and 1% S.

Steel. Pig iron is converted into other forms of iron mainly by altering the carbon content. Above 1000°C the carbide Fe_3C is formed which is soluble in iron at low concentrations. The carbide has a carbon atom surrounded by a near regular trigonal prism of iron atoms. Pure iron has a melting point of 1539°C, reduced to 1015°C (the eutectic temperature) by 4.3% carbon (Fig. 3.9). Iron crystallises in three forms: α-iron, bcc,

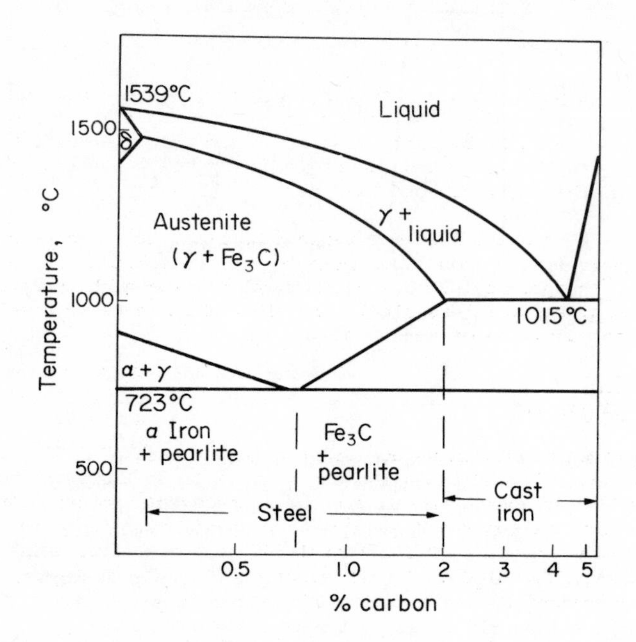

Fig. 3.9. Phase diagram for Fe-C.

<910°C; γ-iron, fcc, 910-1400°C; and δ-iron, bcc, 1400-1539°C. The solid solution of γ-Fe and Fe_3C (<2%C) called "austenite" is a tough form of iron (steel) while α-Fe + Fe_3C (<2%C) is "pearlite", which is more malleable, because its structure consists of alternating layers of α-iron and Fe_3C. Austenite converts to pearlite by slow cooling but rapid cooling "freezes" the steel in the austenite form called "martensite". Above 2%C the iron is no longer steel and the carbide becomes unstable decomposing, in part, to produce graphite;

$$Fe_3C \rightarrow 3Fe + C. \tag{3.42}$$

Remaining Fe_3C occurs as coarse particles, and the result is pig or cast iron which is hard and brittle.

Steel is produced by blowing oxygen through or over molten pig iron. At 1600°C the reaction;

$$2C + O_2 \rightarrow 2CO, \tag{3.43}$$

has a favourable free energy (ΔG^o = -552 kJ). Methods used to bring about the oxidation include the open hearth furnace, rotor process, the basic oxygen furnace and electric arc furnace.

In order to improve the qualities of steel it can be annealed. To prevent its oxidation an atmosphere of CO_2 and CO is used in order to give the ratio $p_{CO}/p_{CO_2} > 1$ to ensure the reaction;

$$CO_{2(g)} + Fe_{(s)} \rightarrow FeO_{(s)} + CO_{(g)} \quad (3.44)$$

lies on the left.

Other materials in pig iron are also removed by oxidation. Silicon and manganese are oxidised and the molten oxides float on the iron as slag. The oxides of S and P are unstable with respect to the formation of FeO, at the temperatures of steel making, and therefore have to be converted to slag with the use of a basic oxide such as CaO.

There are a number of special steels whose properties depend on the amount of other materials in the iron. For example 2.5% Si makes the steel elastic and can be used in springs, 0.4 - 1.6% Mn gives steel a high tensile strength, 12-15% chromium gives stainless steel, and the cutting edge of steel is maintained by the addition of tungsten.

Extraction of Aluminium

In order to produce aluminium using carbon a temperature of 2000°C is required (see Fig. 3.7), which is too expensive. It is therefore necessary to use another energy source and to work at a lower temperature. An electrolytic method called the Hall-Héroult process, discovered independently by Hall (U.S.A.) and Héroult (France) in 1866, uses cryolite, Na_3AlF_6, as a non-aqueous solvent for Al_2O_3. A 2-8% solution of Al_2O_3 in Na_3AlF_6 has a melting point of 960-980°C.

The principal aluminium ore, bauxite, contains aluminium oxide (30-70%) with ferric oxide and silica as the other main components. The oxide is purified by making use of the differing acidic and basic properties of the oxides (Bayer process).

$$\left.\begin{array}{l} Al_2O_3 \\ \text{(amphoteric)} \\ Fe_2O_3 \\ \text{(basic)} \\ SiO_2 \\ \text{(acidic)} \end{array}\right] + 30\%\ NaOH \xrightarrow[460K]{800kPa} AlO^-_{2(aq)} + SiO^{2-}_{3(aq)} + Fe_2O_{3(s)} \quad (3.45)$$

The insoluble materials are removed by sedimentation, and purfied alumina precipitates as $Al_2O_33H_2O$ (or $Al(OH)_3$) by treating the solution with CO_2 and $Al_2O_33H_2O$ seed crystals. The silica remains in solution. Non-aqueous conditions are required for the electrolysis so the oxide is first heated to 1500 K to give αAl_2O_3.

The aluminium oxide is then fed into the cell (Fig. 3.10) containing cryolite.

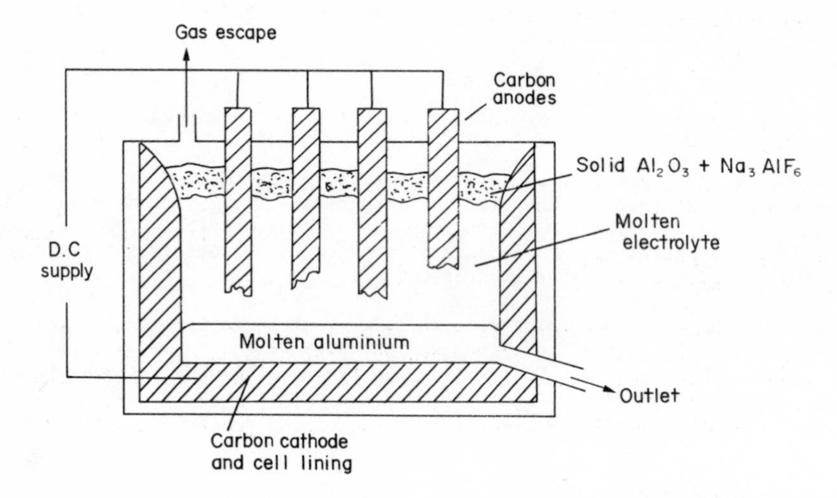

Fig. 3.10. A schematic diagram of the Hall-Héroult cell for the production of aluminium.

The inputs to the cell to produce 1000 kg of Al from 4000 kg of ore (assuming a 50% Al_2O_3 content) are given in Fig. 3.11. The electrochemical reactions are;

$$Al^{3+} + 3e \rightarrow Al_{(l)} \quad \text{(cathode)} \tag{3.46}$$

$$O^{2-} - 2e \rightarrow \tfrac{1}{2}O_2 \quad \text{(anode)} \tag{3.47}$$

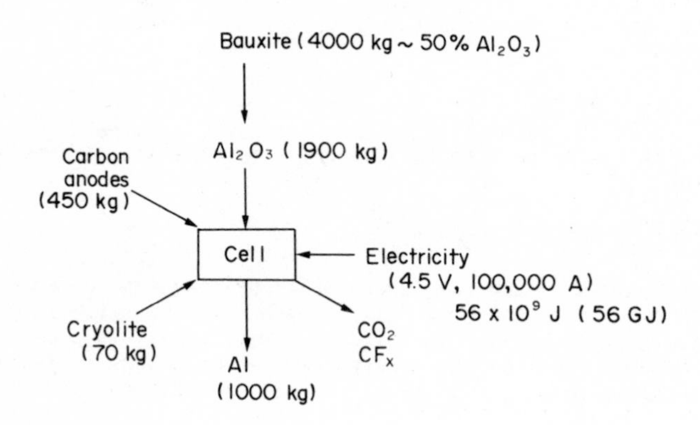

Fig. 3.11. Inputs to produce 1000 kg of aluminium

which add up to;

$$Al_2O_3 \xrightarrow{\text{electrolysis}} 2Al + \frac{3}{2}O_2,\ \Delta G_{1260\ K} = 1255\ kJ. \qquad (3.48)$$

The oxygen reacts with the carbon anodes;

$$\frac{3}{2}O_2 + \frac{3}{2}C \rightarrow \frac{3}{2}CO_2,\ \Delta G_{1260\ K} = -603\ kJ. \qquad (3.49)$$

This reaction, which consumes the carbon anodes, thermally assists in the production of aluminium;

$$Al_2O_3 + \frac{3}{2}C \rightarrow 2Al + \frac{3}{2}CO_2,\ \Delta G_{1260\ K} = 652\ kJ. \qquad (3.50)$$

Hence, indirectly, carbon reduces the oxide, and just a little more than half of the expected energy is required. However, energy is necessary to produce the anodes, but is less than the above saving in power.

The major energy inputs in the production of 1000.kg of aluminium are: (a) the Bayer process, 44 GJ, (b) dehydration of $Al_2O_33H_2O$, 9 GJ, (c) electrolysis, 168 GJ assuming 33% efficiency in the handling of electrical power, and (d) carbon anode production, 14 GJ. This gives a total of 235 GJ making aluminium production expensive in the use of power (see Table 3.11).

Refinement of Metals

Metals are sometimes needed in a high state of purity, and have to be refined. The treatment of pig iron to make steel is an example. We will discuss electrolytic refinement of copper and zone refinement of silicon.

Refinement of copper. Impure copper is refined by making it the anode of a cell with pure copper as the cathode. The cell, which runs at 50-60°C, contains dilute $H_2SO_4/CuSO_4$ as the electrolyte, and over 2-3 weeks a 4.5-5.0 kg cathode will increase to 100-125 kg. Metal impurities in the copper anode either go into solution, if their reduction potentials are less than for copper; or drop on the cell floor as a sludge as the anode disappears, if their reduction potentials are greater than for copper. For example consider copper with Ag and Ni as impurities. The reduction potentials are; Cu(+0.34 V), Ag(+0.80 V) and Ni(-0.25 V). If the cell is run so that the anode reaction is;

$$Cu_{(s)} \rightarrow Cu^{2+}_{(aq)} + 2e \qquad (3.51)$$

then the Ni will also go into solution, but not the Ag. If the cathode reaction is;

$$Cu^{2+}_{(aq)} + 2e \rightarrow Cu_{(s)}, \qquad (3.52)$$

the Ni^{2+} will not be discharged. The electrolyte has to be changed as the impurities build up in solution, and because some impurities may tend to be discharged with the copper. The latter is more likely to happen at high concentrations of the impurity metals. The anode sludge is a good source of rare and precious metals, such as Ag, Au and the platinum metals.

Refinement of Silicon. It is necessary to obtain very pure silicon (1 part in 10^{10}) for use in the transistor industry. This is achieved by zone refinement. A bar of silicon is allowed to move slowly through a small circular induction heater (Fig. 3.12a), melting the silicon, which then solidifies as the bar moves out of the heater. Since the impurities are generally more soluble in the liquid phase they accumulate in the melt as it moves along the bar leaving behind purified silicon. The process can be repeated a number of times, always in the same direction. If we assume that an impurity X and silicon form a solid solution the phase changes are represented by the section of a phase diagram given in Fig. 3.12b. If the concentration of the impurity is C_x, as the liquid cools to temperature T_A at point A solid separates with composition C'_x (point B), i.e. the solid is now more pure and the liquid as a consequence becomes richer in X. On the next treatment the solid with concentration C'_x melts and on cooling at temperature T_B gives an even purer solid with C''_x as the concentration of X. The purified silicon is then doped by adding controlled amounts of material such as boron and arsenic to give p and n type semiconductors respectively.

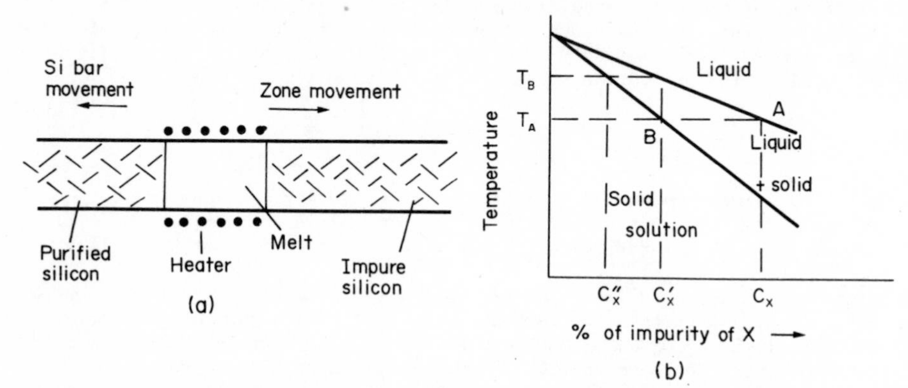

Fig. 3.12. Refinement of silicon, (a) diagram of technique, (b) part of a silicon-impurity phase diagram.

EXTENDING MINERAL RESOURCES

A way of maintaining, or extending, mineral resources is to investigate new methods of extraction where low grade ores can be used without too great an expenditure in energy. Another approach is to search for new materials for specific tasks. The recycling of materials is also of increasing importance, as are methods of protecting materials from corrosion.

Discovery of New Ore Bodies

The search for new mineral resources continues, particularly for ore bodies whose presence is not obvious from a study of the earth's surface. Geochemical and geophysical techniques (with a sound understanding of geology) are the principal methods employed. Geochemical analytical results of trace elements in rocks, sediments and plants can indicate ore bodies. Studies of the dispersion of elements around an ore body can provide information of use in later exploration. For example, around the Robinson mining area in

Nevada, U.S.A., 2000 samples were analysed for 30 elements. The results, when analysed, produced a pattern of contours of element concentration extending out from the copper ore body. It was discovered that thallium and indium had moved furtherest from the ore, and therefore their presence elsewhere may suggest concealed ore deposits.

An area for exploration that is still in its infancy is the sea bed. Manganese nodules are one example of a mineral discovered on the ocean floor. This extension of mineral exploration (including oil), raises political problems of ownership and environmental problems which need to be resolved before developments get too far advanced. The Arctic and Antarctic are also two areas that need careful consideration.

New Methods of Ore Processing and Extraction

Improved methods of ore processing and metal extraction are being investigated for a variety of reasons. These include the use of new ores or low grade ores, pollution control, replacement of an expensive reagent and small scale processes for handling small ore deposits.

Titanium dioxide for use as a pigment is obtained either from rutile by chlorination to give $TiCl_4$ from which TiO_2 is obtained;

$$TiO_{2(s)} + C_{(s)} + 2Cl_{2(g)} \rightarrow TiCl_{4(\ell)} + CO_{2(g)}, \quad (3.53)$$

$$TiCl_4 + O_2 \xrightarrow{650-750^{\circ}C} TiO_2 + 2Cl_2, \quad (3.54)$$

or by treating ilmenite with sulphuric acid. The problem with the latter process is the disposal of the $FeSO_4 7H_2O$. Chlorination of ilmenite has been proposed but this uses up expensive chlorine in producing $FeCl_3$

$$2FeTiO_3 + 3C + 7Cl_2 \rightarrow 2FeCl_3 + 2TiCl_4 + 3CO_2. \quad (3.55)$$

However, this problem has been overcome by separating the $FeCl_3$ from the $TiCl_4$, by fractional condensation, and recovery of the Cl_2 by the reaction;

$$4FeCl_3 + 3O_2 \rightarrow 2Fe_2O_3 + 6Cl_2. \quad (3.56)$$

Both low grade copper and uranium ores have been enriched *in situ* by leaching. Copper oxide ores (Emerald Isle mine Arizona, U.S.A.), after fracturing and while still in the mine, have been leached with acid solutions and the metal recovered from solution by cementation;

$$Cu^{2+}_{(aq)} + Fe \rightarrow Fe^{2+}_{(aq)} + Cu. \quad (3.57)$$

Uranite ores have been leached *in situ* by circulating carbonate/bicarbonate solutions, under oxidising conditions, into the ore strata. Oxidation to U(VI) is achieved and the uranium which dissolves is then recovered using ion exchange techniques.

Mercury has been long used to concentrate gold by amalgamation. This has two disadvantages, pollution and the limited mercury resources. Activated charcoal successfully absorbs gold, which is then separated by flotation. The gold is recovered from the charcoal either by smelting or using cyanide leaching solutions.

In order to obtain the most from small deposits of zinc and lead ores, small scale extraction methods need to be developed. This rules out conventional pyrometallurgical methods. The U.S. Bureau of Mines has developed chlorination of the sulphide ores to give $PbCl_2$, $ZnCl_2$ and S.

$$PbS/ZnS + 2Cl_2 \rightarrow PbCl_2 + ZnCl_2 + 2S. \quad (3.58)$$

The lead chloride is crystallised from brine solutions and the fused salt electrolysed to obtain lead and recover the chlorine.

$$Pb^{2+} + 2e \rightarrow Pb \quad \text{(cathode)} \quad (3.59)$$

$$2Cl^{-} - 2e \rightarrow Cl_2 \quad \text{(anode)} \quad (3.60)$$

Zinc chloride is obtained at a later stage from the original solution, as it is more soluble than $PbCl_2$.

Recycling of Metals

Interest in, and the practice of, recycling metals fluctuates with social, political and economic conditions. Prior to, and during, World War II metal recycling was a significant industrial process, and the amounts of Cu, Pb and Al recycled as a percentage of the total consumption were 44%, 39% and 28% respectively. However, the proportion of recycled metals has decreased, probably as a result of the cost and effort of processing, and optimism over resources. The energy inputs for purifying scrap metal (Table 3.15)

TABLE 3.15 Recycling of Metals - Energy Requirements

Metal	Energy Requirement as % of energy to extract metal from ores*	Metal	Energy Requirement as % of energy to extract metal from ores*
Steel	45 - 50	Aluminium	4.5
Titanium	20	Magnesium	2
Copper	10		

* Does not allow for energy consumption in collection and separation.

indicate the advantage of using scrap over ore. The percentage figures are based on data which excludes the energy component in getting the scrap ready for processing, though this is considered not to a significant energy factor.

The various methods of producing steel make use of scrap iron. The basic oxygen furnace can use 30% scrap, the open hearth furnace 40% and the electric arc furnace 98%. The growing popularity of the basic oxygen method (U.S.A. production in 1977; basic oxygen furnace 63%, electric furnace 22%, open hearth 16%), a process that generates its own scrap, may be a factor in reduced recycling of iron (from 60 to 40% between 1972-74).

Two problems in recycling metals are physical separation of materials, and dealing with alloys. Iron, because of its ferro-magnetism, is one of the easier metals to separate, and the colour of copper is an advantage in separating it from other metals. Aluminium used in cans may be contaminated with Pb and Sn from solder, iron used as a dispersion hardener and magnesium as a hardener. Any glass present will give Si at the temperatures of molten aluminium. These and other metals, such as Zn and Cu reduce the quality of the recycled aluminium. More research is necessary to find ways of removing impurities. One cannot, for example, oxidise the aluminium to Al^{3+} in order to remove the trace metals because this would "throw away" the energy originally spent in getting the aluminium from its ores. Steel cans are a significant source of scrap iron, and in 1972, 700 x 10^6 cans were collected in the U.S.A., and used for cementation extraction of copper, equation (3.57) above. Tin and labels etc. are removed by firing to 900 K which does not impair the reaction. Approximately 2 kg of iron are necessary to obtain 1 kg of Cu. The large surface area of the cans aids in the process.

The future must see an increase in the recycling of metals and materials. More research effort is needed to find economically viable methods. Recovery of copper from waste Cu, can be achieved by making use of the relative stabilities of Cu(I) and Cu(II) in different chemical environments. Acetonitrile, a by-product of the production of acrylonitrile;

$$\underset{\text{propylene}}{CH_3CH{=}CH_2} + NH_3 + O_2 \xrightarrow{\text{catalyst}} \underset{\text{acrylonitrile}}{CH_2{=}CHC{\equiv}N} + HCN + \underset{\text{acetonitrile}}{CH_3C{\equiv}N}, \qquad (3.61)$$

coordinates with Cu(I) stabilizing the oxidation state. In the presence of acetonitrile and acid, Cu metal (waste Cu) is oxidised to Cu(I) by Cu(II);

$$Cu_{(s)} + Cu^{2+}_{(aq)} + 2CH_3CN \xrightarrow{H^+} 2[Cu(CH_3CN)_2]^+. \qquad (3.62)$$

The Cu(I) complex dissolves in aqueous acetonitrile, but if the acetonitrile is distilled off (in the absence of O_2), the above reaction is reversed depositing purified Cu. The recovered CH_3CN and $Cu^{2+}(aq)$ can then be used over again. Other metals, such as zinc, will dissolve and remain in the aqueous acid solution.

Corrosion and its Prevention

The life-time of metals may be extended by preventing corrosion, the reverse of the extraction process. The formation of metal oxides is a favourable process (Fig. 3.7), though for most metals it is kinetically slow at ambient temperatures. There is an economic cost associated with corrosion, one estimate is that approximately half of the amount of metals extracted has been lost through corrosion. The loss is irreversible, as the oxidation products become thinly spread over the face of the earth. Two types of corrosion occur, the reaction of a metal with a gas, called tarnishing, and the reaction of a metal with an aqueous solution called corrosion.

Tarnishing. A common example of tarnishing is the darkening of silver-ware due to the formation of Ag_2S.

$$2Ag + H_2S \rightarrow Ag_2S + H_2 \qquad (3.63)$$

The rate at which metals tarnish increase with the temperature, but the tendency for oxide formation drops (ΔG^o becomes less negative). Three different rate laws can occur for gaseous (oxygen) tarnishing, parabolic, linear and logarithmic (Fig. 3.13). If the oxide film completely covers the

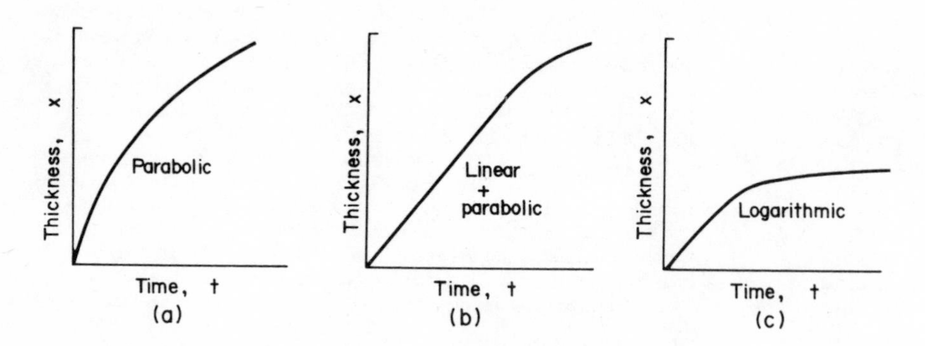

Fig. 3.13. Rate of growth of oxide films during tarnishing, (a) parabolic, (b) linear + parabolic and (c) logarithmic.

metal further tarnishing will be diffusion controlled. That is, as the thickness (x) of the oxide film increases, its rate of formation decreases. Hence we can write;

$$\frac{dx}{dt} = \frac{A}{x} \tag{3.64}$$

where A is a constant. Integration gives;

$$x^2 = 2At + \text{constant}, \tag{3.65}$$

a parabolic relationship (Fig. 3.13a). As the rate of diffusion increases with temperature, the rate of supply of O_2 could be the determining factor, i.e. a linear relationship between x and t;

$$x = Bt + \text{constant, (Fig. 3.13b).} \tag{3.66}$$

However, as the film thickens diffusion may again become the controlling factor and the parabolic rate takes over (Fig. 3.13b). Finally if the oxide film is a poor ionic or electron conductor so that diffusion is negligible the growth in x will follow a logarithmic law,

$$x = c \log[Dt + \text{constant}], \text{ (Fig. 3.13c).} \tag{3.67}$$

A mechanism by which the oxide film grows is shown for zinc oxide in Fig. 3.14. Zinc oxide, a non-stoichiometric oxide, has an excess of zinc ions $Zn_{1+\delta}O$. The zinc ions produced at the metal/oxide interface;

$$Zn_{(s)} \rightarrow Zn^{2+}_{(s)} + 2e \tag{3.68}$$

diffuse to the oxide/gas interface via the interstitial sites (in zinc oxide one half of the tetrahedral cavities are used). The electrons also move in the same direction (negative-type conduction) and reduce the O_2 absorbed at the oxide/gas interface

$$\tfrac{1}{2}O_{2(g)} + 2e \rightarrow O^{2-}_{(s)} \tag{3.69}$$

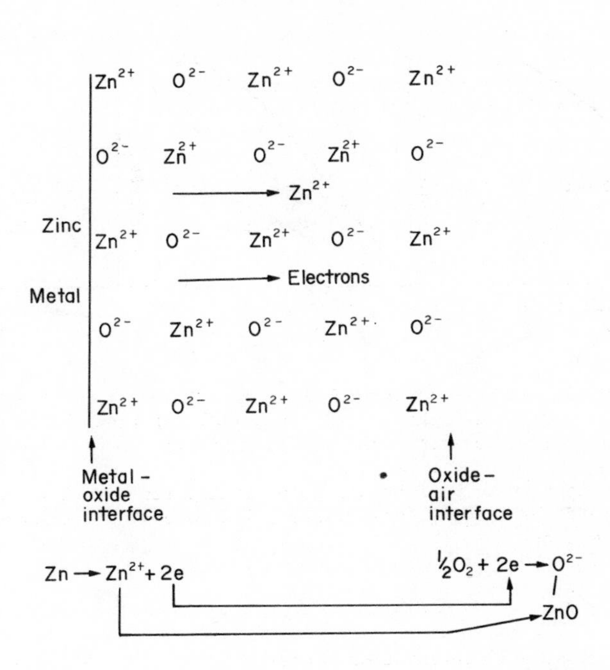

Fig. 3.14. Growth of zinc oxide film on the surface of zinc.

The oxides Al_2O_3 and Cr_2O_3 follow a logarithmic growth rate, Al_2O_3 has few lattice defects, and Cr_2O_3 is a poor ionic conductor. Hence the oxide film effectively reduces further corrosion. This is a reason for the wide use of Al as a structural metal, and the use of Cr in stainless steels and for plating.

Corrosion. Electrochemical reaction of a metal in contact with an aqueous solution is called corrosion. This is a major problem for iron, as the final product does not protect the metal from further reaction. Irrespective of this, iron is widely used because it is plentiful, relatively cheap to extract and has a great number of uses as a metal and in alloys. Iron corrosion is represented by the following processes. The reaction;

$$Fe_{(s)} - 2e \rightarrow Fe^{2+}_{(aq)}, \ (+0.44\ V), \tag{3.70}$$

is anodic. The reduction reaction at the cathode, which can be a good distance from the anode, could involve a number of species, but the most common is oxygen;

$$O_2 + 2H_2O + 4e \rightarrow 4OH^-_{(aq)}, \ (+1.23\ V). \tag{3.71}$$

The overall reaction is;

$$2Fe_{(s)} + O_2 + 2H_2O \rightarrow 2Fe^{2+}_{(aq)} + 4OH^-_{(aq)}, \ (+1.67\ V), \tag{3.72}$$

(see Fig. 3.15). The process is assisted by low pH, high temperatures, and electrolytes which improve the electron and ionic conduction. Salt spray provides a convenient electrolyte. Areas of strain and impurities in the iron can act as points for the anode reaction.

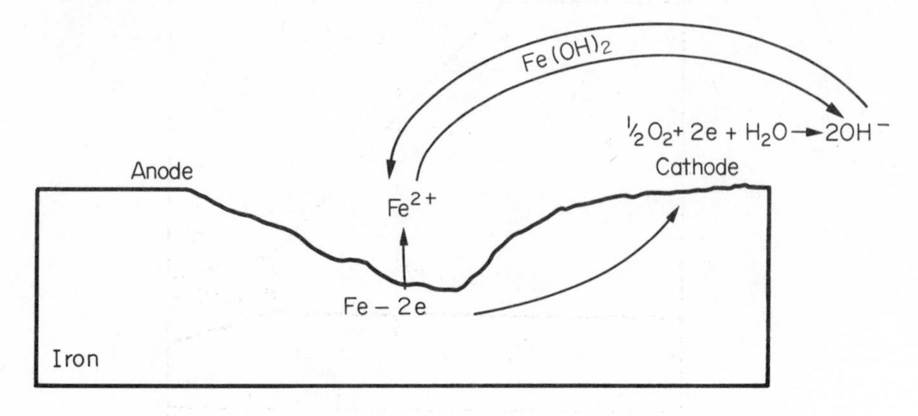

Fig. 3.15. The corrosion of iron

The primary product of corrosion, $Fe(OH)_2$, reacts further in alkaline conditions;

$$4Fe(OH)_2 + O_2 + 2H_2O \rightarrow 4Fe(OH)_3, \qquad (3.73)$$

and the ferric hydroxide loses water to give hydrated ferric oxide, red-brown rust;

$$2Fe(OH)_3 \xrightarrow{H_2O} Fe_2O_3.nH_2O. \qquad (3.74)$$

Zinc undergoes similar corrosion, except that the $Zn(OH)_2$ reacts with CO_2 to give $Zn(OH)_2ZnCO_3$, which protects the metal from further corrosion.

The corrosion of iron is conveniently summarised in an E_h - pH diagram, Fig. 3.16. The primary product Fe^{2+} is prevented from forming by raising the potential (anodic protection) or lowering it (cathodic protection) or increasing the pH (>9). Steel is protected from corrosion when in contact with water by raising the pH. Steel used for reinforcing concrete does not corrode as the pH > 12.

Galvanic corrosion occurs when two metals are connected in an electrolyte. The reason is because the standard reduction potentials of the metals are different. For copper (E^o = +0.34) and iron (E^o = -0.44), iron, the anode, will corrode;

$$Fe_{(s)} - 2e \rightarrow Fe^{2+}_{(aq)}, \quad E = +0.44 \text{ V}, \qquad (3.75)$$

and copper acts as the cathode for the reaction;

$$O_2 + 2H_2O + 4e \rightarrow 4OH^-_{(aq)}. \qquad (3.76)$$

Copper also catalyses the reduction of oxygen.

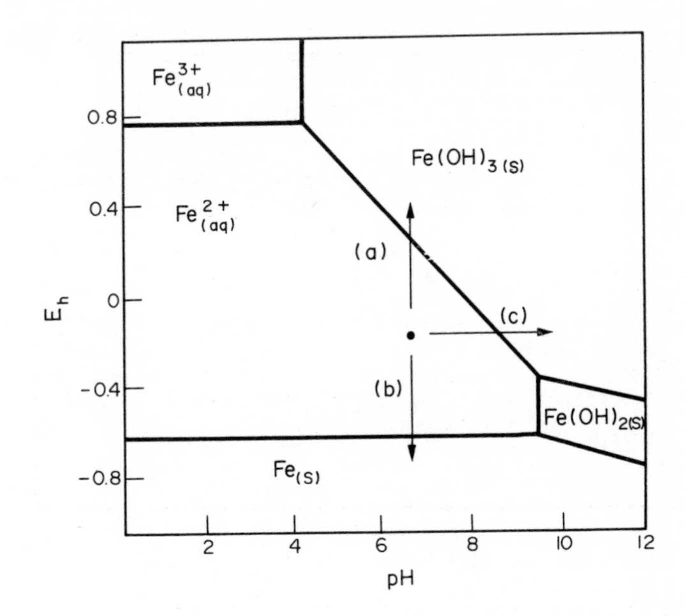

Fig. 3.16. E_h - pH diagram for the Fe/H_2O system in relation to the prevention of corrosion. Movement along (a) i.e. raised potential is anodic protection, along (b) i.e. lower potential is cathodic protection, and along (c) is pH protection.

For some metals (e.g. Al, Cr) the oxides adhere firmly to the metal protecting it from further corrosion, and making the metal "passive". Within the pH range 4 to 9, Al_2O_3 renders Al passive, but outside these limits, the oxide, which is amphoteric, will dissolve. The passivity of a metal is affected by ions. For example chloride ions can replace oxide ions to give corrosion called pitting.

Prevention of corrosion. Cathodic protection of iron is achieved by lowering the potential on the metal so that it becomes the cathode, this can be done directly by applying a potential from a d.c. source. Coating iron with another metal, such as zinc (in the galvanising of iron), provides protection from the $Zn(OH)_2ZnCO_3$ coating. However, if the iron is exposed (as long as the area is not too great) the zinc will corrode preferentially, this is not the case for tin coatings. It is not necessary to coat the iron in some cases; blocks of Mg, called the sacrificial anode, when linked to iron, will corrode preferentially.

Chemical inhibitors added to water will retard the electrode processes. For example, deposits of $CaCO_3$ on the inside of iron pipes inhibit the cathode reaction, raising the pH with NaOH has the same effect. Iron phosphate or chromate deposits act as anode reaction inhibitors.

A process called anodising forms a protective film of Al_2O_3 on Al. The metal is deliberately made the anode of a cell and a film of Al_2O_3 allowed to grow. In either boiling water or $Na_2Cr_2O_7$ solutions the film becomes non-porous. One finds that aluminium cooking vessels used for cooking in oils show signs of corrosion while those used for cooking in water do not.

Various inorganic and organic paint coatings are a means of keeping O_2 and/or water away from the metal surface. This is satisfactory only if the surface is completely covered. If a hole appears corrosion will occur and can continue underneath the paint. The addition of zinc compounds to paint assists in inhibiting both the cathodic and anodic reactions.

NON-METALLIC RESOURCES

We will conclude this chapter with a brief outline of non-metal resources, and some of the more important methods of extraction. In contrast with the metals most non-metals are used in compounds rather than as free elements. We will discuss some of the more important compounds in Chapter 6. Carbon resources will be discussed in Chapter 4.

Resources

The major non-metal resources are listed in Table 3.16. The formation of a number of these resources has already been discussed (Chapters 2 and 3). Over 90% of the world's phosphate comes from marine sedimentary

TABLE 3.16 Non-Metallic Resources

Element	Crustal abundance (ppm)	Principal mineral resource
B	10	Borosilicates, borates, e.g. $Na_2B_4O_7 4H_2O$
C	270	Coal, petroleum, limestone, shells, living matter
N	46 78.1% (atmosphere)	Atmosphere, KNO_3(India), $NaNO_3$ (Chile)
P	1420	Phosphate rock, e.g. $Ca_5(PO_4)_3F$, $Ca_5(PO_4)_3OH$
O	46.6% (crust) 20.9% (atmosphere)	Atmosphere, oxides, silicates
S	480	Elemental S, volcanic and bacterial; sulphides, sulphates, e.g. $CaSO_4$
F	650	Fluorospar, CaF_2
Cl	310 (crust) 1.9% (sea water)	Sea water, Evaporites.
Br	1.6 (crust) 65 (sea water)	Sea water
I	0.3 (crust) 0.05 (sea water)	Iodates, e.g. $Ca(IO_3)_2$ (Chile)

deposits produced from the weathering of granites. The liberated phosphates found their way into the oceans, and could well have been concentrated by marine organisms in bones, shells and tissue and redeposited. Finally, the more soluble $CaCO_3$ was leached away. Other important deposits arise as residual deposits from the weathering of limestone and alkaline igneous rocks containing phosphate. Small deposits on coral atolls are of bio-marine origin.

Extraction of the Halogens

Chlorine is the most important halogen, with a wide range of uses. The major sources are evaporites and sea-water (Table 2.2). Production of NaCl from sea-water accounts for approximately 29% (39 x 10^9 kg annually) of the total amount of salt produced. The salt is obtained by solar evaporation. Water of density 1.03 g l^{-1} is run into concentrating ponds and is held while mud and silt settle, and the density rises to 1.16 g l^{-1}. During this time Fe and Ca carbonates precipitate and the volume of water has been reduced by 80%. The concentrated sea-water is removed to new ponds where calcium sulphate precipitates, first as gypsum $CaSO_4 2H_2O$ and then as anhydrite $CaSO_4$. The brine is held until the density rises to 1.21 g l^{-1} and then moved to harvesting ponds where the NaCl crystallises out. When the density has risen to 1.26 g l^{-1} (4% of the original water now remains), the liquor is removed before other materials, such as $MgSO_4$, $4KCl.4MgSO_4.11H_2O$ (rainite), $KCl.MgCl_2.6H_2O$ (carnallite) and $MgBr_2$ crystallise. These materials are significant sources of magnesium (60% of total production) and bromine (27% of total production).

The halogens (F, Cl, Br) are obtained from halide salts by oxidation

$$2X^-_{(aq)} - 2e \rightarrow X_{2(g,\ell,s)} \qquad (3.77)$$

From factors that contribute to the oxidation potential (Table 3.17) the endothermic heat of reaction can be calculated. The low potentials of bromide and iodide explain why chemical oxidising agents can be used, but for fluoride and chloride electrical oxidation is necessary. The high negative potential for fluorine is a consequence of the high hydration energy of F^- and low F-F bond energy.

Fluorine. Since fluorine reacts rapidly with water it is produced by electrolysis of an anhydrous mixture of KF/HF (~1:2) which is conducting and has a low melting point (72°C). The cell walls are often made of monel metal (Ni/Cu alloy ~2:1), which is resistant to attack by fluorine. Mild steel is used for the cathode and carbon for the anode. The electrochemical reactions are;

$$2H^+ + 2e \rightarrow H_{2(g)} \quad \text{(cathode)} \qquad (3.78)$$

$$2F^- - 2e \rightarrow F_{2(g)} \quad \text{(anode)} \qquad (3.79)$$

producing H_2 and F_2, which react explosively and need to be kept apart.

Chlorine. Chlorine is produced by electrolysing brine solutions, and is normally associated with the production of caustic soda, the "Chloralkali" cell. Two cell types are used, the diaphragm cell and the flowing mercury

TABLE 3.17 Extraction of the Halogens

$$X^-_{(aq)} - e \xrightarrow[E]{\Delta H} \tfrac{1}{2}X_{2(g)}$$

$\downarrow -\Delta H_{hydration}$

$X^-_{(g)}$ — $-\tfrac{1}{2}$(bond energy)

$\downarrow$ - Electron affinity

$X_{(g)}$

	F^-	Cl^-	Br^-	I^-
$-\Delta H_{hydration}$	460*	385	351	305
- Electron affinity	333	348	340	297
½(Bond energy)	-79	-121	-96	-76
Sublimation			-15	-30
ΔH	714	612	580	496
E(Volts)	-2.87	-1.36	-1.07	-0.54

* values in kJ mol^{-1}

cell. In both cases Cl_2 and H_2 are produced, but the cells differ in the way they keep the two gases apart.

An asbestos diaphragm keeps the gases apart in the diaphragm cell, the reactions are;

$$2Cl^-_{(aq)} - 2e \rightarrow Cl_{2(g)} \quad \text{(graphite anode)}, \qquad (3.80)$$

and

$$2H^+_{(aq)} + 2e \rightarrow H_{2(g)} \quad \text{(iron cathode)}. \qquad (3.81)$$

Sodium hydroxide concentrates in the cell, hence the overall reaction is;

$$Na^+_{(aq)} + Cl^-_{(aq)} + H_2O \rightarrow \tfrac{1}{2}Cl_2 + \tfrac{1}{2}H_2 + OH^-_{(aq)} + Na^+_{(aq)} \qquad (3.82)$$

In the mercury cell sodium metal is discharged rather than hydrogen. The metal amalgamates with the mercury cathode, which is allowed to react with water to give H_2 and NaOH.

$$2Cl^-_{(aq)} - 2e \rightarrow Cl_{2(g)} \quad \text{(carbon anode)}, \qquad (3.83)$$

$$\left.\begin{array}{l} Na^+_{(aq)} + e \rightarrow Na \\ Na + Hg \rightarrow Na/Hg \end{array}\right] \quad \text{(mercury cathode)}. \qquad (3.84)$$

The next step may be considered as a secondary cell,

$$Na/Hg - e \rightarrow Na^+_{(aq)} + Hg, \quad \text{(amalgam anode)}, \qquad (3.85)$$

and

$$2H^+ + 2e \rightarrow H_{2(g)}, \quad \text{(graphite or iron cathode)}. \qquad (3.86)$$

The mercury is recycled back to the cell and sodium hydroxide solution is left.

Some comparative data for the two cells are given in Table 3.18. The diaphragm cell uses saturated brine solutions, while the mercury cell requires dry salt to make the brine, but produces purer NaOH. However, the

TABLE 3.18 Operational Data for Chloralkali Cells

	Diaphragm cell	Mercury cell
Current (amp)	10,000 - 60,000	25,000 - 300,000
Output (kg Cl_2/day)	300 - 2,000	1,000 - 10,000
Cell voltage (V)	3.75 - 3.85	4.25 - 4.50
Energy consumption (kJ/kg Cl_2)	~10.8 x 10^3	~12.6 x 10^3
Energy efficiency (%)	~58	~47
Cell temperature (^oC)	95	65 - 70
Graphite consumption (kg/1000 kg Cl_2)	2.7 - 3.6	2.0 - 2.8
Anode life (days)	220 - 310	150 - 450
Salt consumption (kg NaCl/kg Cl_2)	1.8	1.7

loss to the environment of around 0.1 - 0.2 kg of mercury per 1000 kg of Cl_2 produced has led to serious pollution and health problems. The diaphragm cell has the more favourable energy consumption.

For the diaphragm cell

$$Cl^- + H_2O \rightarrow OH^- + \tfrac{1}{2}Cl_2 + \tfrac{1}{2}H_2, \quad \Delta G_{298\ K} = 211\ kJ \qquad (3.87)$$

and for the mercury cell

$$Na^+ + Cl^- \xrightarrow{H_2O/Hg} Na/Hg + \tfrac{1}{2}Cl_2, \quad \Delta G_{298\ K} = 318\ kJ \qquad (3.88)$$

The free energy per kg of Cl_2 produced is 5.9 x 10^3 kJ and 8.9 x 10^3 kJ respectively. That is 33% less energy is required for the diaphragm cell reaction, but in practice the energy required is approximately 14% less. The standard reversible cell potentials are 2.19 V and 3.3 V for reactions (3.87) and (3.88), but higher voltages are used, due in part to the high overvoltage of Cl_2 at the anode in both cells and H_2 at the cathode of the diaphragm cell.

Bromine and Iodine. Both bromine and iodine are obtained using chemical oxidising or reducing agents. Bromide from brines remaining after the removal of NaCl (bitterns) is oxidised with Cl_2 to give bromine.

$$2Br^-_{(aq)} + Cl_{2(g)} \rightarrow 2Cl^-_{(aq)} + Br_{2(g)} \qquad (3.89)$$

The bromine, blown out as vapour, reacts with SO_2 and is then retreated with Cl_2;

$$Br_2 + SO_2 + 2H_2O \rightarrow 2HBr + H_2SO_4, \quad \downarrow Cl_2 \quad Br_2 + 2HCl \tag{3.90}$$

Alternatively Br_2 can be added to sodium carbonate solution to give bromide and bromate;

$$3CO_3^{2-} + 3Br_2 \rightarrow 5Br^- + BrO_3^- + 3CO_2, \tag{3.91}$$

which, on acidification, regenerates bromine.

$$5Br^- + BrO_3^- + 6H^+ \rightarrow 3Br_2 + 3H_2O \tag{3.92}$$

The main source of iodine is iodate, which is reduced with bisulphite to iodide,

$$2IO_3^- + 6HSO_3^- \rightarrow 2I^- + 6SO_4^{2-} + 6H^+ \tag{3.93}$$

The iodide is then reacted with sufficient mother liquor to liberate iodine;

$$5I^- + IO_3^- + 6H^+ \rightarrow 3I_2 + 3H_2O. \tag{3.94}$$

Iodine can also be obtained from brine using Cl_2 or NO_2^- as the oxidising agents.

CHAPTER 4

Energy Resources

Matter and energy are the two principal resources of our Universe. Over recent years, energy has been highlighted as the major problem of the world - viz. the extent of the resources, and how future needs can be met.

The Einstein equation demonstrates the link between matter (mass, m) and energy (E);

$$E = mc^2, \tag{4.1}$$

where c is the velocity of light. The equation states that the total matter and energy in the Universe is fixed. For processes normally encountered (other than nuclear processes) two laws are applicable: the conservation of matter and the conservation of energy.

In this chapter we will discuss the world's energy resources, their formation and use.

WHAT IS ENERGY?

Energy occurs in many forms, some of which are interconvertible. A major use of energy is to produce work. This is embodied in the *first law of thermodynamics*;

$$\Delta U = q - w, \tag{4.2}$$

where ΔU is the change in internal energy of a system, q the heat supplied and w the work done *by* the system. U is a state function, i.e. it is independent of the route by which the change occurs. If this was not so, then by altering the route, energy could be created or destroyed, and there would be no energy problem.

Two principal types of work are electrical work (i.e. EI watts, where E is the potential drop across a resistance R carrying a current I), and pressure-volume work (particularly for gases). If the pressure P is held constant (e.g. atmospheric pressure), the usual situation, the work is;

$$w = PdV, \qquad (4.3)$$

where dV is the change in volume. Therefore, the heat at constant pressure (q_p) from equation (4.2) is;

$$q_p = \Delta U + PdV. \qquad (4.4)$$

This defines ΔH, the *enthalpy* change, the common way of expressing chemical energy. Enthalpy is connected with temperature by the expression;

$$\Delta H = C_p \Delta T, \qquad (4.5)$$

where C_p is the heat capacity of the system at constant pressure. When ΔH is negative the process is exothermic, and when positive, endothermic.

The tendency is for useful energy to be lost for a number of reasons, a major one being inefficient energy interconversions. For example, approximately 25% of the combustion energy of petrol is used moving a car, the rest is "lost" as heat in the engine and exhaust. This energy can not be recovered to move the car.

The one unifying theme in the cooling of a hot meal and the melting of a block of ice is a change in the randomness or disorder of the system. The measure of the disorder is called *entropy* (S). Change in entropy (ΔS) equates with the difference in the entropies of the two states undergoing the change. If $\Delta S_{universe} > 0$ the change is spontaneous and there is an increase in the entropy of the universe (the change does not need to involve energy). In terms of heat transfer;

$$\Delta S = \frac{q_{rev}}{T}, \qquad (4.6)$$

where q_{rev} is the heat absorbed in a reversible process. This is summarised in the *second law of thermodynamics*, the entropy of the universe increases in a spontaneous process, and does not change in a reversible process;

i.e. $$\Delta S_{univ.} \geqslant 0, \qquad (4.7)$$

or $$\Delta S_{univ.} = \Delta S_{system} + \Delta S_{surrounding} \geqslant 0. \qquad (4.8)$$

It is this aspect that relates to an energy crisis, as energy sources are, in time dissipated (but not destroyed), into less useful forms as the entropy increases. A similar problem was discussed in Chapter 3 with respect to extraction of iron from ores and then dissipation as rust, this also corresponds to increasing entropy.

The requirement $\Delta S_{univ} \geqslant 0$ does not apply to the terms ΔS_{sys} and ΔS_{surr} (equation 4.7), only to their difference. For the spontaneous reaction;

$$2H_{2(g)} + O_{2(g)} \rightarrow 2H_2O_{(1)} \qquad (4.9)$$
$$\text{(3 moles gas)} \rightarrow \text{(2 moles liquid)}$$

ΔS is negative, a decrease in entropy. Hence a new function indicating the spontaneity of an isolated reaction is required. This is obtained by linking enthalpy and entropy in the Gibbs free energy G;

i.e. $$G = H - TS, \qquad (4.10)$$

and for a change in the free energy of the system;

$$\Delta G_{sys} = \Delta H_{sys} - T\Delta S_{sys} \tag{4.11}$$

The spontaneity of a reaction is given by the sign of ΔG, i.e.;

$$\Delta G < 0 \quad \text{feasible (spontaneous) reaction,} \tag{4.12}$$

$$\Delta G > 0 \quad \text{unfeasible (non-spontaneous) reaction,} \tag{4.13}$$

and $$\Delta G = 0 \quad \text{at equilibrium.} \tag{4.14}$$

The source of chemical energy is the chemical bond, the attraction of the nucleus of one atom for the valence electrons of another. From bond energies it is possible to determine the value of ΔH for a chemical reaction (see Chapter 5).

Numerous units have been used for energy and power. The SI units are joule (J)(kg m^2 s^{-2}) for energy and watt (W)(kg m^2 s^{-3} or J s^{-1}) for power. The relationships between these units and some of the more common non-SI units are given in Table 4.1.

TABLE 4.1 Energy and Power Units

To convert from	To	Multiply by	To convert from	To	Multiply by
Energy					
kilojoules	joules	1000	joules	kilojoules	0.001
calories	joules	4.184	joules	calories	0.2390
ergs	joules	1×10^{-7}	joules	ergs	1×10^{7}
Btu's	joules	1.055×10^{3}	joules	Btu's	9.479×10^{-4}
kilowatt-hrs	joules	3.6×10^{6}	joules	kilowatt-hrs	2.778×10^{-7}
foot - lbs	joules	1.356	joules	foot - lbs	0.7376
Power					
joule s^{-1}	watts	1	watts	joule s^{-1}	1
kilowatts	watts	1000	watts	kilowatts	0.001
ergs s^{-1}	watts	1×10^{-7}	watts	ergs s^{-1}	1×10^{7}
Btu's hr^{-1}	watts	0.2930	watts	Btu's hr^{-1}	3.4129
ft-lbs min^{-1}	watts	0.0226	watts	ft-lbs min^{-1}	44.27
horsepower	watts	7.457×10^{2}	watts	horsepower	1.341×10^{-3}

ENERGY RESOURCES AND CONSUMPTION

The fossil fuels; coal, oil and natural gas, are the world's major and readily available energy sources. Since they are non-renewable, this poses a problem for future energy supplies.

Figures, based on a number of estimates of the world's fossil fuel reserves, are given in the 4th and 5th columns of Table 4.2. The data is approximate because of difficulties in estimating quantities accurately. Also, the important quantity is the reserves that can be mined economically rather than the total present. Estimates of the world consumption of fossil fuels (at about 1976) are given in columns 2 and 3. The yearly consumption of coal, oil and natural gas is close to 250 x 10^{18} J, i.e. 0.05% of optimistic estimates of reserves (mainly coal) and 1.4% of estimated recoverable and proved reserves. Of the three fuels, the reserves of coal are greatest, viz. 88% of the total recoverable and proved fossil fuel reserves. Inclusion of tar sands and shale oil increase the reserves significantly, but also push up the cost of energy.

The growth in energy consumption (Fig. 4.1) highlights the significant contribution made by oil. The growth from 1890 to the 1970's has been

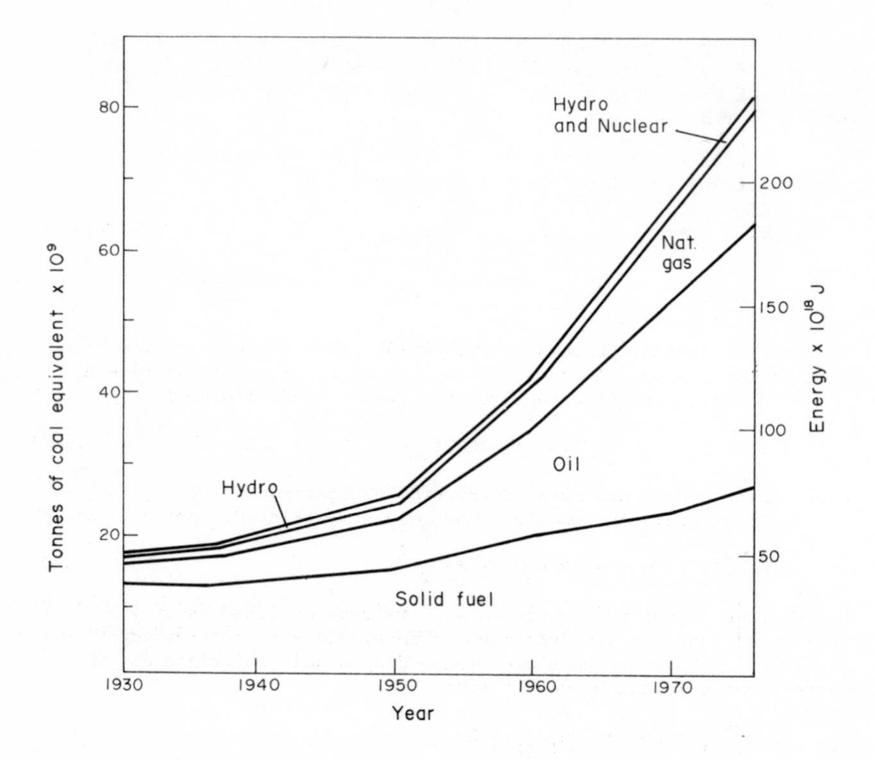

Fig. 4.1. Growth in world energy consumption 1930-1976. Source U.N. Statistical Year Books (1 tonne coal ≡ 7.3 barrels oil ≡ 750 m^3 gas)

around 6.9% a year, while for coal the rate has been less, 4.4% yr^{-1} (1860-1910), 0.75% yr^{-1} (1910-1945) and 3.6% yr^{-1} (from 1945). The proportional fall off in coal is dramatically demonstrated in Fig. 4.2. A small, but increasing use of hydro-, and more recently nuclear, power is also clear from Fig. 4.2.

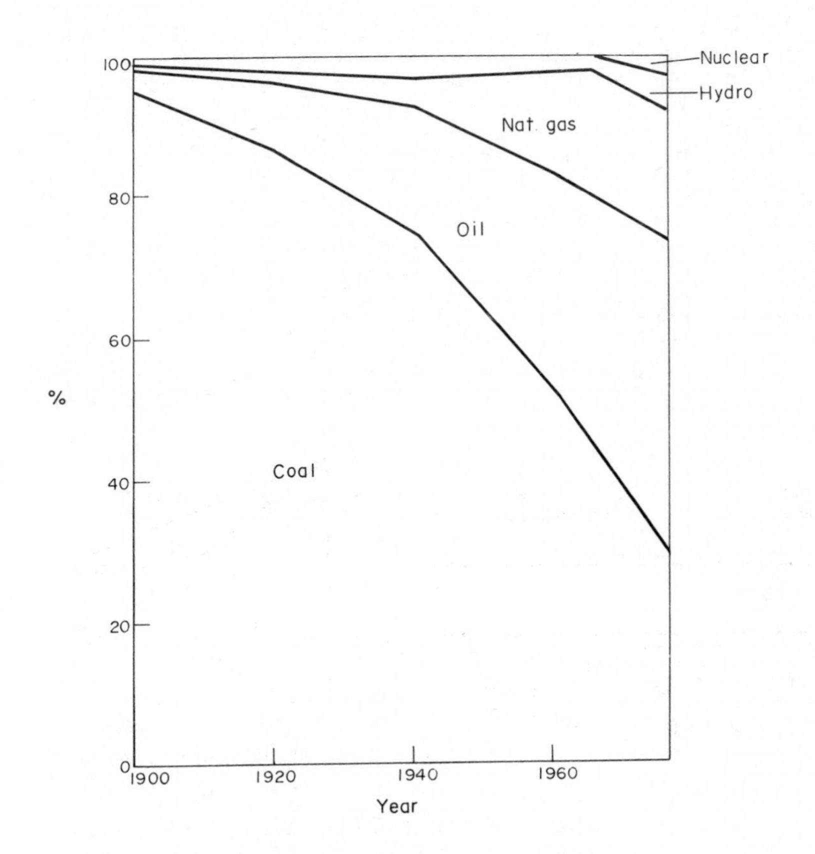

Fig. 4.2. Percentage composition of world fuel use, 1900 - 1978

In the USA 35% of the energy goes to industry, 34% to transport, 17% for residential and 14% for commercial use. Oil contributes significantly in each of these sectors, 99% of transport energy, 49% of commercial and residential energy and 39% of industrial energy. These figures highlight the vulnerability of lifestyles built on oil supplies. A compact summary of the flow of energy in the United States society (1975) is given in Fig. 4.3. Similar diagrams can be constructed for other nations, while there may be differences in the basic energy sources, the end use of the energy is probably similar, for Western nations. Energy losses are high, and in this case, only 44% of the energy produced is used usefully. The greatest losses are in the generation and transmission of electricity (68%) and in transport (72%).

Other energy sources make minor contributions to the world's total energy. However, a few renewable sources, could be more exploited (Table 4.3). Hydro power uses about 9% of the potential water power, estimated at 9.5 x 10^{18} J (per year). Tidal power has a potential of 2.0 x 10^{18} J (per year) and geothermal energy 2 x 10^{8} J to 20 x 10^{18} J (per year). Direct use of

TABLE 4.2 Production and Reserves of Fossil Fuels

Fossil Fuel	Production/Consumption[1] (yearly 1976)		Reserves[1,5]			
	Quantity[2]	Energy[2] (Joules)	Quantity[2]	Energy[2] (Joules)	Major Sources[3]	Life time[4] (years)
Coal	2700×10^6 t	78×10^{18}	Reserves 7.64×10^{12} t 8.8×10^{12} t 9.9×10^{12} t Recoverable 0.44×10^{12} t 0.57×10^{12} t	 220×10^{21} 253×10^{21} 285×10^{21} 12.7×10^{21} 16.4×10^{21}	USSR(56%),USA(19%), Asia(9%),Canada(8%), W. Europe(5%)	Estimate from 300 to 800
Oil	2874×10^6 t or 21×10^9 barrels	124×10^{18}	Reserves 2000×10^9 barrels Proved 500×10^9 barrels to 642×10^9 barrels	 11.8×10^{21} 2.95×10^{21} to 3.79×10^{21}	Middle East(55%),Nth Africa(13%),USSR(12%), USA(6%),China(3%), Venezuela(2%)Other Sth & Cent.America (2.5%),Canada(1.6%), Nth Sea Countries(1%)	50 to 60
Natural gas	1250×10^9 m^3	48.1×10^{18}	Reserves 200×10^{12} m^3 to 280×10^{12} m^3 Proved 53×10^{12} m^3 to 69×10^{12} m^3	 7.7×10^{21} to 11×10^{21} 2.04×10^{21} to 2.66×10^{21}	USSR(34%),Middle East (18%),Nth America (17%),Africa(9%), W. Europe(9%),Asia(6%)	40 to 100
Tar Sand			0.3×10^{12} barrels to 4.2×10^{12} barrels	1.77×10^{21} to 25×10^{21}	Canada, Venezuela, USSR	

Shale oil	17×10^{12} barrels known	100×10^{21}	USA, Africa, Asia, Sth America
	0.72×10^{12} barrels (quality 110-450 l $tonne^{-1}$)	4.2×10^{21}	
	1500×10^{12} barrels known	8900×10^{21}	
	4.6×10^{12} barrels (quality 20-110 l $tonne^{-1}$)	27×10^{21}	

1 Figures given are based on many different estimates.

2 Conversions from one quantity to another are based on energy content. The conversions used in producing the table are:

1 tonne coal $\equiv 28.8 \times 10^9$ J
$\equiv$ 0.77 tonne oil
$\equiv$ 7.3 barrels of oil
$\equiv$ 750 m^3 gas

1 tonne oil $\equiv$ 7.3 barrels (av)
$\equiv 43.2 \times 10^9$ J

1 barrel $\equiv 5.9 \times 10^9$ J

1 m^3 gas $\equiv 38.5 \times 10^6$ J

These are based on statistics from UN and British Petroleum and the oil industry units.

3 Percentages vary from estimate to estimate and can differ by up to 5% in some cases.

4 Life time figures are very approximate.

5 Resource - total amount estimated to be recoverable for the benefit of man, based on knowledge and reasonable conjecture. Reserves - total amount recoverable in terms of economic and operational feasibility. Proved reserves - amount reasonably certain to be produced in the future on current economic and operational conditions from deposits established on known geological and engineering data.

Sources: U.N. Statistical Yearbook 1978; S.S. Penner and L. Icerman, Energy Vols I,II(1974-5) Addison-Wesley, Reading; Royal Society London, Energy in the 1980's, 1974.

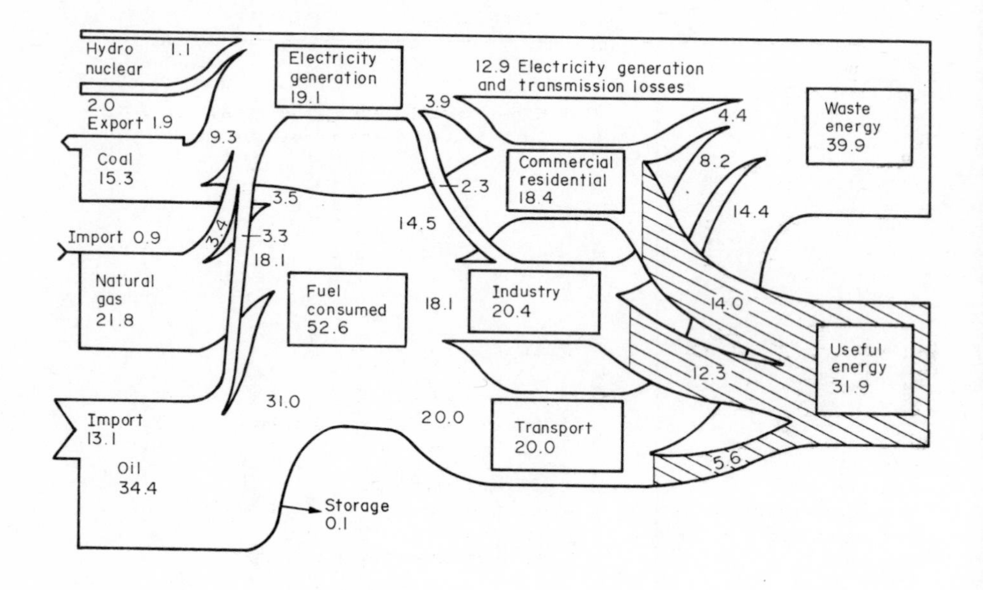

Fig. 4.3. Production and consumption of energy in USA (1975). Figures are in 10^{18} J.

solar energy is under intense investigation. For a 10% conversion of solar energy (at the earth's surface) by photovoltaic cells, and for a solar power of 12.5×10^6 J m^{-2} day^{-1}, an area of 17,000 km^2 is needed to produce in one year, the same energy as the current yearly consumption of coal.

A gram of ^{235}U or ^{239}Pa, has an energy potential of 82×10^9 J. Hence reserves of uranium (Table 3.5) indicate this is a reasonably plentiful source of energy. Better still, is the energy from fusion. A litre of water contains 1×10^{22} deuterium atoms (0.033 g), and their fusion would provide 7.95×10^9 J. The deuterium in 1 km^3 of seawater (10^{12} litres) would provide (potentially) 7.95×10^{21} J. These figures look promising, but the calculations assume no losses, no problem in realising the nuclear energy and ignore the energy required to initiate the fusion and get the raw material.

FOSSIL FUELS

The fossil fuels formed from organic materials millions of years ago. The oldest coals, anthracites, were laid down around 380×10^6 years ago, and some of the youngest lignites 2×10^6 years ago. Oil and natural gas are, in general, younger. Coal and oil are still forming, but slowly relative to present day consumption. One estimate puts coal formation at 9×10^6 tonnes, oil at 0.25×10^6 tonnes and natural gas at 0.15×10^6 tonnes, per year.

TABLE 4.3 Sources of Energy

Source	Comments
Fossil Fuels	
Coal, Oil, Natural Gas, Tar Sands, Shale Oil	Non-renewable energy sources, see Table 14.2 for details on reserves and consumption
Nuclear	
Fission, Breeder reactors, Fusion	Large potential source of energy, see text.
Solar	
Solar collectors	Renewable resource
Solar cells	See text. For direct conversion to electricity
Photosynthesis	Approx. 1.3×10^{21} J used per year.
Water	
Tides	A local resource
Oceans thermal gradient	Heat conversion to electricity
Hydroelectric	See text.
Hydrogen	From decomposition of water.
Earth	
Geothermal, Hydrothermal, Winds	Localised energy resources
Storage	
Fuel cells, Batteries	Chemical energy stored until required.
Wastes	
Waste heat	Using waste heat, e.g. from power generation.
Solid wastes	Combustion to produce electricity.
Other	
Magnetohydrodynamic	Production of electricity when a hot ionized gas passes perpendicular to a magnetic field.

Coal

Coal may become the major fuel of the future, as oil supplies fall off. Coal was used as a source of energy as early as 110 BC in China.

Formation of coal. Coal seams originated from dense vegetation growing in swampy areas. Little is known about the chemical mechanisms involved, but some broad features are understood (Fig. 4.4). The energy available from the combustion of coal was collected by photosynthesis. When the vegetation died bacterial aerobic decay produced humic materials. The process then became anaerobic, and peat formed, as water and sediments covered the vegetation and excluded oxygen. Decay stopped, as the bacteria died from lack of oxygen, and in time, as the pressure from the overlying sediments

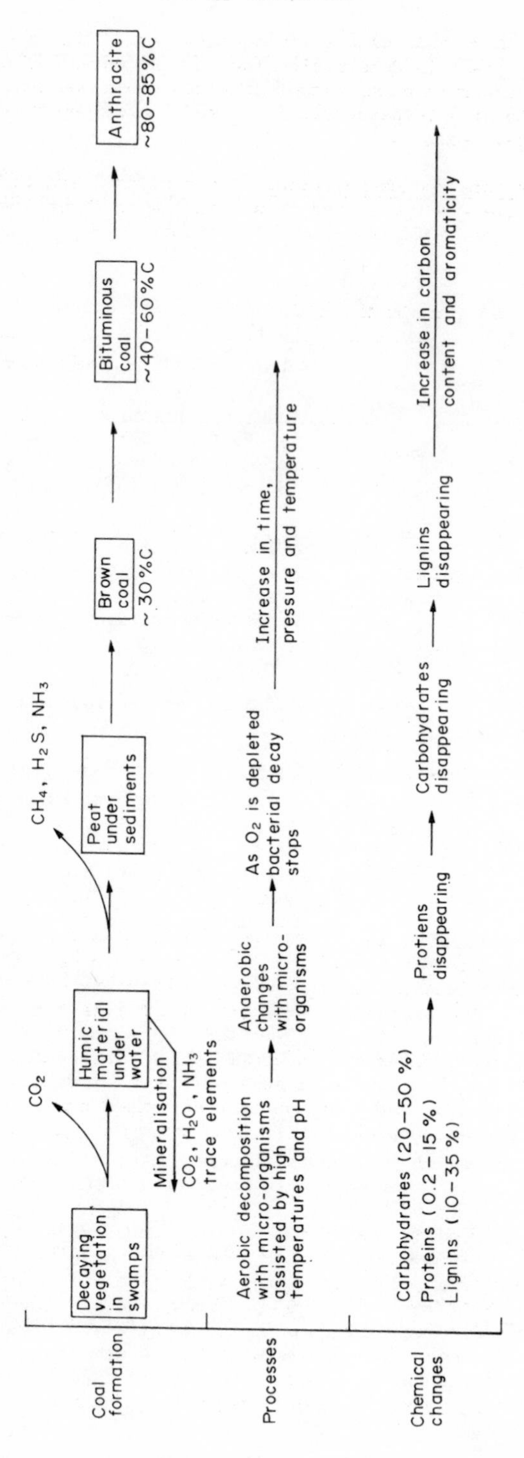

Fig. 4.4. The formation of coal

and temperature increased the peat changed to coal. The time and temperature are probably important in determining the carbon content of coal. Free carbon does not occur in coal, except in a few seams deeply buried which have been subjected to high temperatures. A 3 to 5 meter depth of peat gives about a 30 cm coal seam.

Coal types, structure and composition. Types of coal and their composition are listed in Table 4.4. The more carbon in the coal the greater its energy content, and the more "graphite-like" its structure.

TABLE 4.4 Types of coal[a]

Coal	%C	%ash	%volatiles	%moisture	Energy[b] 10^9 J $tonne^{-1}$
Anthracite	82	9	5	4	30.5 - 37.5
Bituminous					
Low volatiles	66	12	20	2	25.8 - 35.2
Med. volatiles	64	10	23	3	25.8 - 35.2
High volatiles	46	4	44	6	25.8 - 35.2
Sub-bituminous	40	9	32	19	18.8 - 28.1
Lignite	30	5	28	37	12.9 - 18.8

[a] Composition figures are not exact but indicate the general range of values.

[b] For comparison the energy content of peat $\approx 13 \times 10^9$ J $tonne^{-1}$ and wood $10 - 14 \times 10^9$ J $tonne^{-1}$.

Coal, a carbonaceous rock, contains ultrafine capillaries and inorganic materials such as calcite and pyrites. The structure is a three dimensional arrangement of condensed rings, some containing the hetero-atoms O, S and N (Fig. 4.5). Approximately 50% of the sulphur is organic, as thiophenes and sulphides, and 50% as FeS_2. The pyrite may have formed as hydrothermal deposits, after the anaerobic stage, when sulphate was reduced to S^{2-} and S_2^{2-}. Levels of some of the minor components of coa are listed in Table 4.5. Many of the concentrations are low, but since a number of the elements remain in the coal ash, they are concentrated 5 to 10-fold.

Uses of coal. Coal may be burnt for energy, but in general, this is inefficient and polluting. Powdered coal, through which air is passed so that it has the appearance of a liquid (fluidised), burns more efficiently and cleanly. The data in Fig. 4.3 (for the USA) reveals that 61% of coal is used to generate electricity, and 50% of the electricity comes from coal combustion. This is rather inefficient due to energy losses in the

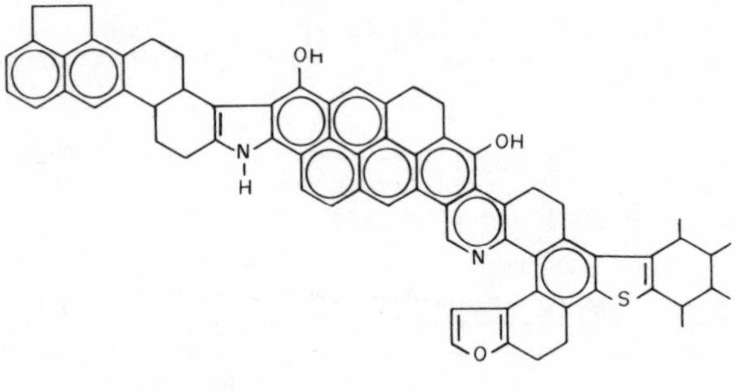

Fig. 4.5. A portion of coal structure

TABLE 4.5 Some Minor Constituents of Coal (>50 mg kg^{-1})

Element	Mean %	Range %	Element	Mean mg kg^{-1}	Range mg kg^{-1}
Al	1.0	0.3 - 3.0	Ba	200	1 - 3000
Ca	0.15	0.05 - 0.37	Cl	500	10 - 8000
Fe	0.8	0.05 - 4.3	F	80	1 - 480
H	5.0		Mn	50	3 - 900
K	0.3	0.005 - 0.65	Na	400	100 - 6000
Mg	0.2	0.01 - 0.35	P	130	6 - 4000
N	1.5		Sr	150	20 - 1000
O	6.4	0.1 - 28.0	Ti	500	200 - 1800
S	1.5	0.1 - 12.0	Zn	50	3 - 300
Si	3.0	0.5 - 11.0			

generation and transmission of electricity (approximately 67%). A more efficient use of coal for producing electricity is magnetic hydrodynamics (MHD). Hot combustion gases (2400°C), which have been ionized, are passed through a strong magnetic field (50,000 gauss) to generate electricity directly.

The future for coal probably lies in its conversion to gaseous and liquid fuels. This means the H to C ratio has to be increased from ~0.3 for anthracite and ~0.8 for bituminous coal to around 1.9 for petrol and 4 for methane. The hydrogenation is achieved using either H_2O or H_2, the conversion routes are shown in Fig. 4.6.

Town gas supply, coke and coal tar are made by conventional pyrolysis, a process being replaced by different methods of pyrolysis or other ways of producing gas or liquids. The reactions appear simple, if coal is just considered as carbon. However, the reactions are more involved because of coal's complex structure, and other constituents in it. The reactants for gas production are coal, oxygen and steam;

$$2C + O_2 \rightarrow 2CO, \text{(combustion)}, \Delta H = -214 \text{ kJ} \quad (4.15)$$

$$2CO + O_2 \rightarrow 2CO_2, \text{(combustion)}, \Delta H = -561 \text{ kJ} \quad (4.16)$$

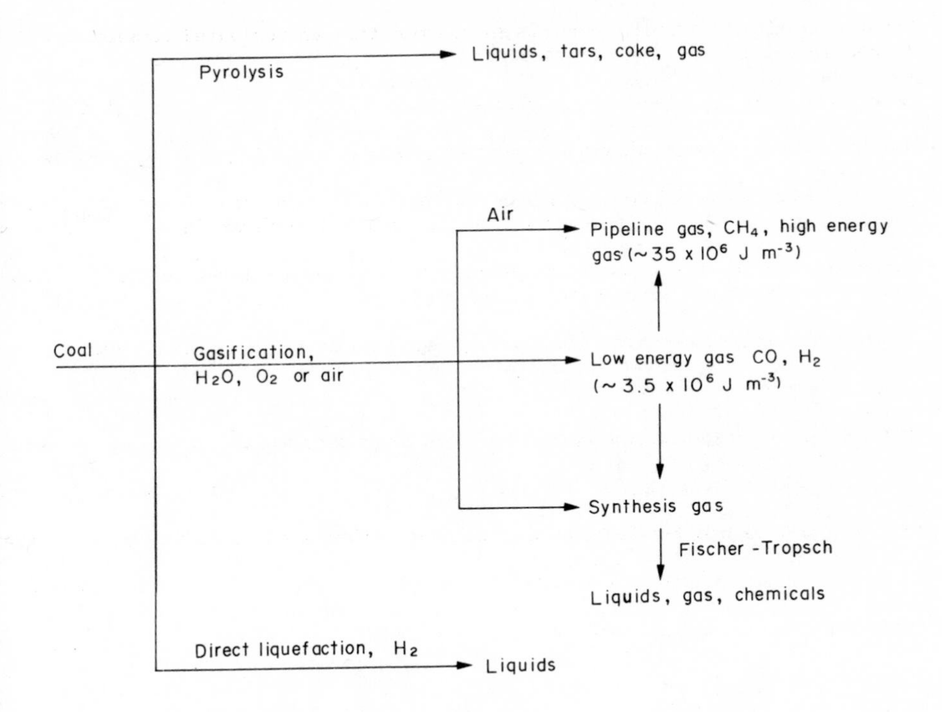

Fig. 4.6. Routes to the conversion of coal to other energy sources

$$C + H_2O \rightarrow \underset{\text{(water gas)}}{CO + H_2}, \text{ (gasification)}, \Delta H = 131 \text{ kJ} \quad (4.17)$$

$$CO + H_2O \rightarrow CO_2 + H_2, \text{ (water gas shift reaction)}, \Delta H = -29 \text{ kJ}. \quad (4.18)$$

The hydrogen produced can be used to make methane;

$$C + 2H_2 \rightarrow CH_4, \Delta H = -75 \text{ kJ}, \quad (4.19)$$

or

$$CO + 3H_2 \rightarrow CH_4 + H_2O, \Delta H = -142 \text{ kJ}. \quad (4.20)$$

The sum of reactions (4.17), (4.18) and (4.19) is;

$$2C + 2H_2O \xrightarrow{900^{\circ}C} CH_4 + CO_2, \Delta H = 27 \text{ kJ}. \quad (4.21)$$

Half the carbon is wasted in this reaction, however the methane has twice the energy content of the original coal and is cleaner burning.

Liquid fuels, such as hydrocarbons and methanol, are produced from a mixture of CO and H_2, called synthesis gas, by adjusting the CO to H_2 ratio with appropriate reactions. Alkanes can be obtained by the Fischer-Tropsch reaction;

$$nCO + (2n+1)H_2 \rightarrow C_nH_{2n+2} + nH_2O, \qquad (4.22)$$

which for CH_4 has $\Delta H = 206\ kJ\ mol^{-1}$. Alkenes and alcohols can also be produced;

$$nCO + 2nH_2 \rightarrow C_nH_{2n} + nH_2O, \qquad (4.23)$$

$$nCO + 2nH_2 \rightarrow H(CH_2)_nOH + (n-1)H_2O. \qquad (4.24)$$

Methanol is formed by the reaction;

$$CO + 2H_2 \rightarrow CH_3OH,\ \Delta H = -91\ kJ. \qquad (4.25)$$

A pressure of 20000 - 30000 kPa, and temperature of 350°C are required, using a Zn/Cr oxide catalyst, while for a copper based catalyst 5000 kPa and 250°C are employed.

Direct liquefaction of coal is achieved by hydrogenation;

$$nC + (n+1)H_2 \xrightarrow[\substack{450^{\circ}C \\ >10000\ kPa}]{catalyst} C_nH_{2n+2} \qquad (4.26)$$

In one process coal is dissolved in a solvent, for example anthracene oil dissolves 80% of coal at 400 - 500°C. The solution is then hydrogenated to give a range of useful products.

Oil

The convenience of oil, its greater energy content than coal (~1.5 on a weight basis), are reasons for its extensive use as a world fuel. It is also a major source of organic, and some inorganic, chemicals.

Formation and extraction. Oil is a younger fossil fuel than coal, its originates from dead micro-organisms and plants buried under sediments. Compaction of the sediments squeezed the oil into porus rocks (e.g. sandstones and carbonate rocks) where, it is stored, if trapped (Fig. 4.7). Trapping

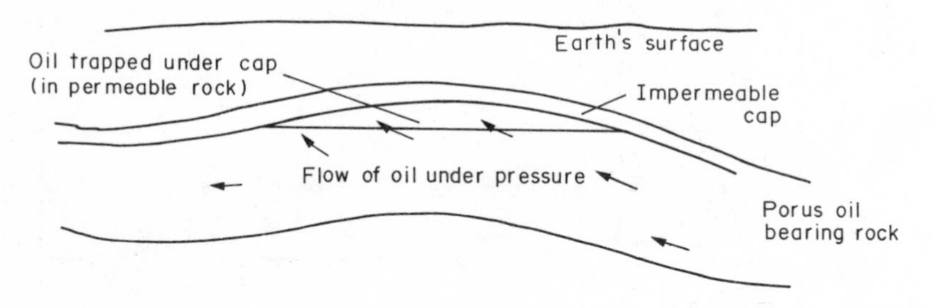

Fig. 4.7. A typical oil trap

and oil formation go together, otherwise the oil will escape. Only about 0.1% of the oil has been retained in the earth. The processes in the change from plant and animal material to hydrocarbons is not well understood. A breakdown (natural cracking) of complex bio-organic molecules to low molecular weight paraffins has occurred, and the older the oil the greater the proportion of low molecular weight materials. The changes occurred below 100°C, high temperatures associated with metamorphic changes would have destroyed the oil. Attempts to simulate oil formation in the laboratory have not been altogether satisfactory. Questions over the influence of bacteria, catalysts (e.g. aluminosilicates), temperature and time have yet to be resolved. The ultimate product of breakdown, methane, is thermodynamically the most stable hydrocarbon.

Approximately 1/3 to 1/2 of the oil in wells is recovered at present. Methods to increase the recovery are being developed. One technique, chemical flooding, uses surfactants, injected into the wells to reduce the oil-water interfacial tension, and so displace more oil.

Oil processing. A major objective of oil refining is to produce petrol and utilise the oil fraction outside the petrol range. Fractional distillation separates out the various fractions as shown in Fig. 4.8. Approximately 20%

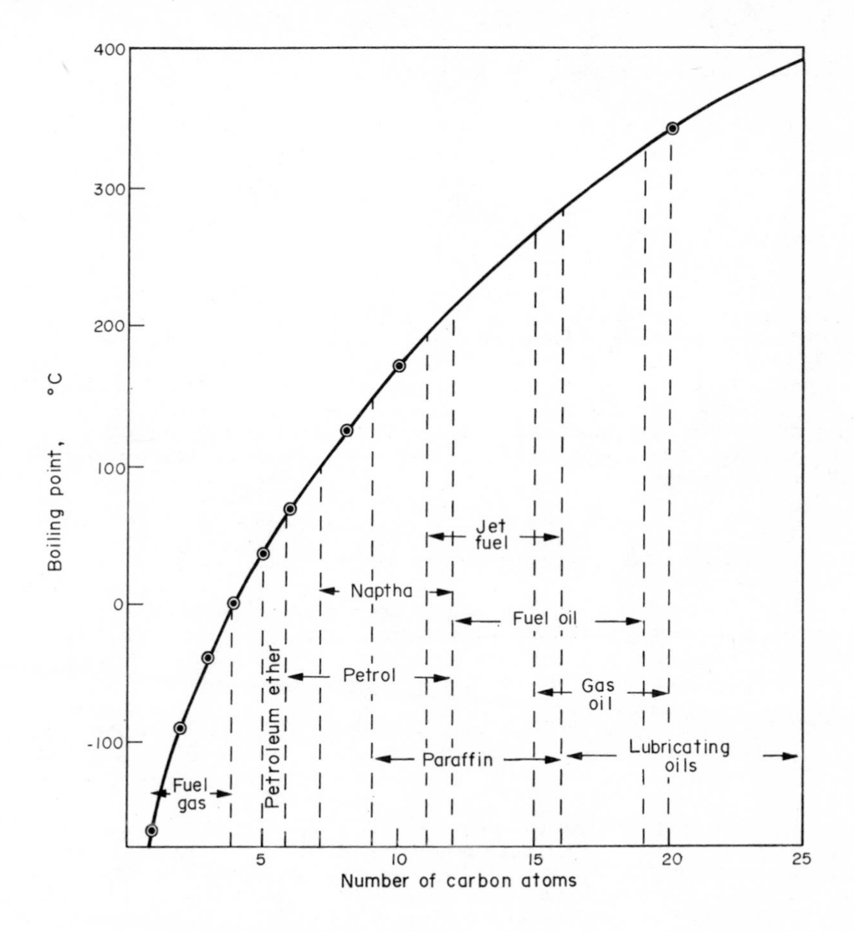

Fig. 4.8. Products of fractional distillation of petroleum

of crude oil is petrol, which may be increased to 40 - 45% by refining processes. The techniques are *cracking* - breaking up of big molecules, *catalytic reforming* - producing aromatics and high octane components; *alkylation* - increasing the proportion of branched chain paraffins (which increases the octane rating), and *polymerisation* - building up bigger molecules.

Cracking can be achieved thermally at 1000 - 1100 K, or by using a catalyst such as potassium on SiO_2/Al_2O_3 at 725 K.

$$C_nH_{2n+2} \rightarrow C_nH_{2n} + H_2, \qquad (4.27)$$

$$C_{m+n}H_{2(m+n)+2} \rightarrow C_mH_{2m} + C_nH_{2n+2}, \qquad (4.28)$$

e.g. $$C_{15}H_{32} \rightarrow C_7H_{14} + C_8H_{18}. \qquad (4.29)$$

Cracking can also be achieved using hydrogen;

$$C_{m+n}H_{2(m+n)} + 2H_2 \xrightarrow[\text{pressure}]{\text{high}} C_mH_{2m+2} + C_nH_{2n+2}. \qquad (4.30)$$

Alkylation can be represented by the equation;

$$RH + H_2C{=}C\begin{matrix} R' \\ R'' \end{matrix} \xrightarrow[\substack{\text{acid} \\ \text{catalyst} \\ 20^\circ C}]{\text{strong}} R - CH_2 - CH\begin{matrix} R' \\ R'' \end{matrix}, \qquad (4.31)$$

for example;

$$CH_3 - \overset{\overset{CH_3}{|}}{C} = CH_2 + CH_3 - \overset{\overset{CH_3}{|}}{CH} - CH_3 \xrightarrow[\text{or HF}]{H_2SO_4} CH_3 - \underset{\underset{CH_3}{|}}{\overset{\overset{CH_3}{|}}{C}} - CH_2 - \overset{\overset{CH_3}{|}}{CH} - CH_3. \qquad (4.32)$$

Catalytic reforming is achieved with catalysts such as MoO_3 on Al_2O_3 or Re/Pt on clay, and at high pressure (1500 - 2000 kPa) and temperature (500 - 600°C). For example;

$$\underset{\text{n-hexane}}{C_6H_{14}} \rightarrow \underset{\text{cyclohexane}}{C_6H_{12}} + H_2 \rightarrow \underset{\text{benzene}}{C_6H_6} + 3H_2. \qquad (4.33)$$

Many reactions require catalysts, which are readily poisoned by some of the materials in oil (Table 4.6) such as sulphide and vanadium. The sulphur occurs as thiophenes, N and O as heteroatoms and some metals, either as salts of carboxylic acids or in porphyrin complexes.

The residue of oil distillation, and heavy crude oil, pose the same problems as the conversion of coal to lighter hydrocarbons. The H:C ratio has to be increased from 1.4 to >2.

At present about 10% of the oil produced is used in the manufacture of petrochemicals. It could be that this proportion will increase in the future, especially as alternative fuel sources are developed, which could replace oil.

TABLE 4.6 Some Constituents of Oil (mg kg^{-1})

Element	Concentration	Element	Concentration	Element	Concentration
Ca	15	K	5	Na	4
Cl	39	N	7000 - 20,000	Ni	10
Cu	1.3	O	6 - 44,000	V	50
Fe	25	S	15,000	Zn	30

Natural Gas

Natural gas forms in the same way as oil, and they generally occur together, with either the gas dissolved in the oil or lying above it. The gas, mainly CH_4, has a number of uses in addition to a primary fuel. One approach is to reform the CH_4 to give CO and H_2, synthesis gas (reverse of reaction (4.20)) which can itself be used as a fuel, or to make methanol or liquid hydrocarbons as described for coal. The reforming reaction is endothermic;

$$CH_4 + H_2O \rightleftharpoons CO + 3H_2 \quad H = 142 \text{ kJ} \tag{4.34}$$

and is carried out at 700 - 900°C and 500 - 4000 kPa pressure with a NiO_2 or Al_2O_3 catalyst.

Synthetic petrol processes have been developed using natural gas, to first make methanol (reactions (4.34) and (4.25)). Then with zeolite catalysts (three dimensional cage aluminosilicates) methanol is converted to paraffins

$$nCH_3OH \rightarrow C_nH_{2n} + nH_2O \tag{4.35}$$

where n is predominantly 5 to 10. Catalysts are also available that avoid the formation of the methanol.

NUCLEAR ENERGY SOURCES

Fossil fuels will probably continue to dominate the energy scene for the rest of this century. However, alternative energy sources have to be developed; some, such as nuclear and solar energy, will be world wide in their impact, others, such as tidal and geothermal will be localised. In this section we will consider nuclear energy.

Nuclear Energy

Relative to coal and oil, nuclear energy is costly, because of stringent construction requirements, and the expense and problems of processing and storing used fuel. At present about 10% of the electricity generated in the USA is from nuclear power. Estimates of the ultimate energy available from different fuels suggests that thermal nuclear power, using ^{235}U, will add little to our present resources, but breeder reactors, and possible fusion processes, will be significant sources of energy (Table 4.7). The first nuclear reactor was built at the Argonne Laboratory, USA, in 1942. However, nature may have provided the conditions for natural reactors around 1800 x 10^6 years ago. Discoveries in Gabon, Africa, have revealed at least

TABLE 4.7 Ultimate Energy Available from Some Fuels[a]

Energy Source	Energy ($J \times 10^{21}$)	Energy Source	Energy ($J \times 10^{21}$)
Coal	220 - 285	Thermal Reactor	1
Oil	11.8	Breeder Reactor(^{249}Pu)	200
Natural Gas	7.7 - 11	Breeder Reactor(^{232}Th)	100
Tar Sands and Oil Shale	100 - 125	Fusion	500 - 12,500

[a] The energies given are probably not realizable for most sources.

six fossil reactor sites.

Nuclear Reactions

The nuclei binding energy curve (Fig. 1.2) shows that fusion of light nuclei and fission of heavier elements to lighter ones are both exothermic. Fission is used for producing power, but as yet fusion is in the experimental stage.

Uranium-235, captures a neutron and undergoes fission releasing energy of the order of 82×10^9 J g^{-1}. Typical fission reactions are;

$$^{235}_{92}U + ^{1}_{0}n \rightarrow ^{139}_{56}Ba + ^{94}_{36}Kr + 3^{1}_{0}n, \quad (4.36)$$

and

$$^{235}_{92}U + ^{1}_{0}n \rightarrow ^{133}_{51}Sb + ^{99}_{41}Nb + 4^{1}_{0}n. \quad (4.37)$$

There are around 30 pairs of fission products within the range Z = 30(Zn) to Z = 63(Eu). Fission initiated by slow, or thermal neutrons, produces more neutrons. To maintain the chain reaction the neutrons produced need to be slowed down, or moderated; and control is achieved by capturing the excess neutrons. Adequate ^{235}U must be present to achieve the critical mass, which means enriching the natural form of uranium containing 0.71% ^{235}U. Since the amount of ^{235}U available is quite small, around 6800 metric tons are readily available, uranium supplies could run out by 2000 A.D.

The uranium sources can be extended by other nuclear reactions, in breeder reactors. The principal isotope of natural uranium (^{238}U, 99.27%), captures a neutron to initiate the process;

$$^{238}_{92}U + ^{1}_{0}n \rightarrow \underset{(23.5m)}{^{239}_{92}U} \xrightarrow{-\beta^-} \underset{(2.35d)}{^{239}_{93}Np} \xrightarrow{-\beta^-} \underset{(24,390\ yrs)}{^{239}_{94}Pu} \quad (4.38)$$

The ^{239}Pu captures fast neutrons and undergoes fission in the same way as ^{235}U. Breeder reactors therefore extend the life time of uranium as a fuel by a factor of 100. Thorium-232 can also be used in the same way.

$$^{232}_{90}Th + ^{1}_{0}n \rightarrow \underset{(22m)}{^{233}_{90}Th} \xrightarrow{-\beta^-} \underset{(27d)}{^{233}_{91}Pa} \xrightarrow{-\beta^-} \underset{(1.62 \times 10^5\ yrs)}{^{233}_{92}U} \quad (4.39)$$

The ^{233}U produced also undergoes fast neutron initiated fission. One gram of ^{238}U provides, through ^{239}Pu fission, 21×10^9 J. If the neutrons produced in a ^{235}U reactor are not moderated, then ^{239}Pu is produced (equation 4.38) providing more fuel than is used, hence the term "breeder".

Nuclear Reactors

Nuclear reactors are planned to maintain, yet control the fission reaction, and to use the energy to produce steam for generating electricity. The ^{235}U in a thermal reactor has been enriched from the natural abundance of 0.71% up to 5%. Enrichment is achieved by passing UF_6 through diffusion barriers. The slightly lighter $^{235}UF_6$ passes through the barrier more readily than $^{238}UF_6$ by a factor of 1.004;

$$\text{i.e.} \quad \sqrt{\frac{\text{RMW} \; ^{238}UF_6}{\text{RMW} \; ^{235}UF_5}} = \sqrt{\frac{352}{349}} = 1.004. \qquad (4.40)$$

A three-fold enrichment requires 275 diffusions;

i.e. $\quad 3 = (1.004)^n$, hence $n = 275$.

Breeder reactors require a mixture of ^{238}U and ^{239}Pu or highly enriched ^{235}U (~45%), because fast neutron capture is not as efficient as for slow neutrons. The fuel is in containers, called cladding, which have low neutron capture and good mechanical strength.

Moderation of fast neutrons is achieved with graphite, or H_2O or D_2O. No moderation is used in breeder reactors. Control rods, made of Cd, or B, or B_4C, which have a high neutron capture cross section, are used to control fission by raising or lowering them into the core.

The heat produced is removed with a coolant, H_2O, or D_2O or CO_2 or He in thermal reactors and molten Na or K/Na alloy in breeder reactors. The coolant is either passed through a heat exchanger and the heat transferred to water to produce steam, or used directly as steam, for generating electricity. A second cooling system is required to cool the water from the steam. A schematic diagram is given in Fig. 4.9 and a few details of some reactors are given in Table 4.8.

Containment of radioactivity. The containment of radioactivity is a major aspect of nuclear power technology, for example, the core is well shielded. The primary cooling water becomes radioactive from irradiation of corrosion products (especially from the heat exchanger) that get into the water, and sometimes leakage from the fuel rods. The sodium used in breeder reactors becomes quite active as ^{24}Na (15 hr) and ^{22}Na (2.6 yr) are formed. Radiolysis of water produces H_2 and O_2 which need to be removed.

After a period spent fuel rods are replaced. Because of their high radioactivity they are placed in cooling ponds for up to six months, by which time many of the short lived isotopes have decayed to low levels. The cooling water becomes hot, and active if corrosion occurs. Radiolysis also takes place. The spent fuel also contains heavier elements, e.g. ^{239}Pu, ^{240}Pu, ^{241}Am and ^{245}Cm.

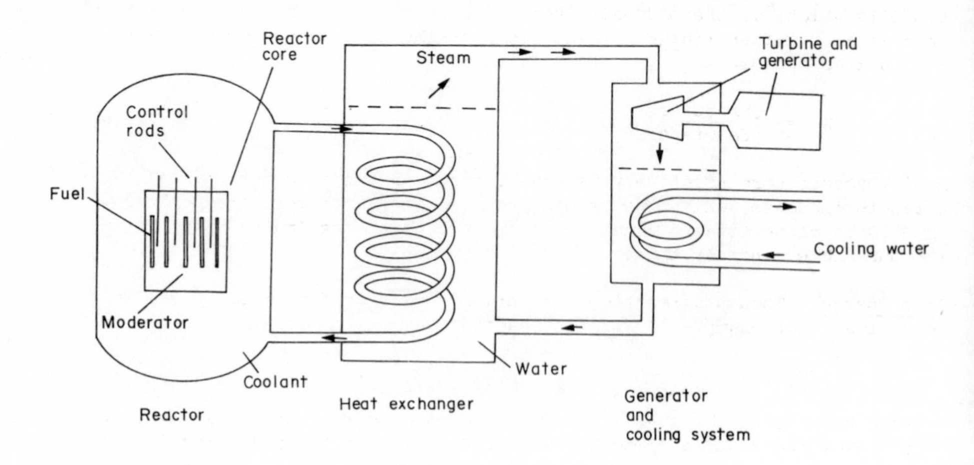

Fig. 4.9. A schematic diagram of a nuclear reactor

The fuel is processed to separate out fission products, uranium and ^{239}Pu. The material, including the cladding, is dissolved in nitric acid, containing nitrite ions, to give $UO_2(NO_3)_2$ and $Pu(NO_3)_4$. The nitrite (as an oxidising and reducing agent) ensures the plutonium is in the tetravalent oxidation state;

$$6Pu^{3+} + 2NO_2^- + 8H^+ \rightarrow 6Pu^{4+} + N_2 + 4H_2O, \tag{4.41}$$

$$PuO_2^{2+} + NO_2^- + 2H^+ \rightarrow Pu^{4+} + NO_3^- + H_2O. \tag{4.42}$$

The uranium and plutonium salts form adducts with tri-n-butyl phosphate (TBP) and become soluble in organic solvents, such as kerosene. By solvent extraction the U and Pu are separated from the fusion products. Plutonium is then removed from the organic layer, and separated from uranium, by adding Fe^{2+} which reduces the element to Pu^{3+};

$$Fe^{2+} + Pu^{4+}_{(org)} \rightarrow Fe^{3+} + Pu^{3+}_{(aq)}. \tag{4.43}$$

The plutonium is then recovered as $Pu(NO_3)_4$ using nitrite, while uranium is recovered from the TBP adduct by lowering the nitrate concentration in the equilibrium;

$$UO_2^{2+}{}_{(aq)} + 2NO_3^- + 2TBP \rightleftharpoons UO_2(NO_3)_2 2TBP_{(org)} \tag{4.44}$$

moving it to the left. The cycling of the fuel is shown schematically in Fig. 4.10.

The long half lives of some of the fission products raise problems over storage. For the first 300 to 400 years ^{90}Sr, ^{137}Cs and ^{244}Cm are the main problems; beyond that time longer lived isotopes such as ^{93}Zn (9.5×10^5 yrs),

TABLE 4.8 The Features of Some Nuclear Reactors

Reactor type	Fuel	Moderator	Control Rods	Coolant
Thermal				
Magnox	Uranium metal in Mg alloy cladding	Graphite	Boron	CO_2 under pressure
CANDU	Uranium metal in stainless steel cladding	Heavy water (D_2O)	Boron in steel	Heavy water under pressure
Gas cooled (GCR)	UO_2 enriched to 2.5% ^{235}U in zirconium alloy cladding	Graphite	B_4C	CO_2 under pressure (or He)
Pressurised water (PWR) also (LWR)	Uranium metal enriched 2 to 5% ^{235}U in stainless steel cladding	Water[a] at high pressure	Ag/In/Cd or B_4C	water
Boiling water[b] (BWR)	UO_2 enriched to $\geq 1.5\%$ ^{235}U	Water under pressure	B_4C	water
Breeder				
Dounreay (DFR)	Uranium metal enriched to 45.5% ^{235}U in niobium cladding	None	B_4C	molten Na/K alloy
Prototype (PFR)	Uranium and plutonium oxides	None	B_4C	molten Na

a water refers to H_2O unless specified as heavy water D_2O.

b BWR uses steam from coolant directly to drive the turbines.

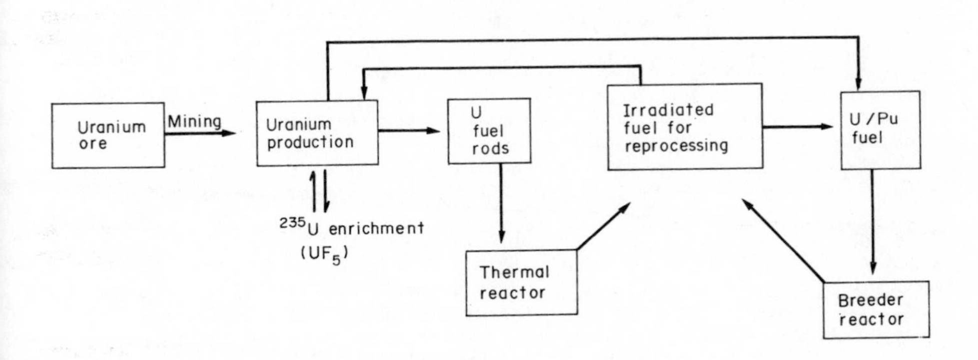

Fig. 4.10. The cycling of uranium in the nuclear energy industry.

^{99}Tc (2.13 x 10^5 yrs), ^{135}Cs (2.3 x 10^6 yrs), ^{239}Pu (24390 yrs), ^{240}Pu (6540 yrs), ^{241}Am (433 yrs) and ^{245}Cm (8.5 x 10^3 yrs), require safe storage for centuries. Initially the products are stored in solution in tanks, which need to be cooled. For long term storage solids (e.g. fused into glass) are better, because of the smaller bulk. Ultimate storage suggestions include trenches on the ocean bed, old mine shafts, injection into rocks, the Antarctic ice cap and outer space. All have drawbacks. The problem is a serious one that has to be faced as more and more fission products are produced. It is essential that the future be made safe from radioactive contamination of the environment.

Fusion

As described in Chapter 1, the stars are giant fusion reactors. Nearly all of the earth's energy stems from fusion reactions occurring in the sun. Achievement of these reactions on earth, promises a plentiful supply of energy, with less associated radioactivity. Some of the reactions that could be used are (the energy evolved per gram of reactants);

$$^2_1H + ^2_1H \rightarrow ^3_2He + ^1_0n, \quad -79.5 \times 10^9 \text{ J g}^{-1} \qquad (4.45)$$

$$^2_1H + ^2_1H \rightarrow ^3_1H + ^1_1H, \quad -96.2 \times 10^9 \text{ J g}^{-1} \qquad (4.46)$$

$$^2_1H + ^3_1H \rightarrow ^4_2He + ^1_0n, \quad -338.9 \times 10^9 \text{ J g}^{-1} \qquad (4.47)$$

Deuterium (^2_1H) is plentiful (1 atom in every 6700 atoms of hydrogen), and so fuel for reactions (4.45) and (4.46) is assured. Tritium (^3_1H) for reaction (4.47) is produced from ^6_3Li or ^7_3Li;

$$^6_3Li + \underset{\text{(slow)}}{^1_0n} \rightarrow ^3_1H + ^4_2He, \quad -62.8 \times 10^9 \text{ J g}^{-1} \qquad (4.48)$$

$$^7_3Li + \underset{\text{(fast)}}{^1_0n} \rightarrow ^3_1H + ^4_2He + ^1_0n, \quad \underset{\text{(endothermic)}}{33.5 \times 10^9 \text{ J g}^{-1}} \qquad (4.49)$$

Tritium is radioactive, and radioactivity will develop in the system from capture of neutrons. Therefore the fusion process is not radioactivity free.

The reactions require ignition temperatures around 4 x 10^7 to 10^8 K, and then 24.5 x 10^6 K to sustain the fusions. This raises the problem of the containment of the materials. At the high temperatures the material is ionic and may be contained by magnetic fields. Laser induced reactions are a possible way forward. As yet, the technology requires further development before useful power can be produced, but it is an approach worth pursuing.

Wide scale use of nuclear power is uncertain because of problems relating to radioactivity and its containment, and the fear that sometime an accident will produce a reactor meltdown. The power debate is linked with the use of nuclear armaments. Both fission - the atomic bomb-, and fusion - the hydrogen bomb-, processes have been used with wide scale and far reaching destruction. Many problems need resolving before unlimited use of nuclear energy is embarked upon.

SOLAR ENERGY

Solar energy is of central importance to the world's energy system. Fossil fuels are stored solar energy, life depends on photosynthesis and solar energy can be collected and utilized (Table 4.9). The sun's energy flux is

TABLE 4.9 Forms and Uses of Solar Energy

Form of solar energy	Utilization
Fossil fuels	Combustion and conversion of coal, oil, natural gas.
Direct thermal collection	Heating and cooling of buildings. Solar furnaces.
Indirect thermal collection	Wind energy. Hydroelectric power. Ocean thermal currents. Solar photovoltaic conversion. Photo electrolysis.
Collection by photosynthesis	Biomass production and conversion (terrestrial and marine). Agriculture and forest residues.

4.2×10^9 J m^{-2} min^{-1}, with maximum intensity in the visible (400 - 700 nm) range (λ_{max} = 483 nm)(Fig. 2.3).

Approximately 0.002% of this energy is intercepted by the earth and its atmosphere so it receives a flux around 8.4×10^4 J m^{-2} min^{-1}. Only a portion of this reaches the earth due to reflection and scattering by the atmosphere.

Solar Energy Balance

Approximately 5.4×10^{24} J enters the earth's upper atmosphere each year. Around 30% is reflected and scattered back into space by clouds and particulate matter. Nearly one half (47%) is used in heating the atmosphere (19%), and earth's surface (28%)(Table 4.10). Photosynthesis uses 0.02% of the solar energy, still a considerable amount of energy, i.e. 1×10^{21} J $year^{-1}$. Solar energy is diffuse, i.e. spread over a wide area, it is intermittent and, at present, harnessing the energy is expensive.

Uses of Solar Energy

There are two main approaches to collection of solar energy. The absorbed energy, producing electron excitation, is used for heating or the absorbed energy initiates chemical changes giving materials which may store the energy (e.g. plants).

Direct collection of solar energy. Solar energy collected directly is mainly used for warming or cooling buildings. Black collectors, covered with glass, trap the infrared radiation. For intense heating, sunlight

TABLE 4.10 Solar Energy Budget

Fate of Solar Energy	Approximate Energy $J \times 10^{21}\ yr^{-1}$	Percent
Reflected by clouds	1620	30
Heating atmosphere and earth's surface	1026	19
Used in evaporation of water	1242	23
Used in generating winds, ocean currents	11	0.2
Used in photosynthesis	1	0.02
Total (approximate)	5400	100

has to be focussed with mirrors, which also need to track the sun. Solar furnaces, such as the one at Odeillo in France, can, for an incident energy of 950 Wm^{-2}, produce an average energy of 1600 W cm^{-2} (1600 J cm^{-2} s^{-1}) and an average temperature of 3825°C. Because of the intermittent nature of solar energy it would be advantageous to store it. The most common method is to use water. Another approach is to use materials with high heats of fusion, and a transition temperatures 10-15 K above ambient. Sodium carbonate decahydrate has a fusion temperature of 305-309 K, and an enthalpy of fusion of 267 J g^{-1}. However, 1 to 2 tonnes would be required to store 2.7 to 5.3×10^8 J day^{-1}, the amount necessary to keep a house warm in freezing weather.

Temperatures up to 100°C are reached in solar ponds with a black bottom and containing a brine solution lying below water.

Photosynthesis. Natural photosynthesis is one of the most important processes in the world. The information in Fig. 4.11 portrays the central position of photosynthesis.

The solar energy collectors in plants are the green chlorophylls (and other pigments). There are a number of chlorophylls, each has a porphyrin nucleus with magnesium at the centre, but differ in the nature of substituents on the porphyrin nucleus (Fig. 4.12). The electronic absorption spectrum of the molecules have bands associated with $\pi \rightarrow \pi^*$ transitions. A low-energy band, around 700 nm (position varies with type of chlorophyll), is the reason for the green colour, and is the transition that captures the solar energy. An absorption band around 430 nm, and high energy absorptions of other pigments, such as carotenoids, protect the plant from radiation damage, and the energy can be transferred to the chlorophyll system involved in the photosynthesis.

Normally energy absorbed in this way is rapidly lost by fluorescence or through heat to the surroundings. The Mg-chlorophyll has a low-lying triplet excited state, which can receive an electron from the excited singlet state by crossing over. The electron in the triplet state has a sufficiently long life time for the energy to be used in chemical reductions, rather than being lost through phosphorescence or heat.

The chemical reactions of photosynthesis are complex, but basically water is decomposed producing dioxygen, and carbon dioxide is reduced to carbohydrate.

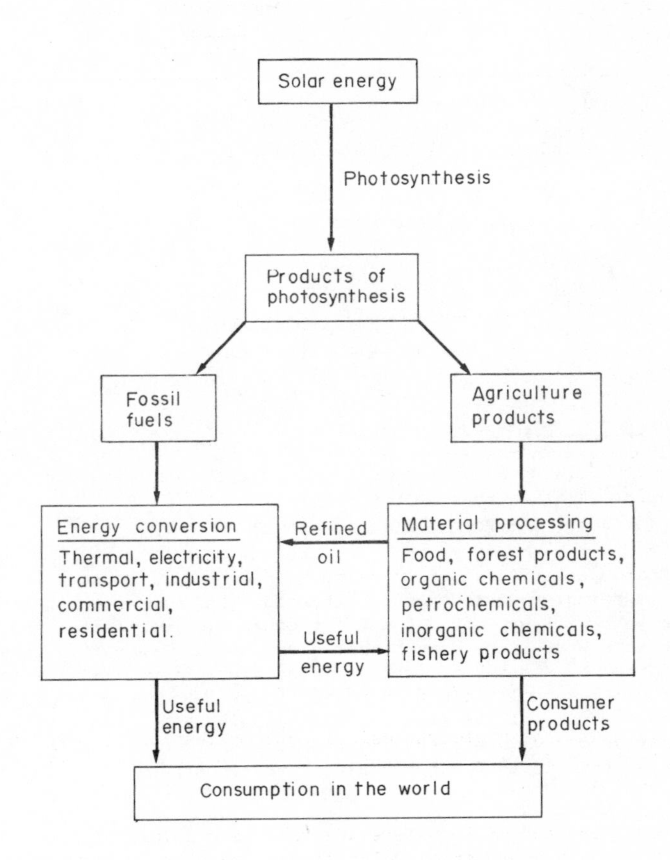

Fig. 4.11. Direct and indirect products of photosynthesis

The energy required to produce one mole of $C_6H_{12}O_6$ is 2830 kJ. In a possible mechanism (Fig. 4.13) two photo-acceptors are proposed. The solar energy is used to decompose water and provide the electrons required for the reduction of nicotinamideadenine dinucleotide phosphate (i.e. NADPH). The energy for converting CO_2 into carbohydrate comes from the ATP, in the "dark reaction" which does not depend on solar energy. The electron flow is in one direction (Fig. 4.13), because of the spatial separation of the two parts of the system. A dinuclear manganese complex in photosystem(II) utilizes the Mn(II)/(IV) couple. The average photosynthesis efficiency is <1%, but under optimum conditions, i.e. sufficient CO_2 and H_2O, warm temperatures and adequate sunlight, the efficiency can reach 10-12%.

The photosynthesis system maintained in reducing conditions has also been used to produce hydrogen. An estimated yield is 9 moles of H_2 day^{-1} m^{-2}.

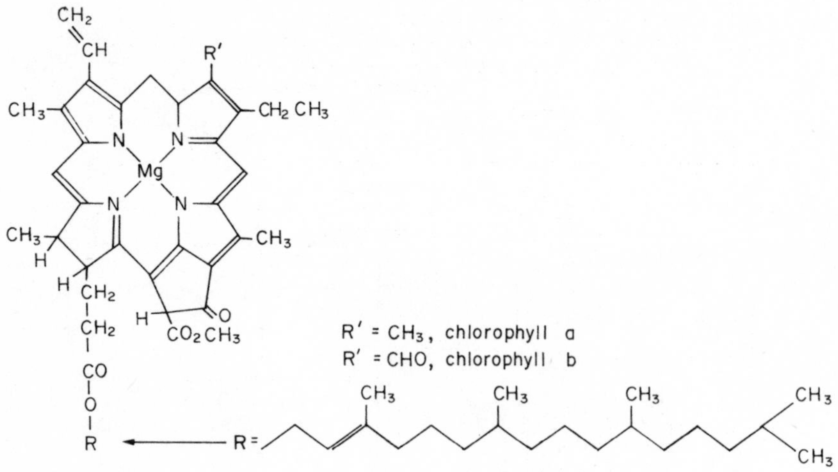

R′ = CH_3, chlorophyll a
R′ = CHO, chlorophyll b

Fig. 4.12. Structure of chlorophyll

Biomass. Photosynthetic material can be used as a fuel source, for example, sugar cane is being grown to produce ethanol. Ten tonnes of sugar per hectare per year will provide 5 tonnes of ethanol. Hydrocarbon producing plants such as rubber trees *Heva brasiliensis* and the *Euphorbia* species are being considered as sources of fuels. Trees are, potentially , a good source of fuel, less polluting than fossil fuels. However, if plant material is used, more effort will be required to replenish it. Simple biomass digesters may become viable energy sources for small communities. When plant material is allowed to decay anaerobically methane is produced.

$$n\{CH_2O\} \rightarrow \frac{n}{2}CO_2 + \frac{n}{2}CH_4 \qquad (4.50)$$

Synthetic photosynthesis. A membrane separates the two main biochemical processes in natural photosynthesis. Successful attempts have been made to construct membranes which, with the right materials, in sunlight produce O_2 and H_2. A phospholipid membrane with sensitizers (chlorophyll-like) on each side has been studied. Electrons are stripped from water giving O_2, by the capture of solar radiations and with manganese compounds. The electrons reduce H^+ to H_2 using iron containing compounds, on the other side of the membrane.

Solar cells. Photocells have been in use as light meters and in control systems for many years. However, they are inefficient. Arising out of the USA space program, silicon photovoltaic solar cells, have been developed, which have 10-15% efficiency. A cell consists of linked *n*- and *p*-type semiconductors. If a photoelectron moves across the junction before combining with a positive hole (Fig. 4.14) an electric current will flow. Spatial separation of the two semiconductors ensures the electron flow is in one direction. The *p*-type semiconductor has to be thick enough to collect the majority of the solar photons, but thin enough to allow the photoelectron a chance to move across the junction. The cell operates at a voltage of

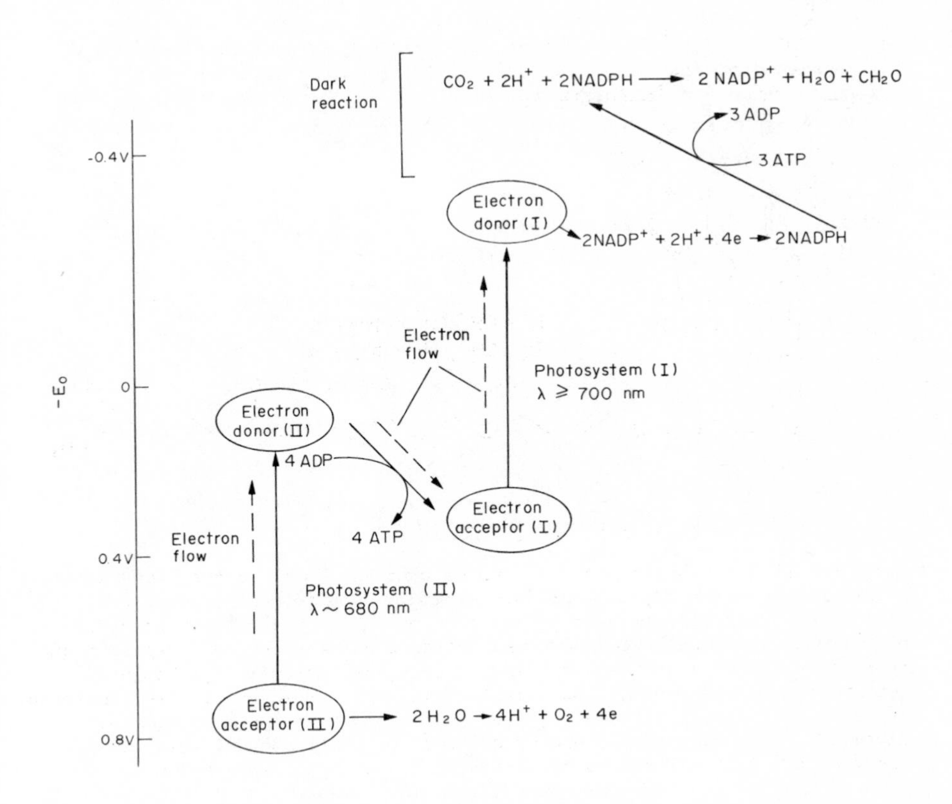

Fig. 4.13. A suggested reaction scheme for photosynthesis

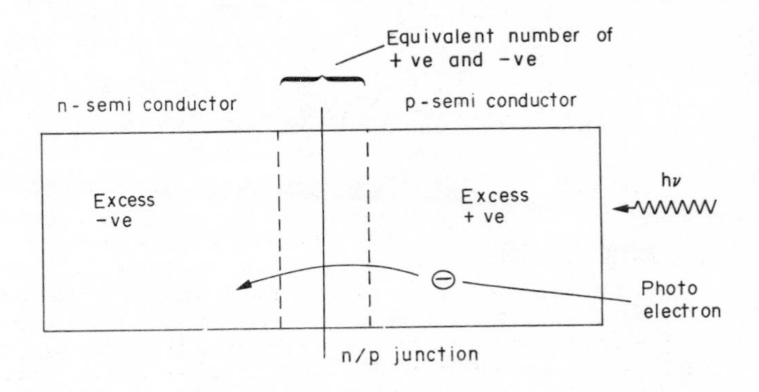

Fig. 4.14. The basic arrangement for a silicon solar cell

0.5 - 0.6 V. Efficiency is reduced by reflection from the cell surface, recombination of electrons and holes, and current leakage across the *n/p* junction.

Photoelectrochemical cells have been investigated using a photosensitive (e.g. semiconductor) anode (Fig. 4.15). Solar energy striking the anode, promotes an electron into the conduction band which can then be transferred to, and involved in, reduction at an inert cathode;

$$e^- + Ox \rightarrow Red^-, \qquad (4.51)$$

e.g. $$e^- + H^+ \rightarrow \tfrac{1}{2}H_2. \qquad (4.52)$$

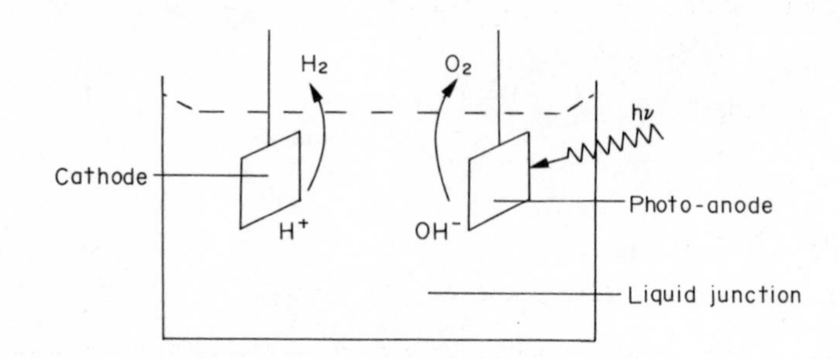

Fig. 4.15. A photoelectrochemical cell

Oxidation occurs at the positive hole on the anode;

$$\oplus + Red \rightarrow Ox^+, \qquad (4.53)$$

e.g. $$2\oplus + OH^- \rightarrow \tfrac{1}{2}O_2 + H^+. \qquad (4.54)$$

It is difficult to produce semiconductors which are corrosion resistant, and with valence band-conduction band energy gaps comparable to the energy of visible radiation. Sensitizing the semiconductors with dyes is a possible solution to the energy gap problem, and the choice of the redox solvent can reduce corrosion. Some semiconductors tried are n-$SrTiO_3$, n-CdS, n-CdSe and n-GaAs. The cell; n-CdSe|1 M Na_2S, 1M-S in 1M NaOH|C, is reported to have an efficiency of 8.4%.

Compounds which undergo photochemical processes are also being studied. Complex ion species, such as $[Ru(bipyridyl)_3]^{3+}$, are involved in the photochemical decomposition of water, and dirhodium complexes in the photochemical reduction of H^+ to H_2. An advantage of these reaction systems is that they are homogenous, which should raise the efficiency.

OTHER FORMS OF ENERGY

Hydrogen

Hydrogen gas, when burnt, produces water and 284 kJ mole^{-1} of energy.

$$H_{2(g)} + \tfrac{1}{2}O_{2(g)} \rightarrow H_2O_{(\ell)}, \quad \Delta H = 284 \text{ kJ mole}^{-1} \qquad (4.55)$$

A mole of CH_4 when burnt evolves 890 kJ, however, on a weight basis dihydrogen is the better fuel producing 142 kJ g^{-1} compared with 55.6 kJ g^{-1} from methane. Hydrogen is a clean fuel, the main pollutants being nitrogen oxides. However, NO_x formation can be minimised, because the low ignition temperature means a low-temperature catalyst can be used.

The supply of hydrogen (in water) is plentiful, but it is expensive to obtain, say by electrolysis. Schemes to decompose water, making use of energy, at present wasted, at power plants have been proposed. One approach uses vanadium catalysts:

$$Cl_2 + H_2O \xrightarrow{700-800^oC} 2HCl + \tfrac{1}{2}O_2, \quad (4.56)$$

$$2HCl + 2VCl_2 \xrightarrow{100^oC} 2VCl_3 + H_2, \quad (4.57)$$

$$2VCl_3 \xrightarrow{700^oC} VCl_2 + VCl_4, \quad (4.58)$$

$$VCl_4 \xrightarrow{100^oC} VCl_2 + Cl_2. \quad (4.59)$$

This adds up to;

$$H_2O \rightarrow \tfrac{1}{2}O_2 + H_2 \quad (4.60)$$

A problem is preventing oxidation of VCl_2 by oxygen.

Photochemical decomposition of water to hydrogen, using Rh and Ru catalysts, may also be a way forward.

The storage of dihydrogen poses problems, because of its low mass. One method, to liquify the gas, requires low temperature technology, as the critical temperature of H_2 is 33 K at 1300 k Pa. Another suggestion is to store the dihydrogen as interstitial hydrides e.g. as TiH_2 or Pd_2H. Hydrogen atoms sit in tetrahedral cavities in the metallic structure.

Storage Batteries

Energy storage using batteries has been, and will continue to be, important but will not provide power outputs >0.5 MW. Batteries, when used, are non-polluting, but their production can involve pollution, as for example, in the lead accumulator industry. Batteries can be constructed in modular form, and they have no moving parts. To be useful batteries must have a high energy density, and the chemical reactions must be rapid and reversible (so that recharging is possible).

Developments in battery technology have been slow, but are expected to increase. The convenience of the lead acid battery in petrol driven cars, where the mass of six cells is not critical, has been a reason for little interest in batteries in the past.

The lead storage battery has a relatively low energy density because of the weight of the battery. The reactions are;

$$\text{anode:} \quad Pb_{(s)} + SO_{4(aq)}^{2-} \rightarrow PbSO_{4(s)} + 2e^- \quad (4.61)$$

$$\text{cathode:} \quad PbO_{2(s)} + 4H^+_{(aq^+)} + SO_{4(aq)}^{2-} + 2e^- \rightarrow PbSO_{4(s)} + 2H_2O_{(\ell)} \quad (4.62)$$

i.e. $$Pb_{(s)} + PbO_{2(s)} + 4H^{+}_{(aq)} + 2SO^{2-}_{4(aq)} \rightarrow 2PbSO_{4(s)} + 2H_2O_{(\ell)}, \quad (4.63)$$

which provides 2 volts. Developments in alkali metal batteries are promising. The Li/TiS_2 battery;

$$Li + TiS_2 \rightarrow \underset{\text{intercalation compound}}{Li/TiS_2} \quad (4.64)$$

has 4 to 5 times the energy density of the lead acid battery. Some of these batteries require a high operation temperature so the metal is liquid. The Na-S battery;

$$2Na_{(\ell)} + 2S \rightleftharpoons Na_2S_2, \ (2.1\ V), \quad (4.65)$$

operates at 600 K, and the Li-S battery at 630 K providing 2.2 V. A Li/Cl_2 battery;

$$2Li_{(\ell)} + Cl_{2(g)} \rightleftharpoons 2LiCl, \quad (4.66)$$

delivers 3.6 volts at 900 K.

In the Fe/Ni cell;

$$Fe + 2NiO(OH) + 2H_2O \rightleftharpoons Fe(OH)_2 + 2Ni(OH)_2, \quad (4.67)$$

the electrolyte is a 25% aqueous KOH solution. The cell delivers 1.4 V at ambient temperatures. The Ni/Cd cell;

$$Cd + NiO(OH) + 2H_2O \rightleftharpoons Cd(OH)_2 + 2Ni(OH)_2, \quad (4.68)$$

is similar, delivering 1.3 V. Some other batteries are listed in Table 4.11.

Fuel Cells

Cells in which fuels are consumed to provide electricity are called fuel cells. The most common is the hydrogen fuel cell, where O_2 or air and H_2 are fed into the cell containing aqueous KOH solution. The reactions are;

at cathode: $$\tfrac{1}{2}O_{2(g)} + H_2O + 2e^- \rightarrow 2OH^-, \ 0.401\ V, \quad (4.69)$$

at anode: $$\left[\begin{array}{l} H_{2(g)} \rightarrow 2H, \\ 2H + 2OH^- - 2e \rightarrow 2H_2O, \ 0.828\ V, \end{array}\right. \quad (4.70)$$

i.e. $$\tfrac{1}{2}O_{2(g)} + H_{2(g)} \rightarrow H_2O_{(\ell)}, \ 1.229\ V. \quad (4.71)$$

Electrode catalysts are also used, e.g. Ni, Ag, platinum metals. The process is reversible by electrolysing the water. Other fuels that can be used are CH_4, N_2H_4 and NH_3.

TABLE 4.11 Batteries

Aqueous		Molten salt	Organic electrolyte	Solid state
Acid	Alkaline			
$Pb-PbO_2$	Ni-Zn	Na-S	$Li-TiS_2$	$Li-TiS_2$
$Zn-Cl_2$(1)	Ni-Fe	$Li/Al-FeS_2$(3)	Li-S	$Cu-TiS_2$
$Zn-Br_2$	Ni-Cd	$Na-SbCl_3$	$Li-V(Fe)S_2$	$Ca-NiF_2$
$Zn-MnO_2$	$Ni-H_2$	$Li-Cl_2$	$Li-NbSe_2$	
(Leclanché cell)	Al-air		$Li-TiO_2$	
	Li-air		$Li-Br_2$	
	Fe-air(2)			
	Zn-air			
	$Zn-Ag_2O$			
	$Cd-Ag_2O$			

Examples	Reaction	Volts	% Efficiency
(1)	$Zn + Cl_2 + 6H_2O \rightleftharpoons ZnCl_2 6H_2O$	2.12	>50
(2)	$Fe + H_2O + \frac{1}{2}O_2 \rightleftharpoons Fe(OH)_2$	1.28	<60
(3)	$4LiAl + FeS_2 \rightleftharpoons 2Li_2S + 4Al + Fe$	1.5	70-80

Source: D.A.J. Rand, *Chemistry in New Zealand*, (1979), 43, 185.

Solid Wastes

Municipal refuse is a possible energy source. Urban refuse consists of approximately 50% paper products and cardboard (by weight), 10% food, 10% glass, 10% metals and 20% other materials. This refuse contains around 60% combustible material, 20% moisture and 20% residue. The energy content (based on the weight of the total refuse) is approximately 12×10^9 J $tonne^{-1}$. If all the refuse in the USA was burnt to generate electricity it would replace the coal used at present. The refuse can also be pyrolysed to give tars, oils and gases. It may be necessary to weigh up the advantages of burning municipal refuse, over recycling the materials in it, because of the energy originally used to produce the materials.

Geothermal

The earth's surface temperature increases approximately 15 to 75°C per km of depth. This heat could be utilized by drilling and circulating water. However, for satisfactory heat transfer it would be necessary to improve the heat conduction of rocks. The best areas to drill would be near known hot spots, such as volcanoes. For this reason geothermal energy will be of local interest rather than world-wide.

ENERGY COSTS

Energy impinges on our lives in areas other than the obvious heating, cooling and transport. For example, certain uncontrolled thermal and chemical reactions, and nuclear reactions are used in explosives. The energy of a hydrogen-bomb explosion is of the order of 4×10^{15} J, about 25 times the energy evolved in a summer thunderstorm.

Energy conservation deals with the returns for energy expended. For example, around 130 kJ are expended pushing a bicycle one passenger km, while a car driven in cities, uses 5310 kJ per passenger km. Energy is important in the overall economics of extraction of elements from their ores (Chapter 3). The rapid growth in plastic containers over glass is because of, in part, the lower energy input. To make a 1 litre PVC bottle, 4 to 5×10^6 J are required while for the same size glass bottle, 10×10^6 J are necessary. But also, it is necessary to weigh up the benefits of recycling, which is superior for glass.

Energy conversions can be wasteful, and effort to improve inefficient conversions is worthwhile. Electrochemical batteries (chemical into electrical energy) have a 70-90% efficiency, whereas the conversion of thermal

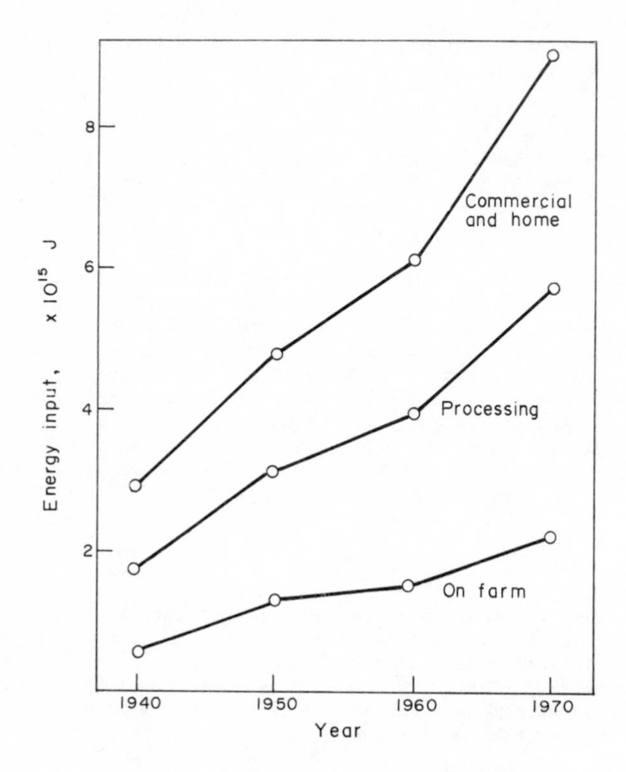

Fig. 4.16. Energy input in production of food, on the farm, processing and preparation

to electrical energy (as in a thermal power station) has an efficiency of around 30%. There are theoretical limits to the efficiency of some conversions, but even so practical efficiencies can be improved.

Food Production

Food production is an area where often more energy is expended than produced. For example, the ratio of energy input to energy output (in the form of food) is >10 for feed-lot beef and far-distant fishing, 5-10 for fish protein concentrate, 2-5 for grass-fed beef and intensive egg production, 1-2 for coastal fishing and milk production and <1 for range-fed beef, low intensity egg production and production of corn, potatoes, soybeans and rice. Fertilizers account for approximately 20% of the total energy used on the farm. While these figures suggest energy wastage, it must be kept in mind that energy usage on the farm (USA) is approximately 4% of the total energy consumption. Also food has more uses to people than just energy. Data presented in Fig. 4.16 shows that the energy expended on the farm (USA) is less than the energy used in processing the food and food preparation. There appears to be more room for energy conservation in the latter two areas.

Fuels have uses other than primary energy sources. As the world's resources (both energy and materials) decrease the relative value of energy sources as energy, or as useful materials, needs to be carefully considered. Energy and materials can not be divorced from each other in future plans for conservation and usage.

PART III

Chemical Changes and Inorganic Materials

CHAPTER 5

Principles of Chemical Change

Chemistry is the study of matter and how one substance can be changed into another. We have considered chemical changes that occur in the environment, and the chemical changes necessary to extract elements from their ores. In this section we will give attention to some important chemicals, how they are made, and how and why they are used. As a preface we will, in this chapter, give a brief description of the basic principles of chemical change.

ENERGETICS

In many reactions reactivity is controlled by the energetics of the chemical reaction, that is the relative thermodynamic stability of the reactants and products. Chemical stability is of particular value when comparing a number of related reactions, for example, the reaction;

$$K_3FeCl_6 + 6KF \rightarrow K_3FeF_6 + 6KCl, \qquad (5.1)$$

proceeds because the lattice energy of K_3FeF_6 is greater than that of K_3FeCl_6. This may be related to the size of the FeX_6^{3-} anions. Chemical reactivity of the anions follows the order;

$$FeF_6^{3-} > FeCl_6^{3-} > FeBr_6^{3-} \qquad (5.2)$$

Reactivity and Free Energy.

If, for a reaction, the enthalpy change $\Delta H < 0$ and entropy change $\Delta S > 0$, then the reaction is feasible. The two terms are linked by the equation;

$$\Delta G = \Delta H - T\Delta S, \qquad (5.3)$$

where ΔG is the free energy change. The free energy is connected with the equilibrium constant by the equation;

$$\Delta G = -RT\ln K. \qquad (5.4)$$

At 298K $\Delta G = -5.71 \times 10^3 \log_{10}K$. (5.5)

If $\Delta G < 0$ then $\log_{10}K > 1$ and the equilibrium lies on the side of the products which, in thermodynamic terms, is a feasible reaction. On the other hand if $\Delta G > 0$ then $\log_{10}K < 1$ and the equilibrium favours the reactants. This arises because

$$K = \frac{[\text{products}]}{[\text{reactants}]}. \quad (5.6)$$

As ΔG becomes more negative K becomes larger and the concentration of the products increase, while the reverse holds when ΔG becomes more positive. When ΔG is close to zero, that is $K \approx 1$, the reactants and products may be at comparable concentration.

Free Energy. Though a reaction is thermodynamically feasible, it may be very slow. For example ΔG for the reaction;

$$SF_{6(g)} + 4H_2O_{(\ell)} \rightarrow H_2SO_{4(aq)} + 6HF_{(aq)}, \quad (5.7)$$

is -460 kJ, but no observable reaction occurs, due to kinetic factors. In other cases the reaction products may protect one of the reactants from further change, as in the corrosion of aluminium.

$$4Al + 3O_2 \rightarrow 2Al_2O_3, \ \Delta G = -3152 \text{ kJ} \quad (5.8)$$

The spontaneous reaction is prevented by the formation of a thin oxide film adhering strongly to the surface of the metal. Reactants may produce different products and the reaction that proceeds will normally have the more favourable free energy change. The combustion of carbon can proceed by two different routes;

$$C_{(s)} + O_{2(g)} \rightarrow CO_{2(g)}, \ \Delta G = -394 \text{ kJ}, \quad (5.9)$$

and $$C_{(s)} + \tfrac{1}{2}O_{2(g)} \rightarrow CO_{(g)}, \ \Delta G = -137 \text{ kJ}. \quad (5.10)$$

At ambient temperatures the first reaction occurs, but the second reaction which is associated with a greater positive entropy change (+89 J K^{-1} compared with +3 J K^{-1} for the first reaction), is the more favourable reaction above 983 K. The reaction of NH_3 with O_2 can proceed by two routes;

$$4NH_{3(g)} + 3O_{2(g)} \rightarrow 2N_{2(g)} + 6H_2O_{(g)}, \ \Delta G = -1303 \text{ kJ}, \quad (5.11)$$

and $$4NH_{3(g)} + 5O_{2(g)} \rightarrow 4NO_{(g)} + 6H_2O_{(g)}, \ \Delta G = -960 \text{ kJ}. \quad (5.12)$$

In this case the less favourable reaction is achieved with the use of a platinum or platinum-rhodium catalyst, which allows for an alternative reaction pathway.

The products of a reaction which has a positive ΔG, may be produced by either increasing the temperature, if $S > 0$, or seeking other reactions with the more favourable free energies, or forcing reactions to proceed by supplying the energy, as in electrolytic processes.

The magnitude of ΔG is clearly dependent on the values of ΔH and ΔS as seen from the free energy data for the formation of H_2O, NH_3 and CH_4 (Table 5.1).

TABLE 5.1 Thermodynamic Data for the Formation of H_2O, NH_3 and CH_4 (kJ)

Reaction (at 298K)	ΔH_f^o	$T\Delta S_f^o$	ΔG_f^o
$H_{2(g)} + \frac{1}{2}O_{2(g)} \rightarrow H_2O_{(g)}$	-242	-13.4	-228.6
$\frac{1}{2}N_{2(g)} + \frac{3}{2}H_{2(g)} \rightarrow NH_{3(g)}$	-46	-29.3	-16.7
$C_{(s)} + 2H_{2(g)} \rightarrow CH_{4(g)}$	-75	-24.3	-50.7

In all cases the entropy change is unfavourable, and only for water is the enthalpy change large. However, in the case of water the situation alters as the temperature is raised, as shown in Figure 5.1a. The unfavourable $T\Delta S$ term becomes greater than ΔH, so that above 4000K ΔG is +ve, and water spontaneously decomposes to H_2 and O_2. A similar diagram for the water gas reaction;

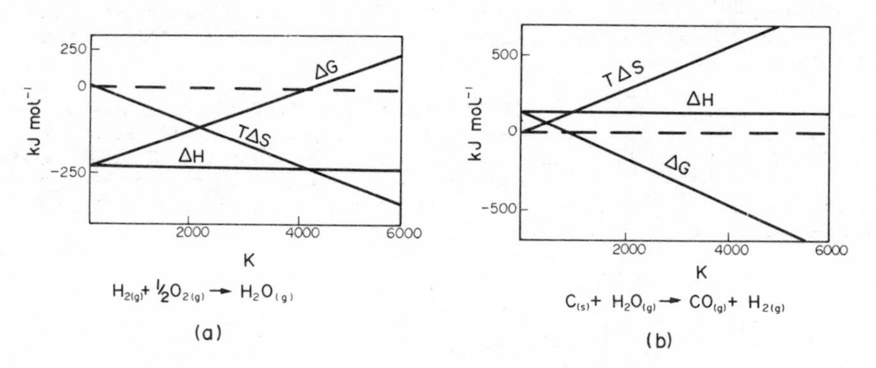

Fig. 5.1. Energy diagrams for (a) the formation of water, and (b) the water gas reaction.

$$C_{(s)} + H_2O_{(g)} \rightarrow CO_{(g)} + H_{2(g)}, \qquad (5.13)$$

indicates how a favourable $T\Delta S$ term makes the reaction feasible above 1000° K (Fig. 5.1b). The signs of ΔH and $T\Delta S$ do not change with temperature and at low temperatures ΔH is the dominant term, while at high temperatures $T\Delta S$ becomes dominant. This is because of the increase in T, rather than any great change in ΔS.

Two important reactions in the use of natural gas (Chapter 4) are steam reforming:

$$CH_4 + H_2O \rightarrow CO + 3H_2, \quad \Delta G = +142 \text{ kJ} \qquad (5.14)$$

and the shift reaction

$$CO + H_2O \rightarrow CO_2 + H_2, \quad \Delta G = -29 \text{ kJ} \qquad (5.15)$$

The first reaction becomes feasible around 800-900 K, and the second reaction is only feasible below 1000 K. The reasons relate to the interplay of ΔH and $T\Delta S$ (Fig. 5.2a, b).

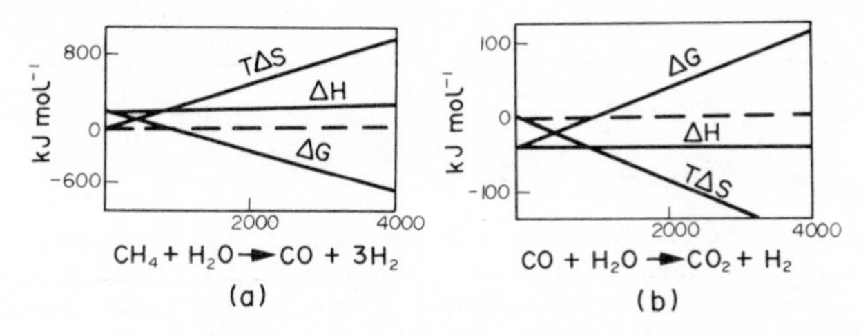

Fig. 5.2. Energy diagrams for (a) the steam reforming of methane, and (b) the shift reaction.

Enthalpy. Since, in the region of ambient temperatures, ΔH dominates, it is not unreasonable to use enthalpy data alone. This is because the term $T\Delta S$ tends to zero as the temperature drops, and ΔS is in Joules while ΔH is normally in kiloJoules. However, it is necessary to be cautious when using just enthalpy data, as at times the entropy term can be significant. The safest time to use just ΔH values is when comparing trends within a series of related reactions.

The enthalpy of a reaction may be estimated from the difference between the energy required to break and reform bonds. Analysis of the formation of H_2O (equation 5.16), NH_3 (5.17) and CH_4 (5.18) shows why the ΔH values given in Table 5.1 differ. The high sublimation energy of carbon, and the strong triple bond in N_2 reduce the overall enthalpy of the formation of CH_4 and NH_3 respectively (figures in kJ).

$$\begin{array}{ccccc} H_{2(g)} & + & \tfrac{1}{2}O_{2(g)} & \rightarrow & H_2O_{(g)} \\ 432 \downarrow & & 249 \downarrow & \nearrow & 2(O\text{-}H) \\ 2H_{(g)} & + & O_{(g)} & & -922 \end{array} \tag{5.16}$$

$$\Delta H_f^o = 432 + 249 - 922$$

$$= -241 \text{ kJ mol}^{-1}$$

$$\begin{array}{ccccc} \tfrac{1}{2}N_{2(g)} & + & \tfrac{3}{2}H_{2(g)} & \rightarrow & NH_{3(g)} \\ 471 \downarrow & & 648 \downarrow & \nearrow & 3(N\text{-}H) \\ N_{(g)} & + & 3H_{(g)} & & -1158 \end{array} \tag{5.17}$$

$$\Delta H_f^o = 471 + 648 - 1158$$

$$= -39 \text{ kJ mol}^{-1}$$

$$
\begin{array}{ccccc}
C_{(s)} & + & 2H_{2(g)} & \rightarrow & CH_{4(g)} \\
714 \downarrow & & 864 \downarrow & \nearrow & 4(C\text{-}H) \\
C_{(g)} & + & 4H_{(g)} & & -1644
\end{array}
\qquad (5.18)
$$

$$\Delta H_f^o = 714 + 864 - 1644$$

$$= -66 \text{ kJ mol}^{-1}$$

Some of the more common enthalpy processes, which are important in inorganic systems are listed in Table 5.2. Such terms are used in the estimation of enthalpies of reactions using a thermochemical cycle, based either on a covalent model or an ionic model. The covalent model approach has been given above for equations (5.16) to (5.18).

TABLE 5.2 Common Enthalpy Processes

Enthalpy	Process	Symbol	Sign of ΔH (as written)
Ionization	$M_{(g)} \rightarrow M^+_{(g)} + e_{(g)}$	I.	+
Electron affinity	$X_{(g)} + e_{(g)} \rightarrow X^-_{(g)}$	E.A.	- (mostly)
Sublimation or Atomization	$M_{(s)} \rightarrow M_{(g)}$	A	+
Lattice	$M^+_{(g)} + X^-_{(g)} \rightarrow MX_{(s)}$	U	-
Solvation	$M^+_{(g)} \rightarrow M^+_{(soln)}$ $X^-_{(g)} \rightarrow X^-_{(soln)}$	ΔH_{soln}	-
Bond	$M\text{-}X_{(g)} \rightarrow M_{(g)} + X_{(g)}$	D	+
Electrode	$M_{(s)} \rightarrow M^+_{(aq)} + e_{(aq)}$ $X_{(g)} + e_{(g)} \rightarrow X^-_{(aq)}$	E^o	

The ionic model can be used for estimating the enthalpy of formation of ionic solids, such as the alkaline earth halides.

$$
\begin{array}{ccccc}
M_{(s)} & + & X_{2(g)} & \overset{\Delta H}{\rightarrow} & MX_{2(s)} \\
\downarrow A & & \downarrow D & & \nearrow \\
M_{(g)} & & 2X_{(g)} & -U & \\
\downarrow I_1+I_2 & & \downarrow -EA & \nearrow & \\
M^{2+}_{(g)} & + & 2X^-_{(g)} & &
\end{array}
\qquad (5.19)
$$

Hence $\Delta H = A + I_1 + I_2 + D - 2EA - U$, and for $MgBr_2$ we have;

$$\Delta H = 149 + 737 + 1450 + 193 - 648 - 2398,$$

$$= -517 \text{ kJ mol}^{-1}.$$

It is clear that the dominant energy term is the lattice energy. Provided data are available, enthalpies of more complex reactions can be obtained, for example the decomposition of a metal halide;

$$MX_{n(s)} \rightarrow MX_{n-1(s)} + \tfrac{1}{2}X_{2(g)}. \quad (5.20)$$

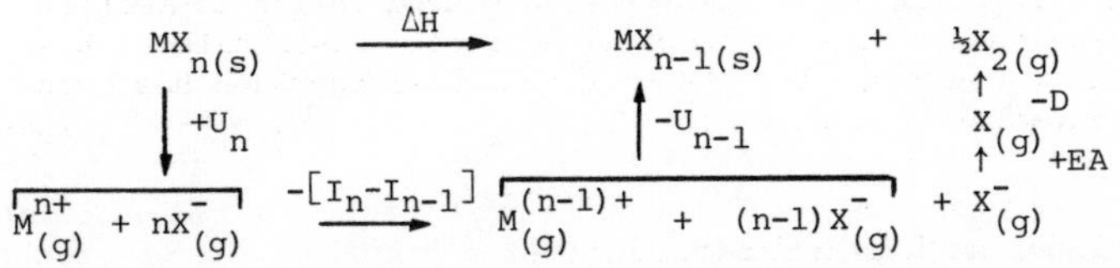

$$\begin{array}{ccccc} MX_{n(s)} & \xrightarrow{\Delta H} & MX_{n-1(s)} & + & \tfrac{1}{2}X_{2(g)} \\ \downarrow +U_n & & \uparrow -U_{n-1} & & \uparrow -D \\ & & & & X_{(g)} \\ & & & & \uparrow +EA \\ M^{n+}_{(g)} + nX^-_{(g)} & \xrightarrow{-[I_n - I_{n-1}]} & M^{(n-1)+}_{(g)} + (n-1)X^-_{(g)} & + & X^-_{(g)} \end{array} \quad (5.21)$$

Hence,

$$\Delta H = U_n - U_{n-1} - \left[I_n - I_{n-1}\right] + EA - D. \quad (5.22)$$

The unknown quantities in this equation are usually the lattice energies. However, if use is made of Kapustinskii's equation;

$$U = \frac{1071\ \nu\ Z_+ Z_-}{r_+ + r_-} \text{ kJ mol}^{-1}, \quad (5.23)$$

then it is possible to estimate ΔH. For the reaction;

$$CrCl_3 \rightarrow CrCl_2 + \tfrac{1}{2}Cl_2, \quad (5.24)$$

$$\Delta H = 5143 - 2425 - [3056 - 1592] + 348 - 121$$

$$= 1481 \text{ kJ mol}^{-1}$$

On the basis of the enthalpy data, the reaction is unfavourable, however the entropy change is favourable and at high temperatures ($\sim 600^{\circ}C$) the reaction occurs.

Entropy. Analysis of the entropies of a number of reactions show that in general, when more gaseous molecules occur in the products than in the reactants the entropy is positive, i.e. favourable. For example;

$$H_2O_{(\ell)} \rightarrow H_{2(g)} + \tfrac{1}{2}O_{2(g)}, \quad \Delta S_{298} = 45 \text{ J K}^{-1} \quad (5.25)$$

$$SiCl_{4(\ell)} + 2H_2O_{(\ell)} \rightarrow SiO_{2(s)} + 4HCl_{(g)}, \quad \Delta S_{298} = 410 \text{ J K}^{-1} \quad (5.26)$$

Reactions in solution are not so readily analysed, as solvent and solute interactions may influence the entropy, as in dissolving magnesium chloride.

$$MgCl_{2(c)} \rightarrow Mg^{2+}_{(aq)} + 2Cl^-_{(aq)} \quad (5.27)$$

$$\Delta S_{298} = S_{Mg^{2+}_{(aq)}} + 2S_{Cl^-_{(aq)}} - S_{MgCl_{2(c)}} \quad (5.28)$$

$$= -118 + 2 \times 55 - 89$$

$$= -97 \text{ J K}^{-1}$$

KINETICS

Whereas thermodynamic data gives information on whether a reaction is feasible or not, it gives no indication of the rate of the reaction. The reaction;

$$2H_{2(g)} + O_{2(g)} \rightarrow 2H_2O_{(g)}, \quad (5.29)$$

has a favourable free energy of -457 kJ, but a mixture of the two gases at room temperature shows no sign of reacting, over a period of years. Yet in the presence of a heat source, such as a spark, reaction occurs explosively. Chemical kinetics is the study of reactions from the point of view of their rates, and kinetic data allows chemists to decide on reaction mechanisms, the relationship between structure and reactivity and how to optimise conditions to produce a desired rate.

Rate Laws

For a reaction, such as;

$$NO + O_3 \rightarrow NO_2 + O_2, \quad (5.30)$$

the rate is given by;

$$-\frac{d[NO]}{dt} = -\frac{d[O_3]}{dt} = \frac{d[NO_2]}{dt} = \frac{d[O_2]}{dt} \quad (5.31)$$

The rate is a function of the concentration of the reacting species NO and O_3;

$$\text{Rate} = k_r[NO]^m[O_3]^n, \quad (5.32)$$

where k_r is the rate constant (which is independent of the concentration of the reacting species) and has units of moles ℓ^{-1} s^{-1}. The order of the reaction is m + n, which in this case is found experimentally as 1 + 1, i.e. a second order reaction. It is also a bimolecular reaction as the reaction involves the collision of two molecules (NO and O_3). The reaction orders (zero, 1, 2 and sometimes 3) generally indicate the number of molecules involved in collisions in the rate determining step. Often, but not always, the order is the same as the molecular stoichiometry of the reactants, i.e. for;

$$2NO + O_2 \rightarrow 2NO_2 \quad (5.33)$$

$$-\frac{d[O_2]}{dt} = k_r[NO]^2[O_2], \quad (5.34)$$

but for;

$$H_2O_2 + 2HI \rightarrow 2H_2O + I_2, \quad (5.35)$$

the Rate $= k_r[H_2O_2][HI]$, (5.36)

indicating the rate controlling step involves *one* molecule of HI reacting with H_2O_2. A mechanism consistent with this is;

$$H_2O_2 + HI \xrightarrow[\text{determining}]{\text{rate}} H_2O + HIO, \quad (5.37)$$

$$HIO + HI \rightarrow H_2O + I_2. \quad (5.38)$$

For a first order reaction, A → products, the rate is;

$$-\frac{d[A]}{dt} = k_r[A], \quad (5.39)$$

which on integration gives; (5.40)

$$\ln\frac{[A]_o}{[A]_t} = k_r t, \quad (5.40)$$

where $[A]_o$ is the initial concentration of A, and $[A]_t$ is the concentration of A at time t. This equation applies to the decay of radio-nuclei.

For a reaction where one of the reactants is in a large excess so that change in its concentration is negligible compared with the others, then the rate is virtually independent of the substance and the reaction has a pseudo order. Such a situation arises when a reactant also acts as the solvent for the materials.

Reaction Rate Theory. In order for a reaction to proceed molecules or ions must collide with sufficient energy and in the correct orientation. Two theories have been developed to quantify these terms, the collision theory and the transition state theory.

The reaction rate in the collision theory expressed in general terms is;

$$\text{Reaction rate} \propto \text{Collision frequency} \times \text{Collision energy} \times \text{Collision geometry} \quad (5.41)$$

For the reaction A + B → products;

$$\text{Reaction rate} = Zn_An_B \times e^{-\Delta H^*/RT} \times P, \quad (5.42)$$

where Z is the frequency of collisions, ΔH^* the activation energy, i.e. the energy to initiate the reaction, and P the probability factor that the molecules will collide with the correct geometry. Since;

$$\text{Reaction rate} = k_r n_A n_B, \quad (5.43)$$

for a bimolecular reaction;

$$k_r = ZPe^{-\Delta H^*/RT} \quad (5.44)$$

For the transition state theory it is assumed that an activated complex is formed in equilibrium with the reactants;

$$A + B \rightarrow AB^* \rightarrow \text{products} \quad (5.45)$$

In this case the reaction rate is expressed as;

$$\text{Reaction rate} \propto \text{Collision frequency} \times \text{Enthalpy of activation} \times \text{Entropy of activation} \quad (5.46)$$

Since an equilibrium is involved the rate will be proportional to the equilibrium concentration of AB*;

$$\text{Reaction rate} \propto [AB^*] \quad (5.47)$$

$$\propto [A][B]K^*, \quad (5.48)$$

where K* is the equilibrium constant. The reaction rate is;

$$\text{Reaction rate} = \nu[A][B]K^*, \quad (5.49)$$

where $\nu(=\frac{kT}{h})$ is the frequency with which AB* converts to products.

$$\therefore \quad \text{Reaction rate} = \frac{kT}{h}[A][B]K^*. \quad (5.50)$$

But,

$$\text{Reaction rate} = k_r[A][B], \quad (5.51)$$

hence

$$k_r = \frac{kT}{h}K^* \quad (5.52)$$

However,

$$K^* = e^{-\Delta G^*/RT} = e^{\Delta S^*/R}e^{-\Delta H^*/RT} \quad (5.53)$$

Therefore

$$k_r = \frac{kT}{h}e^{-\Delta H^*/RT}e^{\Delta S^*/R} \quad (5.54)$$

The transition state theory can be represented diagrammatically (Fig. 5.3). The reaction rate is determined by the magnitude of $\Delta G^*_{forward}$, and since it is positive it becomes apparent why the rate increases with a rise in temperature. For the $H_2 + O_2$ reaction (5.29) a spark provides the energy for $\Delta G^*_{forward}$, while at room temperature there is insufficient energy. Once the reaction is initiated the overall release of energy ($-\Delta G_{reaction}$) maintains the reaction. The free energy of the reaction $\Delta G_{reaction}$ is the difference between the two free energies of activation

$$\Delta G_{reaction} = (\Delta G^*_{forward} - \Delta G^*_{reverse}) \quad (5.55)$$

Figure 5.3 is a compact summary of both the kinetic and thermodynamic features of a reaction. The top section ($\Delta G^*_{forward}$) relates to the rate, and the bottom section ($\Delta G_{reaction}$) indicates the thermodynamic feasibility of a reaction.

Factors Affecting Reaction Rates

An important question in chemistry is, how can a reaction rate be modified, speeded up or slowed down? We will consider some of the factors that influence the reaction rate.

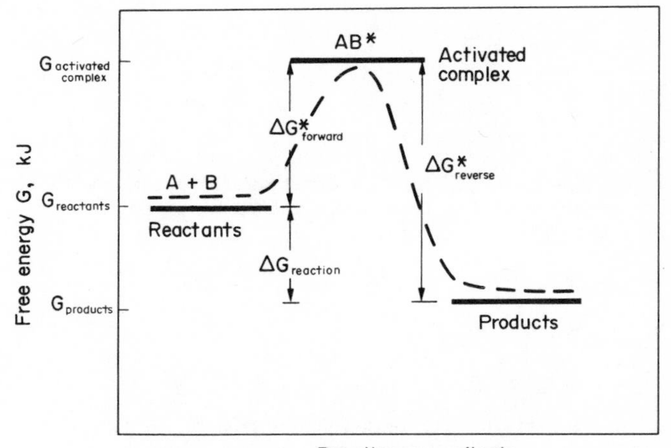

Fig. 5.3. Free energy diagram for a reaction via an activated complex.

The *nature and form of the reacting species* influence the rate, for example the size of bubbles passed through a solution. For large bubbles much of the gas does not come into contact with the other reactants, so a reaction would be speeded up using many small bubbles. The particle size of solid material is important in the same way, the small size providing a greater reaction surface. Since the rate is proportional to the *concentration of the reacting species* higher concentrations will increase the rate, by increasing the frequency of collisions. The influence of *temperature* has already been mentioned in relation to ΔG^*. An increased temperature also will increase the frequency of collisions. Some reactions are influenced by *radiation*, i.e. photochemical reactions. Mixtures of H_2 and Cl_2 do not react in the dark, but react explosively in sunlight, because the light provides the energy for the initial reaction;

$$Cl_2 + h\nu \rightarrow Cl + Cl. \tag{5.56}$$

An important method of changing reaction rates is by the use of *catalysts*. A catalyst is a material which influences the rate, is involved in the reaction, but does not change the overall stoichiometry, and is therefore chemically unchanged at the end of a reaction. A catalysed reaction can be written as;

$$A + B + \text{catalyst} \rightarrow \text{products} + \text{catalyst}, \tag{5.57}$$

and the rate is;

$$-\frac{d[A]}{dt} = k_r[A][B] + k_c[A][B][\text{catalyst}]. \tag{5.58}$$

However, the contribution to the rate of the first term is small compared with the catalysed reaction, and since the catalyst is chemically unchanged we can write;

$$-\frac{d[A]}{dt} = k_r'[A][B], \tag{5.59}$$

where $k_r' = k_c[\text{catalyst}]$. (5.60)

The reaction;

$$2SO_2 + O_2 \rightarrow 2SO_3 + \Delta H, \quad 5.61$$

is feasible, but the rate is slow. The rate will increase with temperature but the reaction becomes less feasible. The catalyst NO_2 speeds up the reaction and a possible mechanism is;

$$SO_2 + NO_2 \rightarrow SO_3 + NO, \tag{5.62}$$

$$NO + \tfrac{1}{2}O_2 \rightarrow NO_2, \tag{5.63}$$

i.e. $SO_2 + \tfrac{1}{2}O_2 \rightleftarrows SO_3$ (5.64)

The process is the basis of the lead-chamber method for producing sulphuric acid. The reaction is probably important in the urban atmosphere where both SO_2 (from coal burning) and NO_2 (from automobile emissions) are present. Such a catalyst is called homogeneous as it is in the same phase as the reactants. Other catalysts are heterogeneous, normally solids, which are in a different phase to the reactants. The contact process for the production of H_2SO_4 makes use of platinum or vanadium pentoxide as catalysts for reaction (5.61). This type of catalyst provides a surface for reactions to occur, and also assists in a favourable geometry for collisions. In the Haber process;

$$N_2 + 3H_2 \overset{\alpha Fe}{\rightleftharpoons} 2NH_3, \tag{5.65}$$

the iron catalyst is probably involved in the following processes;

$$N_2 + \text{catalyst} \xrightarrow{\text{absorption}} N_2\text{—catalyst} \tag{5.66}$$

$$N_2\text{—catalyst} \xrightarrow{\text{bond breaking}} 2N\text{—catalyst} \tag{5.67}$$

Similarly for hydrogen, then;

$$\begin{matrix} N\text{—catalyst} \\ + H\text{—catalyst} \end{matrix} \xrightarrow{\text{bond formation}} NH\text{—catalyst}, \tag{5.68}$$

and so on until NH_3 is produced. Since ΔG and K are state functions they are not changed, as the catalyst provides an alternative route for the reactants to give the products, i.e. the catalyst lowers ΔG^* but does not change $\Delta G_{reaction}$. Enzymes are very important catalysts in living material, as they assist complex reactions to occur at ambient temperatures and pressures, and often in the absence of light. For example respiration;

$$C_6H_{12}O_{6(s)} + 6O_{2(g)} \rightleftharpoons 6CO_{2(g)} + 6H_2O_{(\ell)} \tag{5.69}$$

has a very favourable $\Delta G_{reaction}$ of -3176 kJ, but a high activation energy at ambient temperatures. The reaction is achieved by being broken into a number of steps, each catalysed by a specific enzyme;

$$A \overset{E_1}{\rightleftharpoons} B \overset{E_2}{\rightleftharpoons} C \overset{E_3}{\rightleftharpoons} D \overset{E_4}{\rightleftharpoons} \quad (5.70)$$

circumventing the high energy barrier of the uncatalysed reaction.

CHEMICAL EQUILIBRIA

All chemical reactions should be considered as equilibrium reactions, and the extent that a reaction proceeds is a dynamic balance between the forward and reverse reactions. The equilibrium constant K is a measure of this dynamic equilibrium, and for a reaction such as;

$$aA + bB \rightleftharpoons cC + dD, \quad (5.71)$$

the constant is;

$$K_c = \frac{[C]^c[D]^d}{[A]^a[B]^b}, \quad (5.72)$$

or, more correctly in terms of activities;

$$K_a = \frac{a_C^c\, a_D^d}{a_A^a\, a_B^b} \quad (5.73)$$

The reaction of N_2 and H_2 to give NH_3, in the gas phase, is a good example of an equilibrium reaction;

$$N_2 + 3H_2 \rightleftharpoons 2NH_3, \quad (5.74)$$

and $$K_a = \frac{a_{NH_3}^2}{a_{N_2} a_{H_2}^3} \quad (5.75)$$

or $$K_p = \frac{P_{NH_3}^2}{P_{N_2} P_{H_2}^3}$$

since $a \approx \frac{n}{V} = \frac{P}{RT}$, where $K_p = K_a/(RT)^2$.

The dissociation of acetic acid in water is a good example of an equilibrium reaction in solution;

$$CH_3COOH + H_2O \rightleftharpoons CH_3COO^- + H_3O^+, \quad (5.77)$$

and $$K_a = \frac{a_{CH_3COO^-}\, a_{H_3O^+}}{a_{CH_3COOH}\, a_{H_2O}}, \quad (5.78)$$

or $$K_c = \frac{[CH_3COO^-][H_3O^+]}{[CH_3COOH]}, \quad (5.79)$$

where the activity of water, the solvent, is 1.

The position of equilibrium, involving gases, may be altered by changing the pressure (or volume) if the number of moles of gaseous components differ on either side of the reaction. For the reaction (5.74) the moles of gas change from 4 to 2 and therefore an increase in pressure (or reduction in volume) will drive the reaction to the formation of more ammonia, the equilibrium constant of course remaining constant. This type of event is encompassed in Le Chatelier's principle, which states that if a stress is applied to a system in equilibrium, the position of equilibrium changes in order to absorb the stress. A change in temperature has a similar affect, and if a reaction is endothermic an increase in temperature will favour the formation of the products. An increase in the concentration of one of the species (reactants) or products) will also change the equilibrium position, but of course without affecting the equilibrium constant. Sufficient OH^- ions are present in ammonia dissolved in water,

$$NH_3 + H_2O \rightleftharpoons NH_4^+ + OH^- \quad (5.80)$$

to exceed the solubility product of $Mg(OH)_2$. However, if NH_4Cl is added to the solution reaction (5.80) moves to the left and $Mg(OH)_2$ dissolves.

STRUCTURE AND BONDING AND CHEMICAL REACTIVITY

The factors, thermodynamics, reaction rates and chemical equilibria which influence reaction rates, include a knowledge of structure and bonding in compounds. It is instructive to consider structure and bonding separately, and a few examples will be considered.

Comparative Reactivity of N_2, CO and F_2

The three molecules N_2, CO and F_2 increase in reactivity in the order given. Dinitrogen and carbon monoxide are isoelectronic with strong triple bonds, N≡N (946 kJ mol^{-1}), C≡O (1080 kJ mol^{-1}). Of the two compounds CO is the more reactive forming metal carbonyls readily, and is a useful reducing agent at elevated temperatures. The different reactivity may be accounted for by the different bonding, as illustrated in Fig. 5.4. Because of the different relative energy levels of the 2s and 2p orbitals of oxygen and carbon the bonding orbitals in CO are not symmetrically arranged in the molecule as for N_2. The π_1 orbital of CO is concentrated more on the oxygen and the π_2 (anti-bonding) more on the carbon atom. It is the π_2 orbital that can interact with nucleophiles such as the d-electrons of a transition metal or a X^- group. The lone pair in the π_3 orbital which is localised on the carbon, can act as an electron donor. The tight, symmetrical bonding in N_2 does not provide a reactive centre, lone pairs are not readily available for bonding, and the molecule is not readily polarisable.

The reactivity of difluorine is because of the number of electron pairs localised on each F atom and only a single bond between each atom. The lone pairs on one F atom strongly repel those on the other because of their proximity, due to the small size of the fluorine atoms. This leads to a long (142 pm) and weak (158 kJ mol^{-1}) F-F bond and the greater reactivity.

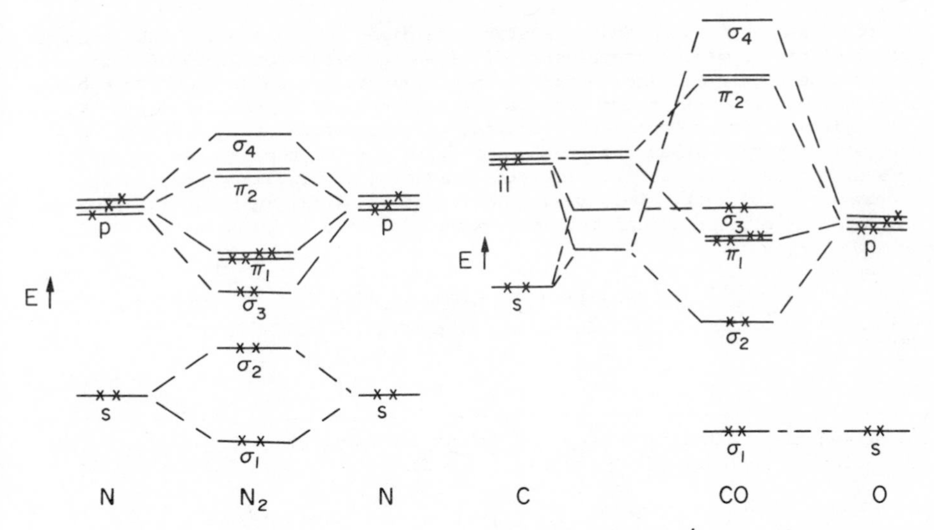

Fig. 5.4. Molecular orbital diagrams for N_2 and CO.

Reactivity of Silicates.

The distinct difference in the reactivity of the oxy-species of carbon (CO, CO_2, HCO_3^- and CO_3^{2-}), and the oxy-species of silicon, (SiO_2 and various silicates), is a consequence of bonding differences. The small C and O atoms, which are comparable in size, achieve their bonding requirements by the formation of σ and p_π - p_π bonds e.g. C≡O, O=C=O. Silicon is significantly larger than carbon and the overlap of Si p_π and O p_π orbitals is poor and does not contribute to the Si-O bonding. In this case Si and O achieve their bonding requirements by polymerisation through oxygen bridging between silicon atoms, or oxygen atoms gaining an electron. This gives rise to the wide range of silicate structures (Fig. 6.10). The Si-O bond is strong and a system of Si-O single bonds is overall more stable than a system of Si=O bonds, the reverse holds for carbon. The polymeric silicate structures are therefore 1, 2 and 3 dimensional covalent lattice structures and are relatively stable and unreactive especially under ambient environmental conditions. This fact, together with the abundance of Si and O in the earth's crust, accounts for the dominance of silicates in rocks and soils.

The Hydrogen-Bond

The hydrogen-bond, while relatively weak, is an important bond in chemistry, and is formed by a hydrogen atom bridging between the elements O, N and F (mainly). The three elements O, N and F are good proton acceptors with lone pairs of electrons which can interact with hydrogen, and when bonded to hydrogen the XH group is a strong proton donor due to the high electronegativity of X(N, O, F). The bond can be represented as;

$$\overset{-\delta}{X} \text{ — } \overset{+\delta}{H} \text{ ----- } Y.$$

↑ covalent bond ↑ hydrogen bond

Some typical hydrogen-bond parameters are given in Table 5.3. The properties of water are a direct consequence of H-bonding, and even at 90°C considerable H-bonding persists. The H-bond is the reason for the wide liquid range of water and therefore its importance as a solvent and reaction medium. In ice H-bonding is fully developed producing a diamond type structure (Fig. 2.7) with a very open structure. When ice melts some of the H-bonds break and H_2O molecules fall into the cavities producing a more dense structure (between 0 and 4°C). This means ice floats on water, an important fact for aquatic life. The H-bond influences the chemistry of fluorine and for

TABLE 5.3 Hydrogen-Bond Parameters

Bond	Compound	Bond energy kJ mol^{-1}	X-H---Y distance pm
F-H---F	$HF_{(g)}$	~29	255
F-H-F	KHF_2	~150	229
O-H---O	$H_2O_{(s)}$	~21	276
N-H---N	melamine	~25	300

example one factor in the low acidity of HF is the relatively strong H-bonds in $(HF)_n$. The H-bond is the mode by which the two strands of DNA are linked together, by joining base pairs. The weakness of the H-bonds must be a factor enabling the two strands to break apart, move and rejoin.

CHAPTER 6

Inorganic Chemicals in Everyday Use

We now will consider some of the more important inorganic chemicals, how they are produced industrially and their use as consumer products. The chemical industry has grown to become a feature of the industrialised developed world and our environment. Some of the major inorganic chemicals and their uses are listed in Table 6.1, where they are grouped according to the principal element in each compound. It is clear that one of the more important chemical industries is fertilizer production. However, its rate of growth from 1963 to 1971 as an industry, does not match the growth rate for synthetic resins, plastic materials and pharmaceuticals (Fig. 6.1 for the United Kingdom). Structural materials, such as cement, pig iron and bricks, top the list of quantities of inorganic materials consumed in the United Kingdom.

THE NITROGEN INDUSTRY

The principal routes to compounds produced in the nitrogen industry are summarised in the diagram

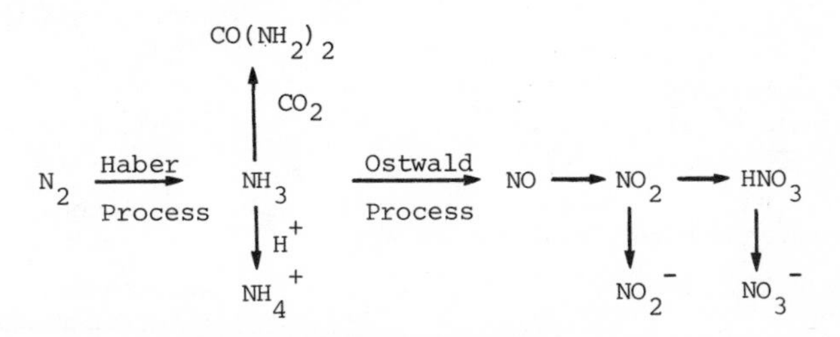

The two main steps are the fixation of atmospheric dinitrogen to give ammonia, and its oxidation to nitric oxide leading eventually to nitric acid. Production of nitrogenous fertilizers, explosives, plastics and synthetic fibres depend on achieving these reactions.

TABLE 6.1 Some Inorganic Chemicals and Their Uses

Element	Compounds	Production(annual)	Some Uses
Nitrogen	NH_3	$14x10^6$ tonnes (USA) 1974	Fertilizer, refrigerant, cleaning agent, intermediate in producing Na_2CO_3, sulpha-drugs, synthetic fibres (eg. nylon) dyestuffs.
	$(NH_4)_2SO_4$		N-fertilizer (and sulphur).
	NH_4NO_3		N-fertilizer, explosives.
	KNO_3		N-fertilizer (and potassium).
	HNO_3	$7x10^6$ tonnes (USA) 1974	N-fertilizer production, N-compounds in explosives, organo-nitrocompounds.
	$(NH_2)_2CO$(urea)		N-fertilizer, intermediate in plastic production.
Phosphorus	P_4, P_n	$1.2x10^6$ tonnes (world) 1976	Production of H_3PO_4, phosphorus sulphides, chlorides & organo-phosphorus compounds, insecticides, flame retardants.
	H_3PO_4	$23.7x10^6$ tonnes(P_2O_5)	Production of fertilizers, food phosphates, detergent phosphates, industrial phosphates, soft drinks (cola), derusting & pickling steel.
	Phosphates Alkali metal phosphates		Soaps, detergents, acid/base buffers.
	$Ca(H_2PO_4)_2$ superphosphate	$24.3x10^6$ tonnes(P_2O_5) (world) 1974/5	Fertilizer.
	$Ca(H_2PO_4)_2$(pure)		Animal feed, food aerator & leavener (baking powder).
	$CaHPO_4$		Toothpaste (abrasive)
	Polyphosphates	$0.8x10^6$ tonnes (USA) 1974	Detergents, in foodstuffs maintain texture & structure, stops discolouration due to metal oxidation.

TABLE 6.1 contd.

Element	Compounds	Production (annual)	Uses
Sulphur	S		Production of H_2SO_4
	H_2SO_4	$4.4-4.7x10^6$ tonnes (UK) 1976 $30x10^6$ tonnes (USA) 1974	Production of fertilizers, paints, detergents, synthetic fibres, wide range of chemicals, in alkylation of petroleum to give high octane petrols, battery acid.
Silicon	Si		Transistors, Si-chips.
	SiC		Abrasive, electrical heating elements.
	Na silicates (soluble)		Soap, detergent builders, catalysts, adhesives, desiccants, pigments..
	Silicates (insoluble)		Glass, cement, stone, also asbestos, talc, zeolites.
	Silicones		Water proofing, oils, greases, rubbers.
Sodium	NaOH	$31x10^6$ tonnes (world) 1974	Pulp & paper, textiles eg. rayon, soap, detergents, Al production, inorganic & organic chemicals, neutralization (acid/base).
	Na_2CO_3	$26x10^6$ tonnes (world) 1974	Glass manufacture, washing, sodium silicates, phosphates.
	$NaHCO_3$		Baking.
	NaOCl		Bleach.
	NaCl		Production of Cl_2, in foodstuffs.

Nitrogen Fixation

The difficulty in fixing dinitrogen arises from the kinetic stability of the N_2 molecule as discussed in Chapter 5, even though the formation of NH_3 from N_2 is thermodynamically feasible. Over 50% of the nitrogen fixed annually (Table 7.1) is achieved biologically with nitrogenase bacteria in soils (e.g. *Clostridium* and *Azotobacter*) or associated with plants, such as legumes *(Rhizobium)*. The fixation by bacteria on plants occurs where the plants grow, therefore it is necessary at times to add nitrogen fertilizers where such plants are scarce or the climatic conditions are such that biological N-fixation is not sufficient for crop needs.

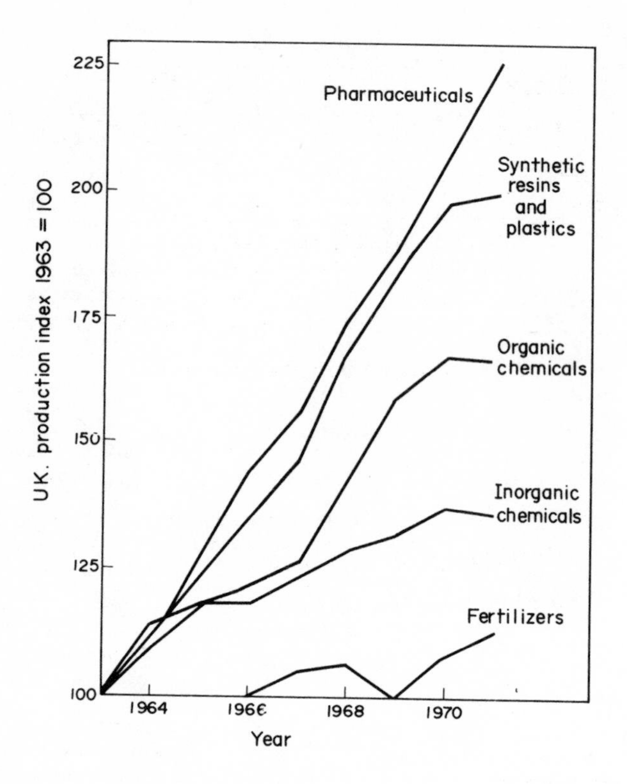

Fig. 6.1. Growth in United Kingdom chemical industry 1963-1971, indexed to 1963=100. Source: B.G. Rubens and M.L. Burstall, The Chemical Economy, Longmans, 1973.

The enzyme in Nitrogenase is made up of two proteins, a Mo-Fe protein (MW ~ 222,000) and a Fe protein (MW = 55,000 - 70,000). When they are combined with ATP, Mg^{2+} and a reducing agent, such as sodium dithionite, dinitrogen is reduced to NH_3. The iron-protein containing the Fe_4S_4 cluster, which is probably reduced by dithionite, is involved in electron transfer reactions with the other protein, where the N_2 is probably bonded to Mo atoms. A very simplified mechanism for nitrogenase action is given in Fig. 6.2. Attempts have been made to find model systems which aid our understanding of the reduction of dinitrogen, and provide a commercial route to ammonia. The discovery of the dinitrogen complex ion $[Ru(NH_3)_5N_2]^{2+}$, in 1965 was the first step in this direction. Reactions such as;

$$[WCl_4P_3] + P + 2N_2 + 4Mg \xrightarrow{THF} cis[W(N_2)_2P_4] + 4MgCl, \quad (6.1)$$

and $$cis[W(N_2)_2P_4] \xrightarrow[CH_3OH]{H_2SO_4} 2NH_3 + N_2 + 4[PH]HSO_4 + \text{oxidised Mo, compounds} \quad (6.2)$$

(where P = Me_2PhP), which give a 90% yield of NH_3 point the way to finding a technology that may compete with the Haber process in the future - i.e. route (a) in Fig. 6.3.

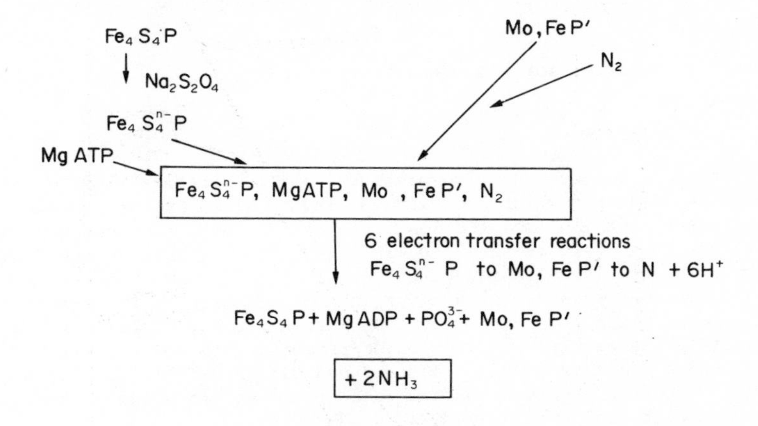

Fig. 6.2. A simplified scheme for the action of nitrogenase. Source: R.L. Richards Chem.Soc. London, Special Publication 36, 1979.

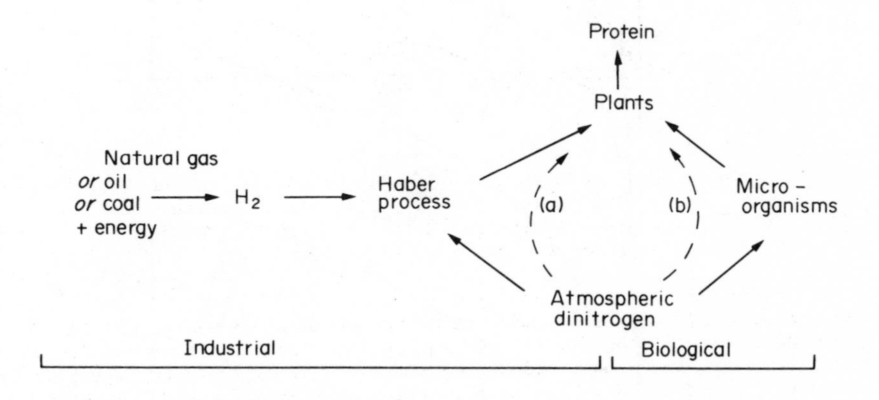

Fig. 6.3. Nitrogen fixation pathways. Source: J. Leigh New Scientist 1976, 385.

Another approach is genetic engineering (route (b) Fig. 6.3), where genes that control nitrogen fixation are isolated and transferred to other plants, such as wheat and potatoes.

Production of Ammonia

The industrial production of ammonia accounts for approximately 26% of the world's nitrogen fixed annually. It is carried out entirely by the Haber process developed early in the 20th century. The rapid development of the process gained impetus from the First World War, when supplies of nitrates from South America were cut off from Germany. Over the six years 1913 to 1918, nitrogen fixed by the Haber process in Germany rose from 1000 to 95,000 tons

The chemical equation for the preparation of NH_3 is;

$$N_{2(g)} + 3H_{2(g)} \rightleftharpoons 2NH_3,\ \Delta G_{298} = -33.4\ kJ \quad (6.3)$$

The reaction is favoured at high pressure, but is kinetically slow and exothermic, two features which necessitate finding a temperature which will give a satisfactory rate but not too low a yield. The effect of pressure and temperature on % yield is shown in Fig. 6.4. Temperatures in the region of 400-500°C and pressures in the range 10×10^3 to 35×10^3 kPa are the conditions normally used. A catalyst is used to speed up the reaction, and is a reason why temperatures below 400°C are not used, because of its low reactivity. The catalyst is iron oxide promoted with K_2O (0.35%) Al_2O_3 (0.84%) and a trace of CuO and then reduced with hydrogen. Dinitrogen is absorbed on to the catalyst surface where nitrogen atoms are produced which then react with dihydrogen (see Chapter 5).

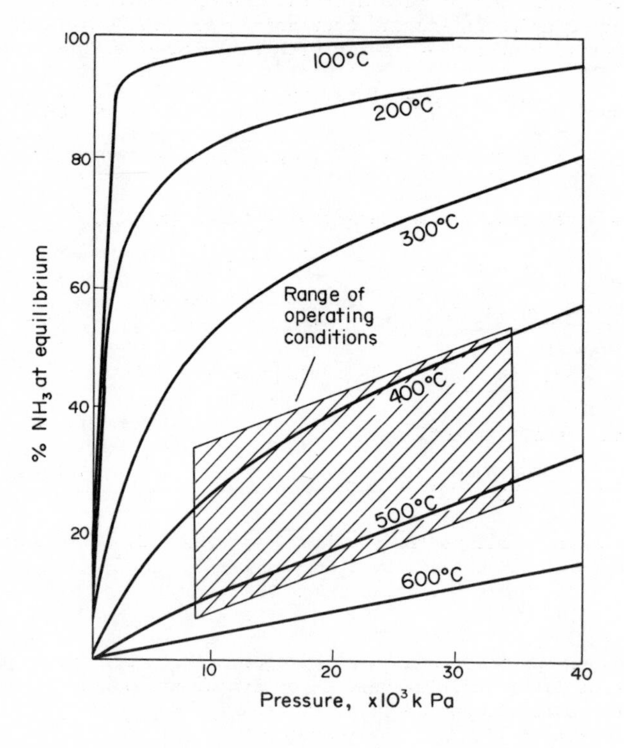

Fig. 6.4. Conditions for the production of ammonia by the Haber process.

The dinitrogen component of the synthesis gas (N_2 and $3H_2$) comes from the air, but the dihydrogen has, over the years, come from a number of sources depending on cost. Initially dihydrogen was produced by treating coal/coke with steam to give water gas (Bosch process);

$$C + H_2O \rightarrow CO + H_2, \ \Delta H = 131 \text{ kJ} \quad (6.4)$$

The energy for this process comes from the reactions;

$$C + O_2 + 4N_2 \rightarrow CO_2 + 4N_2, \ \Delta H = -407 \text{ kJ}, \quad (6.5)$$

and

$$C + \tfrac{1}{2}O_2 + 2N_2 \rightarrow CO + 2N_2, \ \Delta H = -107 \text{ kJ}, \quad (6.6)$$

to give producer gas. The two processes are combined and adjusted to give the right proportion of H_2 and N_2. The overall reaction, which is slightly endothermic is;

$$\underbrace{4N_2 + O_2}_{\text{air}} + 12H_2O + 7C \rightarrow 8\,NH_3 + 7CO_2, \ \Delta H = 38.5 \text{ kJ mol}^{-1}\ NH_3. \quad (6.7)$$

Today the steam reforming of liquid or gaseous hydrocarbons (Chapter 4) is used to produce dihydrogen and the synthesis gas. For example between 700 and 900°C paraffin hydrocarbons react with steam to give H_2;

$$C_nH_{2n+2} + nH_2O \rightarrow nCO + (2n + 1)H_2, \quad (6.8)$$

$$C_nH_{2n+2} + 2nH_2O \rightarrow nCO_2 + (3n + 1)H_2. \quad (6.9)$$

The most common feedstock is methane from natural gas;

$$CH_4 + H_2O \xrightarrow[700\text{-}800^\circ C]{\text{Ni catalyst}} CO + 3H_2, \ \Delta H = 142 \text{ kJ}, \quad (6.10)$$

$$2CH_4 + O_2 + 4N_2 \rightarrow 2CO + 4H_2 + 4N_2, \ \Delta H = -72 \text{ kJ}, \quad (6.11)$$

followed by the 'shift' reaction;

$$CO + H_2O \rightarrow CO_2 + H_2, \ \Delta H = -29 \text{ kJ}. \quad (6.12)$$

By balancing these reactions the synthesis gas ($N_2 + 3H_2$) is produced.

Urea

Urea is being increasingly used as a nitrogen fertilizer, due to its high nitrogen content (46%) and slow release of useful nitrogen. It is normally produced on the same site as ammonia, by reacting NH_3 with CO_2, the product of the shift reaction (6.12). The reaction;

$$CO_2 + 2NH_3 \rightarrow CO(NH_2)_2 + H_2O, \ \Delta H_{298} = -3 \text{ kJ}, \quad (6.13)$$

is carried out at high pressure, around 15 x 10^3 kPa, and temperatures around 180-200°C. It proceeds in two stages, the formation of ammonium carbamate, $OC(NH_2)ONH_4$, which then decomposes to urea and water. In soil urea hydrolyses to ammonium carbonate;

$$CO(NH_2)_2 + 2H_2O \rightarrow (NH_4)_2CO_3, \quad (6.14)$$

which undergoes rapid nitrification, but some loss of nitrogen can occur because the carbonate hydrolyses to ammonia.

Nitric Acid

The oxidation of ammonia can take one of two paths;

$$4NH_3 + 3O_2 \rightarrow 2N_2 + 6H_2O, \ \Delta H = -1270 \text{ kJ}, \quad (6.15)$$

and

$$4NH_3 + 5O_2 \rightarrow 4NO + 6H_2O, \ \Delta H = -950 \text{ kJ}. \quad (6.16)$$

Reaction (6.15) is the favoured, but (6.16) can be achieved with a catalyst of Pt (90%)/Rh (10%) gauze at 800-850°C. While this is an important reaction, leading eventually to nitric acid and nitrates, the reaction uses up the hard-won dihydrogen in the Haber Process. The burning produces H_2O and heat, which is mostly lost to the environment around the plant. The next reaction;

$$2NO + O_2 \rightleftharpoons 2NO_2, \ \Delta H \sim 0, \quad (6.17)$$

is one of the few termolecular reactions and, rather oddly, its rate decreases with an increase in temperature, therefore NO is cooled to around 25°C before reacting with O_2. The NO_2 then dissolves in water giving 61-65% nitric acid;

$$3NO_2 + H_2O \rightleftharpoons 2HNO_3 + NO, \ \Delta H = -117 \text{ kJ}, \quad (6.18)$$

Three-quarters of the nitric acid produced is used to make NH_4NO_3 by direct reaction of 40-60% acid with gaseous ammonia.

THE PHOSPHORUS INDUSTRY

The natural source of phosphorus is phosphate rock, fluoro, chloro or hydroxy apatite $[Ca_3(PO_4)_2]_3CaX_2$, the most common form being fluoroapatite. Around 10^8 tonnes are mined annually (approximately 9.5 x 10^6 tonnes of phosphorus). Large reserves, estimated at 5.7 x 10^9 tonnes, exist, mostly in North Africa . Nearly 90% of rock phosphate is used in the production of phosphate fertilizers, as shown in Fig. 6.5.

Phosphorus

The production of elemental phosphorus is a declining industry, and has high energy inputs (see Fig. 6.6). Phosphate is reduced with carbon in an electric furnace at 1500°C, sand is added to remove Ca as slag.

$$2Ca_3(PO_4)_2 + 6SiO_2 + 10C \rightarrow 6CaSiO_3 + P_4 + 10CO, \ \Delta H = +3.06 \text{ MJ} \quad (6.19)$$

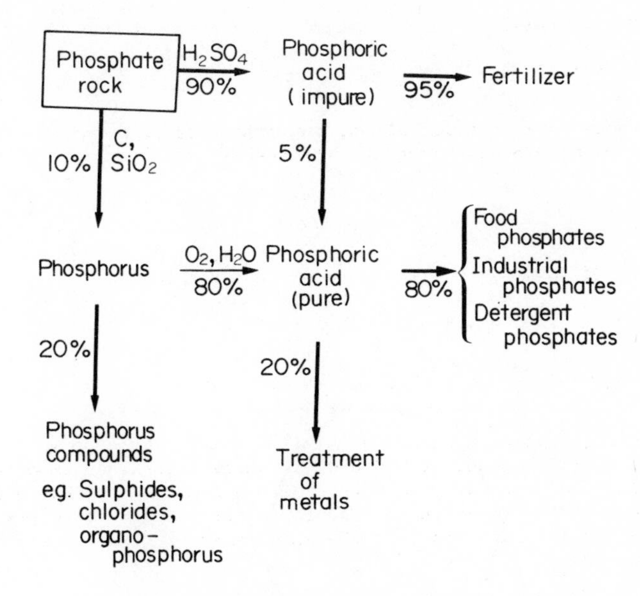

Fig. 6.5. The uses made of phosphate rock. Source: R. Thompson (Editor) The Modern Chemicals Industry, Chem.Soc., London 1977.

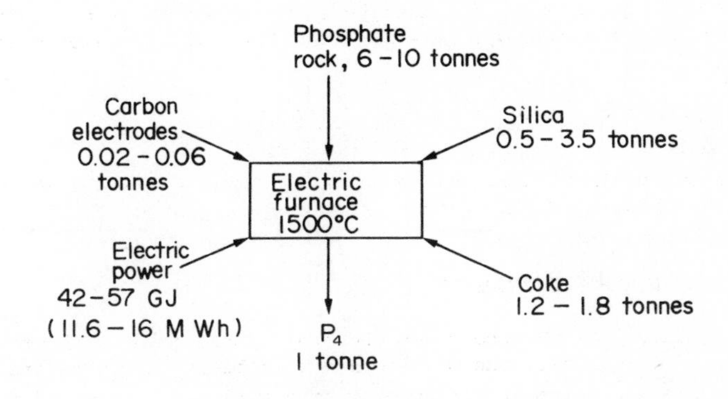

Fig. 6.6. Production of phosphorus.

Superphosphate

Natural phosphate rock is of low solubility, and except for the more acid soils it is not satisfactory as a fertilizer. Treatment with sulphuric acid, under controlled conditions gives mainly $Ca(H_2PO_4)_2$ which is more soluble in water.

$$[Ca_3(PO_4)_2]_3CaF_2 + 7H_2SO_4 \rightarrow \underbrace{3Ca(H_2PO_4)_2 + 7CaSO_4}_{\text{superphosphate}} + 2HF \qquad (6.20)$$

The less soluble $Ca(HPO_4)$, is sometimes produced by treating superphosphate with lime to give 'reverted superphosphate'. This material is also used in animal feeds. Owing to the low solubility of iron, aluminium and calcium phosphates and the restricted pH range over which the $H_2PO_4^-$ ion exists, phosphate fertilizers are most efficient over the pH range 6-7 (Fig. 6.7). These limitations mean that the use of phosphate fertilizers is expensive, and often, wasteful.

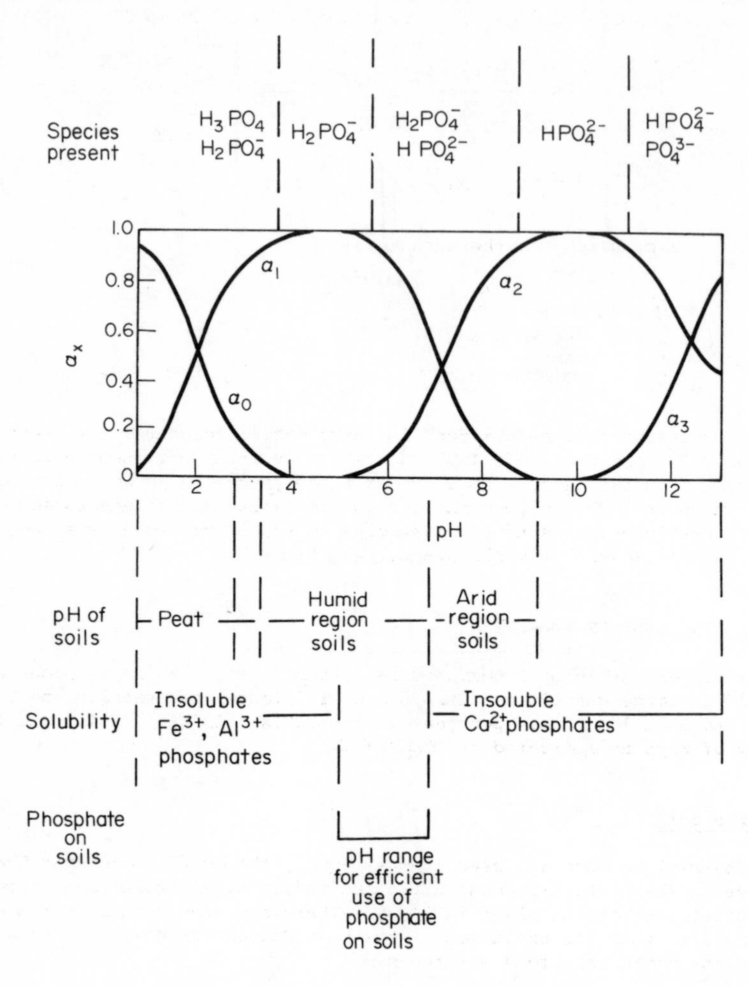

Fig. 6.7. Distribution diagram for phosphate species and relation to soil pH and phosphate solubility in soils.

When excess sulphuric acid is added to phosphate rock, phosphoric acid is produced

$$[Ca_3(PO_4)_2]_3CaF_2 + 10H_2SO_4 + 20H_2O \rightarrow 10CaSO_4 2H_2O + 2HF + 6H_3PO_4 \quad (6.21)$$

Phosphoric acid is used in the "pickling" of steels, where phosphate bonded to the iron produces a thin protective coating.

Sodium Tripolyphosphate

The polyphosphate $Na_5P_3O_{10}$ is prepared from heating together the two salts $NaPO_3$ (metaphosphate) and $Na_4P_2O_7$ (pyrophosphate) at 300-500°C;

$$NaPO_3 + Na_4P_2O_7 \rightarrow Na_5P_3O_{10}, \tag{6.22}$$

or heating Na_2HPO_4 and NaH_2PO_4 at 450°C;

$$2Na_2HPO_4 + NaH_2PO_4 \rightarrow Na_5P_3O_{10} + 2H_2O. \tag{6.23}$$

The $P_3O_{10}{}^{5-}$ ion, which has the structure;

```
   O⁻        O        O⁻
   ||        ||       ||
O = P - O - P - O - P = O
    |        |        |
    O⁻       O⁻       O⁻
```

is used extensively as a "builder" in heavy duty detergents. It reacts with Ca^{2+} and Mg^{2+} ions to form soluble salts and removing permanent hardness. It acts as a buffer, around pH 9-10, at which pH the fatty acids are more readily soluble. The polyphosphate also disperses oils by absorption on to the soil particles, causing the particles to repel because of the negative charge associated with the phosphate.

THE SULPHUR INDUSTRY

The most important sulphur compound is sulphuric acid, which is produced and consumed in large quantities, around 4.6×10^6 tonnes per year in the U.K. (1976) and 30×10^6 tonnes per year in the U.S.A. (1974). It is used in a variety of ways as indicated in Table 6.2.

Sulphuric acid

Two industrial methods are used to make H_2SO_4, the Contact and Lead Chamber processes. The Contact process gives the purer, more concentrated acid, and the capital cost of the plant is lower. Therefore, the method is the more popular, almost to the exclusion of the Lead Chamber process. In both methods the basic reactions are the same;

$$2SO_{2(g)} + O_{2(g)} \rightarrow 2SO_{3(g)}, \tag{6.24}$$

$$SO_{3(g)} + H_2O_{(\ell)} \rightarrow H_2SO_{4(aq)}. \tag{6.25}$$

Sulphur dioxide is obtained from burning sulphur;

$$S_{(s)} + O_{2(g)} \rightarrow SO_{2(g)}, \quad \Delta H = -300 \text{ kJ}, \tag{6.26}$$

TABLE 6.2 Uses of Sulphuric Acid

Use	Comments
Fertilizers*	In production of superphosphate Ammonium sulphate
Paints/Pigments*	In production of TiO_2
Detergents and Soaps*	
Chemicals*	In production of plastics, HF, HCl, Cu, Al and Ba sulphates
Fibres, cellulose film*	
Metallurgy	'Pickling' of steels
Dyestuffs	
Battery Acid	
Oil, petrol	Alkylation reactions

* Major uses of sulphuric acid

roasting metal sulphides, reduction of $CaSO_4$;

$$CaSO_{4(s)} + 2C_{(s)} \rightarrow CaS_{(s)} + 2CO_{2(g)}, \quad (6.27)$$

$$3CaSO_{4(s)} + CaS_{(s)} \rightarrow 4CaO_{(s)} + 4SO_{2(g)}, \quad (6.28)$$

or oxidation of H_2S.

In the contact process pure dry SO_2 and air are passed through a catalyst V_2O_5/Na_2O, initially at 575^oC and later at 450^oC. The optimum conditions for the equilibrium reaction;

$$SO_{2(g)} + \tfrac{1}{2}O_{2(g)} \underset{\leftarrow}{\overset{V_2O_5/Na_2O}{\rightarrow}} SO_{3(g)}, \ \Delta H = -100 \text{ kJ}, \quad (6.29)$$

are around 450^oC (see Fig. 6.8). The reaction is carried out at 575^oC where the oxidation is faster though the yield is only around 80%, but when the temperature is dropped to 450^oC the yield is increased to around 97%. Finally SO_3 is passed up a tower, down which 98% H_2SO_4 trickles, the gas reacts with the water in the acid and more water is added to keep the concentration at 98%

$$SO_{3(g)} + H_2O_{(\ell)} \rightarrow H_2SO_{4(aq)}, \ \Delta H = -200 \text{ kJ} \quad (6.30)$$

The total energy released per mole of H_2SO_4 produced is -600 kJ, which is often used to produce steam to generate electricity, normally more than the plant needs. A workable unit, is a plant which produces around 1600 tonnes per day accounting for 99% of the sulphur input. The other 1% contributes to air pollution giving 32 tonnes per day of SO_2 - not insignificant.

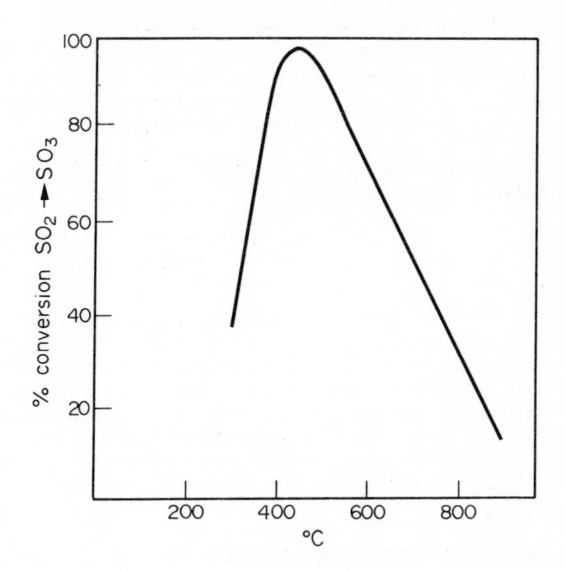

Fig. 6.8. The percentage conversion of SO_2 to SO_3 at various temperatures.

In the Lead Chamber process the catalyst is nitric oxide, discussed earlier in Chapter 5. These reactions are probably more significant in atmospheric pollution chemistry, rather than in the commercial production of sulphuric acid due to the dominance of the contact process.

THE SILICON INDUSTRY

Though not as apparent as for other elements silicon is one of the more important chemical elements in the world. This is partly because much of the industry and technology of silicon compounds has been developed for centuries, such as glass making and building materials. Silicon is the second most abundant element in the earth's crust. A crustal abundance of 27.7 atoms % is the weighted mean of the various materials that make up the crust. The thermodynamic and kinetic stability of the -Si-O-Si- system, extensive polymerisation of silicates, and the abundance, ensures silicates are the principal structural material in the world. In the U.S.A. the amount of building stone mined is in the region of 2×10^6 tonnes per year, of which limestone and granite account for 60%.

Silicon

The amount of elemental silicon used is small, but of increasing importance, because of its use for semi-conductors and transistors and the explosive development in microprocessors. The very pure silicon required (impurities < 1 part in 10^{10}) is obtained by first reducing silica (sand) with carbon (coke) at 1770 K in an electric furnace and converting the silicon to trichlorosilane (or silicon tetrachloride) at 570 K.

$$SiO_{2(s)} + C_{(s)} \xrightarrow{1770\ K} Si_{(\ell)} + CO_{2(g)}, \tag{6.31}$$

$$Si_{(s)} + 3HCl_{(g)} \xrightarrow{570\ K} HSiCl_{3(\ell)} + H_{2(g)}. \tag{6.32}$$

The colourless liquid (bp. 305 K) is then purified by fractional distillation a number of times and then reduced to pure silicon by H_2 or Zn or Mg.

$$2HSiCl_{3(g)} + 2H_{2(g)} \xrightarrow{1300\ K} 2Si_{(s)} + 6HCl_{(g)}. \tag{6.33}$$

The silicon is further purified by zone refining (Chapter 3) and then doped with the requisite amount of a Group (III) element (B, Ga or In) to give a p-type semiconductor, or with a Group (V) element (P, As, Sb or Bi) to give an n-type semiconductor. The doped material is grown as single crystals from which thin plates are cut to be used in constructing p-n junctions and printed circuits for various electronic requirements.

Structural Materials

Stones, gravel and sand, consisting primarily of silica and silicates, are used extensively as building materials either as cut stone or crushed material, sometimes bonded together with cement. Portland cement, which contains Ca and Mg-silicates and aluminates, has an approximate composition, CaO, 64%; SiO_2, 21%, Al_2O_3, 5%; Fe_2O_3, 5%; MgO, 2%; and SO_3, 2%. The world production in the region of 0.5 to 1 x 10^6 tonnes per year is used for binding sand and gravel to make concrete. Cement is obtained by heating limestone, or a dolomitic limestone mixture ($CaCO_3$, 75%, $MgCO_3$, 4%) with clay to 1400°C in a rotary kiln. An approximate equation for the reaction is;

$$\underset{\text{clay}}{CaAl_2Si_2O_{8(s)}} + \underset{\text{limestone}}{6CaCO_{3(s)}} \rightarrow \underbrace{2Ca_2SiO_{4(s)} + Ca_3(AlO_3)_{2(s)}}_{\text{cement}} + 6CO_{2(g)}. \tag{6.34}$$

However, other substances are also produced, such as $CaSiO_3$ and Ca_2AlFeO_5. The kiln clinker which is powdered, reacts when wet producing hydrated silicates and aluminates that crystallise in a rigid mass of interlocking crystals with high mechanical strength.

Bricks, another important building material, are made by firing clay to 1100°C. The rigid material is not influenced by water, as are sun dried bricks. In the firing process water is first removed and at high temperatures the oxides of Si and Al change to mullite, while SiO_2 is converted to the crystobalite structure. Mullite and crystobalite crystallise in a system of interlocking crystals held together by an amorphous glassy material.

$$\underset{\text{kaolinite}}{Al_4Si_4O_{10}(OH)_8} \xrightarrow{900\ K} \underset{\text{alumina}}{2Al_2O_3} + \underset{\text{tridymite}}{4SiO_2} + 4H_2O, \tag{6.35}$$

$$\underset{\text{alumina}}{3Al_2O_3} + \underset{\text{tridymite}}{2SiO_2} \xrightarrow{1300\ K} \underset{\text{mullite}}{3Al_2O_32SiO_2}, \tag{6.36}$$

$$\underset{\text{tridymite}}{SiO_2} \xrightarrow{1740\ K} \underset{\text{crystobalite}}{SiO_2}. \quad (6.37)$$

Glass

Glass is an extremely useful material, due to its structural qualities, transparency to light and low chemical reactivity. It is produced by fusing silica with alkali and alkaline earth metal oxides to give silicates which attain rigidity without crystallising (i.e. a supercooled solution). The structure is amorphous with a random silicate network (Fig. 6.9). Since some Si-O bonds are under more strain than others glass does not have a sharp melting point, but softens over a temperature range. This is a distinct advantage for its working. The reaction;

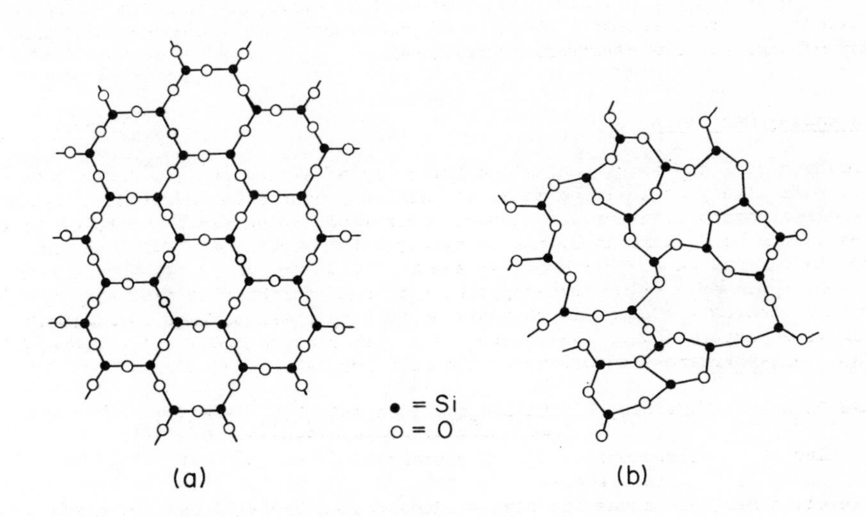

Fig. 6.9. A two-dimensional representation of (a) a crystalline silicate, and (b) a silicate glass.

$$SiO_{2(s)} + Na_2CO_{3(s)} \rightarrow Na_2SiO_{3(\ell)} + CO_{2(g)}, \quad (6.38)$$

when strongly heated gives a water soluble material called water-glass, which is used in fire proofing and some washing powders. The introduction of CaO reduces the solubility and raises the softening temperature to give glass.

$$CaCO_{3(s)} + Na_2CO_{3(s)} + 2SiO_{2(s)} \xrightarrow{1800\ K} \underbrace{Na_2SiO_{3(\ell)} + CaSiO_{3(\ell)}}_{\text{glass on cooling}} + 2CO_2 \quad (6.39)$$

A typical reaction mix is 50% sand, 15% Na_2CO_3, 5% Na_2SO_4, 5% lime and 25% cullet (cullet is recycled glass). This soda-glass accounts for around 90% of all glass made. The properties of glass depend on the additives (Table 6.3), for example CdO gives neutron absorbing glass, while As_2O_3 gives a glass transparent to infrared radiation. Glass is also coloured by the addition of materials, for example, CuO gives red, blue or green, MnO_2 gives purple and CoO gives blue. Glass-making is an energy intensive process, one tonne requiring around 2 to 5 x 10^9 J of energy.

TABLE 6.3 Principal Types of Glasses

Glass name	Constituents	Properties	Uses
Silica	SiO_2 > 96%	good heat resistance high MPt transparent to UV light	lenses, mirrors, high temperature laboratory ware
Soda-lime	70% SiO_2, 15% Na_2O, 10% CaO.	low working temperature	bottles window panes
Borosilicate	60-80% SiO_2, 10-25% B_2O_3.	good thermal shock, high softening temp.	glassware that requires heating
Flint	30-70% SiO_2, 20-60% PbO, 5-20% Na_2O.	high density good refractive index	crystal, optical equipment
Aluminosilicate	70% SiO_2, 15% Al_2O_3.	good chemical resistance and heat resistance	thin film circuitry, ignition tubes

Insoluble Silicates

A number of other insoluble silicates, such as clays, have uses other than in the structural industry, a partial list is given in Table 6.4. The structural chemistry is based on the tetrahedral arrangement of oxygen atoms around silicon (with different Si:O condensation ratios (Fig. 6.10)), and the octahedral arrangement of oxygen around Al. Useful properties are often a consequence of the structure, for example asbestos, mica and talc, have the sheet silicate structure and the slippery feel of talc arises from the sheets slipping over each other.

Zeolites are alumino-silicates with an open framework structure, and have the general composition;

$$M^{n+}_{\frac{x}{n}} (AlO_2)_x (SiO_2)_y \; zH_2O,$$

e.g. chabazite has M=Ca, x=4, y=8, z=13, and faujasite has M=Na, Ca, x=64, y=128 and z=256. They are synthetically produced and are used as molecular sieves, ion exchangers, petroleum cracking catalysts and catalysts for the production of synthetic petrol (equation 4.34);

$$nCH_3OH \xrightarrow[\text{catalyst}]{\text{zeolite}} C_nH_{2n} + nH_2O, \quad (n=5-10), \qquad (6.39)$$

aromatic hydrocarbons + branched aliphatic hydrocarbons.

Structure	Formula	name	Si:O ratio	Examples
= SiO_4	SiO_4^{4-}	Orthosilicate (neosilicate)	1:4	Olivine, $(MgFe)_2SiO_4$ Zircon, $ZrSiO_4$
	$Si_2O_7^{6-}$	Pyrosilicate (sorosilicate)	1:3.5	Thortvetite, $Sc_2Si_2O_7$
	$Si_3O_9^{6-}$	Cyclic silicate (sorosilicate)	1:3	Benitoite, $BaTiSi_3O_9$ Beryl $Be_3AlSi_6O_{18}$ (6 membered ring)
	$(SiO_3^{2-})_n$	Pyoxene (inosilicate)	1:3	Enstatite, $MgSiO_3$ Diopside, $(Ca, Mg)SiO_3$
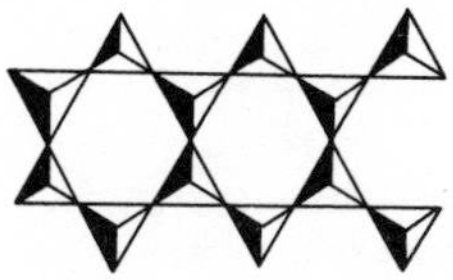	$(Si_4O_{11}^{6-})_n$	Amphibole (inosilicate)	1:2.75	Tremolite, $Ca_2Mg_5(Si_4O_{11})_2(OH)_2$ Hornblende, $(Ca, Na)_{2-3}(Mg, Fe, Al)_5[(Si, Al)_4O_{11}]_2(OH)_2$
	$(Si_4O_{10}^{4-})_n$	Infinite sheet silicate (phyllosilicate)	1:2.5	Biotite, $K(Mg,Fe)_3(AlSi_3O_{10})(OH)_2$ Muscovite $KAl_2(AlSi_3O_{10})(OH)_2$
	$(Si_{4-x}Al_xO_8^{x-})_n$	Framework (tektosilicate)	1:2	Zeolites eg. Stilbite, $NaCa_2(Al_5Si_{13}O_{36})zH_2O$
	$(SiO_2)_n$	Silica (tektosilicate)	1:2	Quartz, chalcedony, agate, jasper, opal

Fig. 6.10. Silicate structures and types.

TABLE 6.4 Some Uses of Insoluble Silicates and Alumino Silicates

Name	Formula	Uses
Clays	aluminosilicates	Building and construction, e.g. bricks, tiles, pipes, cement, glass.
Kaolinite	$Al_4Si_4O_{10}(OH)_8$	Porcelain, china, pottery, filler and white pigment in paper.
Bentonite (Montmorillonite)	$(Al_{3.5}Mg_{0.5})Si_8O_{20}(OH)_4$	Bonding agent in sand moulds, pelletizing iron ore, drilling mud.
Vermiculite	$Mg_6Si_7AlO_{20}(OH)_4$	Horticulture medium, insulation, paints.
Asbestos (Chrysolite)	$Mg_3Si_2O_5(OH)_4$	Building material, fire proofing.
Mica (Muscovite)	$K_2Al_4Si_6Al_2O_{20}(OH)_4$	Electrical insulation, furnace windows.
Talc	$Mg_3Si_4O_{10}(OH)_2$	Cosmetics, crayons, wallpaper.
Zeolites (Chabazite)	$Ca_6Al_{12}Si_{24}O_{72}40H_2O$	Ion exchange, molecular sieves, catalysts in cracking and reforming petroleum.

The open structure allows for movement of small molecules and ions into the lattice hence the molecular sieve and ion-exchange properties. It is possible that the large internal surface area of zeolites is significant in their catalytic properties.

Silicones

Organo silicon-oxygen compounds called silicones have a Si-O backbone framework and organic side-groups (mainly $-CH_3$ and $-C_6H_5$). They are polymeric and have good heat stability due to the strong Si-O bond (452 kJ mol^{-1}) and the organic groups confer water repellancy and non-polar properties to the silicones. Some typical structures are:

```
       CH3      CH3      CH3
        |        |        |
 - O - Si - O - Si - O - Si - O -
        |        |        |
       CH3      CH3      CH3
```

```
        CH3       CH3       CH3
        |         |         |
O  -  Si  -  O  -  Si  -  O  -  Si  -  O  -
        |         |         |
        O         O         O
        |         |         |
O  -  Si  -  O  -  Si  -  O  -  Si  -  O  -
        |         |         |
        CH3       CH3       CH3
```

Silicones are expensive to make, but their particular properties find a number of uses and in the U.S.A. (1969) 6 x 10^6 kg were produced. A typical sequence of chemical reactions in their formation are:

preparation of chlorosilane;

$$CH_3Cl_{(g)} + Si_{(s)} \underset{Cu}{\overset{570\ K}{\rightarrow}} (CH_3)_nSiCl_{4-n(g)}, \qquad (6.41)$$

hydrolysis;

$$m(CH_3)_2SiCl_{2(\ell)} + mH_2O_{(\ell)} \rightarrow [(CH_3)_2SiO]_{m(\ell)} + 2mHCl \qquad (6.42)$$

(m = 3 to 15)

and polymerisation;

$$x[(CH_3)_2SiO]_{m(\ell)} \underset{420\ K}{\overset{KOH}{\rightleftharpoons}} HO[(CH_3)_2OSi]^-_{mx} + K^+. \qquad (6.43)$$

silicone

The silicones range from liquids to solids and are used as water repelling agents, non-stick surfaces, as greases which show little change in viscosity over a wide temperature range, -50 to 150°C, as rubbers and sealants as in the encapsulation of electronic circuits.

ALKALI, ALKALINE EARTH METAL INDUSTRY

We have already discussed the formation of caustic soda (Chapter 3), which is made at the same time as chlorine by electrolysis of brine. The uses of sodium hydroxide are listed in Table 6.1.

Sodium Carbonate

Originally NaOH was made by the Solvay Process, the Na_2CO_3 produced was treated with slaked lime to give NaOH;

$$Na_2CO_3 + Ca(OH)_2 \rightarrow 2NaOH + CaCO_3. \qquad (6.44)$$

The Solvay Process is still used today for making Na_2CO_3, 50% of which is used in the glass industry. The overall chemical reaction is;

$$2NaCl + CaCO_3 \rightleftharpoons Na_2CO_3 + CaCl_2. \qquad (6.45)$$

The reaction, as written, has a tendency to move to the left in solution owing to the low solubility of $CaCO_3$. However, a number of intermediate steps are involved in the preparation. Ammonia is dissolved in a brine solution, through which CO_2, obtained from heating $CaCO_3$, is bubbled to give ammonium bicarbonate, which reacts with the NaCl to give sodium bicarbonate.

$$CaCO_{3(s)} \xrightarrow{1100^{\circ}C} CaO_{(s)} + CO_{2(g)} \quad (6.46)$$

$$CO_{2(g)} + NH_3/H_2O \rightarrow NH_4HCO_{3(aq)} \quad (6.47)$$

$$NH_4HCO_{3(aq)} + NaCl_{(aq)} \rightarrow NaHCO_{3(aq)} + NH_4Cl \quad (6.48)$$

The crystalline $NaHCO_3$ is heated to $150^{\circ}C$ to give sodium carbonate;

$$2NaHCO_3 \rightarrow Na_2CO_3 + CO_2 + H_2O. \quad (6.49)$$

Ammonia is recovered from the NH_4Cl by treatment with lime;

$$2NH_4Cl + CaO/H_2O \rightarrow 2NH_3 + 2H_2O + CaCl_2. \quad (6.50)$$

A natural form of sodium carbonate/bicarbonate, $2Na_2CO_3.NaHCO_3.2H_2O$, called trona, is also a source of sodium carbonate;

$$2Na_2CO_3.NaHCO_3 2H_2O \xrightarrow{heat} 3Na_2CO_3 + CO_2 + 3H_2O. \quad (6.51)$$

Trona is the cheaper source of Na_2CO_3, and when available it is used in preference, for example in the U.S.A., 71% of Na_2CO_3 comes from trona.

Hypochlorites

Both sodium and calcium hypochlorites are widely used as oxidising agents, eg. in laundry bleaches and disinfectants for water systems. They are obtained by chlorination of sodium and calcium hydroxide solutions.

$$2NaOH + Cl_2 \rightarrow NaOCl + NaCl + H_2O \quad (6.52)$$

$$2NaOH + 2Ca(OH)_2 + 3Cl_2 \rightarrow NaOCl + Ca(OCl)_2 + NaCl + CaCl_2 + 3H_2O \quad (6.53)$$

Under refrigeration the products of reaction (6.53) crystallise as $NaCl.NaOCl.Ca(OCl)_2.12H_2O$ from which $Ca(OCl)_2$ can be obtained.

Potassium Salts

Potassium is the third most important mineral nutrient for plants, and is normally applied as KCl. The chloride is obtained from seawater, and also from sylvinite deposits, a physical mixture of 42.7% KCl and 56.6% NaCl. The two chlorides are separated from the ore making use of their different solubilities in hot water (Fig. 6.11). The crushed ore is mixed with a saturated brine solution and heated, the KCl goes into solution while the NaCl remains undissolved. The solution is removed, cooled and the KCl crystallises out. Floatation may also be used, because organic amines pref-

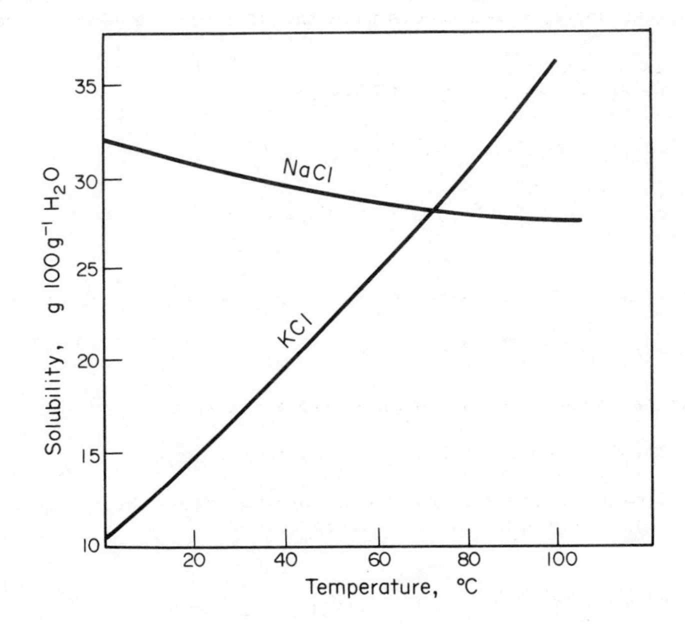

Fig. 6.11. Solubilities of NaCl and KCl in water saturated with respect to the other halides.

preferentially join to KCl crystals, which then attach air bubbles, and so are removed in the froth. Evaporation of brine from the Dead Sea gives a double salt called carnallite $MgCl_2.KCl.6H_2O$ from which KCl can be obtained by careful manipulation of the various liquors produced during the evaporation process.

CONSUMER CHEMISTRY

Many of the chemicals mentioned above have uses in consumer products, either directly or in the making of materials. In order to demonstrate the diversity of use of chemicals, we will consider two common consumer products, cleaning agents (soaps and detergents) and paper.

Soaps and Detergents

Soaps and detergents are a major consumer product in our society, affecting in some way most people's lives. The products range from toilet soaps through to heavy duty detergents, and probably well over 25 x 10^6 tonnes are produced annually in the world. A satisfactory cleaning agent must assist in; the wetting of a material, the removal of dirt particles, keeping the particles dispersed in the water, and cope with the hardness of water. Soaps and detergents can also contain bleaches and whitening (or brightening) agents. However, the principal ingredient is the surfactant (surface acting agent) responsible for the removal of dirt.

Soap surfactants are normally the sodium salts of fatty acids such as stearic acid $CH_3(CH_2)_{16}COOH$, lauric acid $CH_3(CH_2)_{10}COOH$, oleic acid $CH_3(CH_2)_7CH{=}CH(CH_2)_7COOH$ and palmitic acid $CH_3(CH_2)_{14}COOH$. Fats such as tallow or oils such as coconut and palm oil when treated with caustic soda form soaps.

$$\begin{array}{lcccccc} RCOOCH_2 & & & & & & CH_2OH \\ | & & & & & & | \\ RCOOCH & + & 3NaOH & \rightarrow & 3RCOO^-Na^+ & + & CHOH \\ | & & & & & & | \\ RCOOCH_2 & & & & \text{Fatty acid salt, (soap)} & & CH_2OH \\ \text{Fat or oil} & & & & & & \text{glycerol} \end{array} \quad (6.52)$$

The salt consists of two parts, a long hydrocarbon chain, which is water insoluble and hydrophobic (but soluble in organic materials such as oils) and an ionic part $-COO^-Na^+$, which is water soluble and hydrophilic. Some types are; dodecylbenzene sulphonate $CH_3(CH_2)_{11}-C_6H_4-SO_3^-M^+$ (an anionic surfactant), quaternary ammonium salts $C_nH_{2n+1}N^+X(CH_3)_3^-$ (a cationic surfactant) and fatty acid monoglycerides $C_nH_{2n+1}COOCH_2CHOHCH_2OH$ (a non-ionic surfactant).

Surfactants reduce the surface tension of water by reducing the attractive forces between the water molecules. This improves the wetting properties of water. The removal of fats and oils occurs because the hydrocarbon chain can dissolve in the oil (Fig. 6.12) while the ionic end remains in contact with the water. Since the attraction between the ionic section and water is strong, the dirt particle becomes dislodged from the material with agitation. The negative charge surrounding the particles keeps them apart and so the dirt remains in the solution as an emulsion.

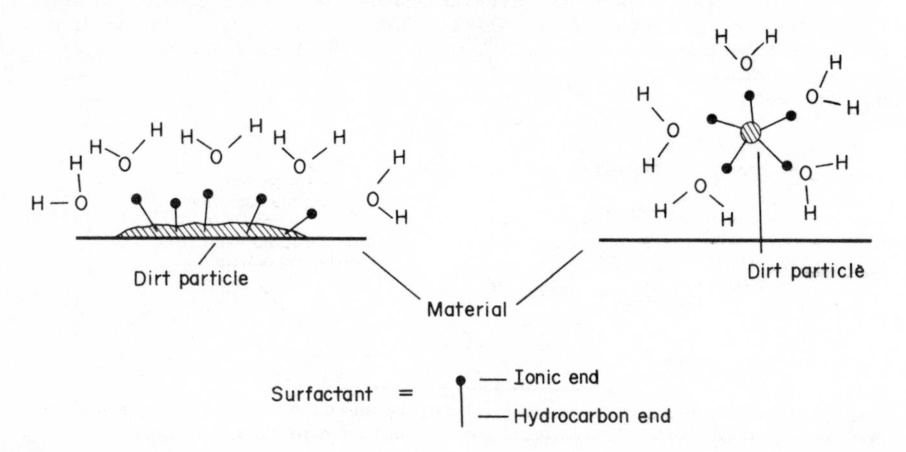

Fig. 6.12. Removal of dirt from material with a surfactant.

Detergents may also contain builders, such as $Na_5P_3O_{10}$, which forms stable compounds with Mg^{2+} and Ca^{2+} ions, removing their interference with the effectiveness of the surfactant. The triphosphate also buffers the solution at pH 9-10, where salt formation of the free fatty acids is more readily achieved. Finally the highly charged phosphate groups which surround a dirt particle aids in dispersing the dirt. Soluble silicates also aid in dispersion and with sodium sulphate they add bulk (fillers) to powdered detergents improving their free flowing quality.

Bleaches such as sodium perborate $NaBO_3 4H_2O$ i.e. $Na^+[HOOB(OH)_3]^- 2H_2O$ are a source of H_2O_2, which oxidise unsaturated organic systems. Unsaturated organic systems are often coloured due to $\pi \rightarrow \pi^*$ type electronic transitions which occur in the near UV and visible regions of the spectrum. The oxidation products, being saturated, are generally colourless. Optical whiteners or brighteners are compounds which absorb UV radiation but fluoresce in the blue region of the visible spectrum. The emitted light compensates for the loss of blue light due to absorption by aging material (the effect of absorption of blue is the appearance of yellow colour) Fig. 6.13. The apparent brightness of a garment is also due to the extra radiation emitted by the fluorescent molecule.

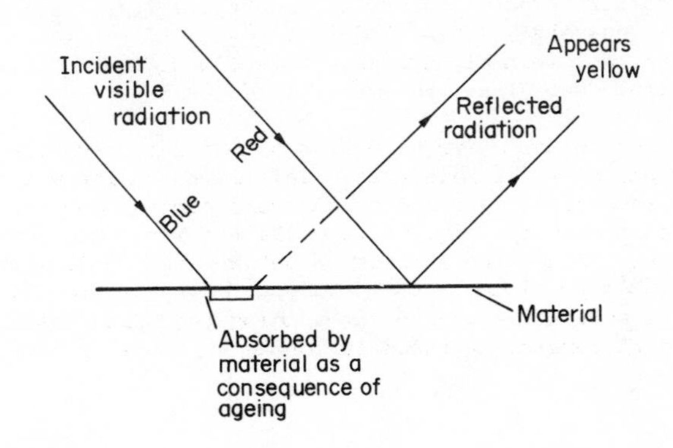

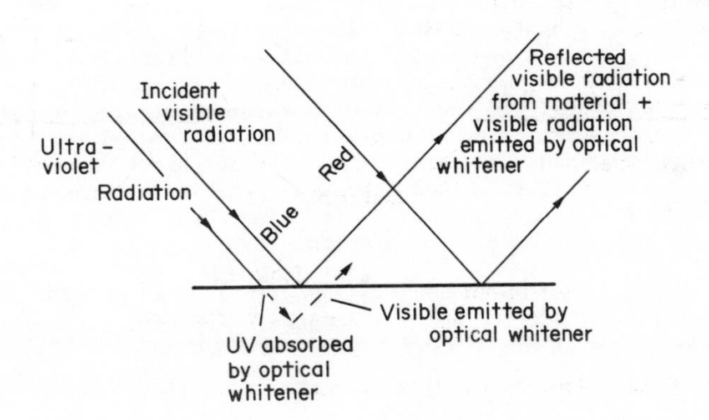

Fig. 6.13. Schematic illustration of the principle of optical whiteners used in washing.

Enzymes which are biological catalysts can be added to detergents to break down protein to water-soluble peptides and amino acids by soaking at ambient temperatures. Proteins are difficult to remove with normal surfactants. Enzymes that live naturally at high temperatures may also have a use in detergents as they will retain their activity in hot water.

Paper

The production of wood pulp and paper is an essential industry in a world where the printed word is pre-eminent. Trees are the resource material and a number of chemicals are used in the processing. In the U.S.A. (1971) 44×10^6 tonnes of pulp were produced from primary fibre and 13×10^6 tonnes of paper material recycled. The industry is a big user of chemicals, and in the U.S.A., 70% of the total consumption of Na_2SO_4 was used in the pulp and paper industry, 17% of the dichlorine, 14% of the caustic soda and 9% of the sodium carbonate. The major categories of paper produced are newsprint, printing and fine paper, industrial papers, sanitary papers and box and construction boards.

Paper is mainly cellulose, polymeric glucose;

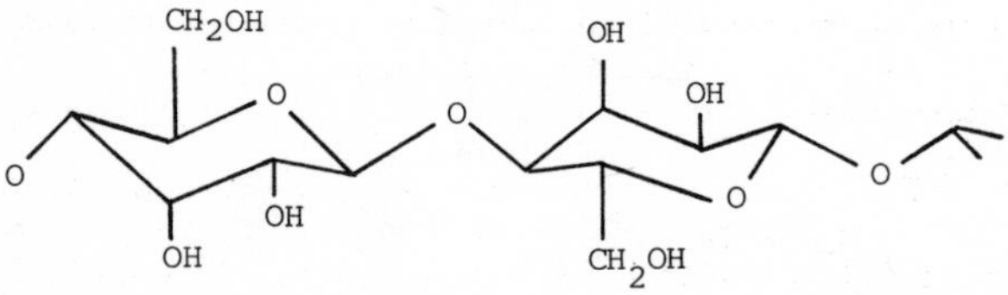

which accounts for approximately 60% of dry wood, by weight. The remainder of wood is approximately 30% lignin and 10% sugars, inorganic salts and resin. The production of pulp and paper has five major steps, mechanical breakdown of wood, removal of lignin and other non-cellulose materials, production of a mat of cellulose fibres, mechanical and chemical treatment to produce different types of paper, and recovery of unspent chemicals. Lignin is not removed from the cheapest paper (newsprint), but is removed for better grades, by one of two main methods, the Kraft (alkaline sulphate) process or the bisulphite process. A third method, the soda process is superseded by the other two. Lignin is removed by making it soluble which means breaking down the polymer and forming water soluble salts.

In the Kraft process wood chips are treated with a NaOH and Na_2S solution at 600 kPa pressure and 170°C for about six hours. A possible reaction is;

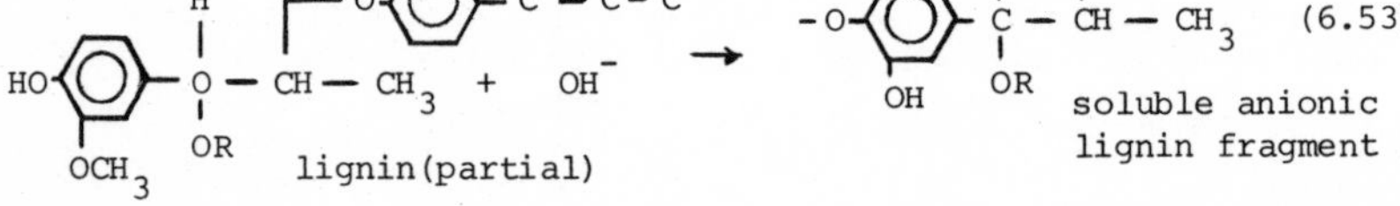

$$\text{lignin (partial)} + OH^- \rightarrow \text{soluble anionic lignin fragment} \quad (6.53)$$

The Na_2S acts as a continual source of OH^- because of the equilibrium;

$$S^{2-} + H_2O \rightleftharpoons SH^- + OH^-. \qquad (6.54)$$

As the OH^- concentration decreases more is produced which keeps the process going. In the bisulphite process the HSO_3^- ion reacts with lignin to produce sulphonic acid groups on the structure, which are soluble in water. The highest grade paper is produced by this method.

When digestion is complete the pulp, mainly cellulose, is filtered off and the liquor is treated to recover unused chemicals. The liquor is evaporated to dryness and excess NaOH which has reacted with CO_2 to give Na_2CO_3 is recovered by the use of lime;

$$\underset{\text{From liquor}}{Na_2CO_3} + Ca(OH)_2 \rightarrow CaCO_3 + \underset{\text{reused}}{2NaOH}. \qquad (6.55)$$

$$CaCO_3 \xrightarrow{\text{heat}} CaO + CO_2; \quad CaO \xrightarrow{H_2O} Ca(OH)_2$$

The sulphide, which has been oxidised to SO_4^{2-} is recovered using carbon,

$$SO_4^{2-} + 2C \rightarrow \underset{\text{reused}}{S^{2-}} + 2CO_2 \qquad (6.56)$$

In the bisulphite process, when magnesium is the counter cation, the residue from the liquor is burnt to give SO_2 which is converted to bisulphite

$$Mg(HSO_3)_2 \xrightarrow{\text{heat}} MgO + H_2O + SO_2 \qquad (6.57)$$

$$SO_2 + H_2O/OH^- \rightarrow HSO_3^- \qquad (6.58)$$

Pulp is also treated with dichlorine and chlorine dioxide to remove last traces of lignin and to bleach. Chlorine dioxide does not degrade cellulose to any great extent, however, HOCl does and formation of it is to be avoided in the chlorination process, hence acid conditions are used. Dichlorine and caustic soda are produced at many pulp and paper mills, from the electrolysis of brine.

Paper is produced from the pulp fibres by forming a mat which is then dried. Various mechanical and chemical treatments are carried out to produce papers with required properties. Talc or kaolinite or TiO_2 are added as fillers, and $Al_2(SO_4)_3$ is added when necessary to change the negative surface charge of cellulose to positive, by the absorption of Al^{3+} ions. This means that both positively and negatively charged chemicals can be absorbed on to the cellulose fibre depending on the pre-treatment.

Because of the large scale use of chemicals, pulp and paper mills are heavy polluters of air and water.

PART IV

Impact on the Environment

CHAPTER 7

Environmental Cycles

In previous chapters we discussed the origin of the chemical elements, how they have been distributed in the earth, and some of the chemical processes occurring in the Earth's crust today. Also we looked at the processes of extraction of some elements from their ores, and some of the chemical changes that are necessary to provide consumer goods. In the discussion it was assumed that no adverse affects other than a depletion of both mineral and energy resources occurred. In the remainder of this book, we will consider some of the chemical consequences to our environment resulting from the activity of man. This will be done in the context of the four major regions on the habitable earth, the atmosphere, the hydrosphere, the biosphere and the lithosphere (solid crust). Movement of materials between the various regions will be discussed first.

ENVIRONMENTAL CYCLES

The objective of constructing element environmental cycles is to indicate the various chemical pathways within the environment. It must be kept in mind, however, that the details are general rather than specific, and therefore have limited use.

Transport of elements within and between the various regions of the earth have been worked out for a number of elements, in particular the major nutrient elements. The types of transport within and between the zones of the physical world are shown in Fig. 7.1. The addition of the biosphere would have complicated the diagram; the biosphere is included when considering individual elements. Many of the transport mechanisms are reversible, though not necessarily quantitatively. For example transfer of soil particles into the atmosphere as an aerosol, is reversed by fallout of the aerosol over a period of time as a function of the size of the particles. The fallout will occur over the lithosphere and hydrosphere, not just the area from which the material originated. The drop in the concentration of NaCl in soil with respect to distance from the ocean suggests strongly that some of the salt originates from sea spray carried inland.

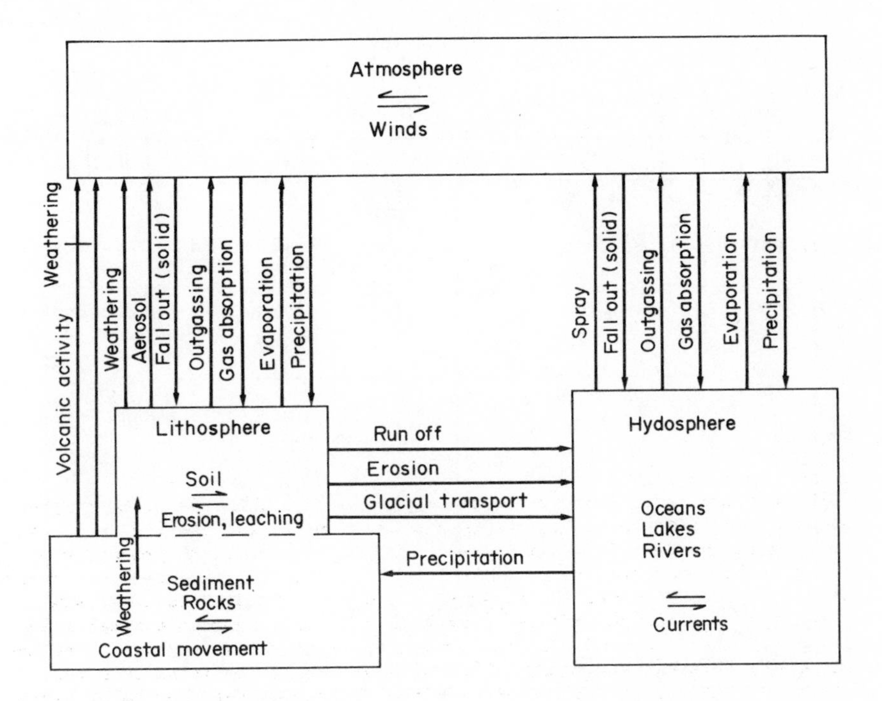

Fig. 7.1. Transport of materials within the physical world

Carbon Cycle

The carbon cycle (Fig. 7.2) is intimately connected with the biosphere. The major carbon pools are: CO_2 in the air, carbon in living and dead material in both the lithosphere and hydrosphere, the inorganic forms of carbon in the hydrosphere, carbonate rocks, and reduced carbon in fossil fuels. Except for carbonate rocks the tendency is for carbon in the absence of air to be in a reduced form (e.g. hydrocarbon), because of anerobic conditions. Four important chemical equilibria exist in the carbon cycle;

$$CO_{2(g)} \rightleftharpoons CO_{2(aq)} \tag{7.1}$$

$$CO_2 + H_2O \rightarrow \text{fixed C in plants} + O_2, \tag{7.2}$$

$$CO_2 + H_2O \leftarrow \text{respiration in animals (and plants)} + O_2, \tag{7.3}$$

$$CO_{2(aq)} + H_2O \rightleftharpoons H^+_{(aq)} + HCO^-_{3(aq)} \rightleftharpoons 2H^+_{(aq)} + CO_3{}^{2-}{}_{(aq)} \tag{7.4}$$

and

$$CaCO_{3(s)} \rightleftharpoons Ca^{2+}_{(aq)} + CO^{2-}_{3(aq)} \tag{7.5}$$

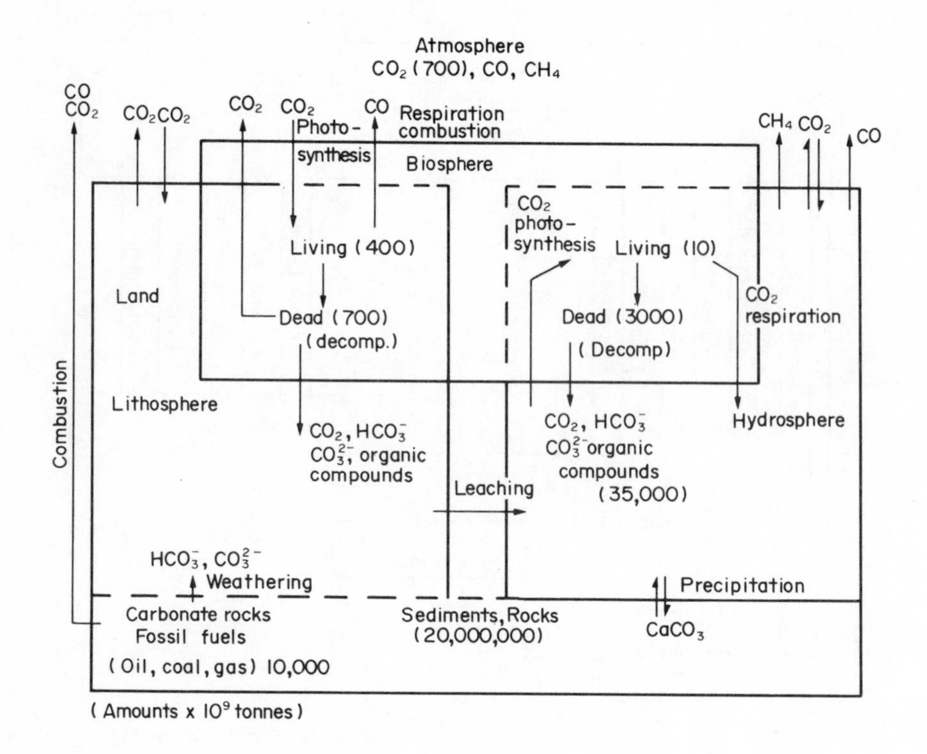

Fig. 7.2. The global carbon cycle

We have already discussed equation (7.2), photosynthesis (Chapter 4), in relation to energy sources. Carbon dioxide provides the world's major environmental source of acid (7.4). The latter equilibrium (7.5), shifts towards the right by virtue of the forms of carbonic acid.

In general carbon in the various areas of the cycle is in balance, though daily (Fig. 7.3) and annual (Fig. 7.4) variations in the CO_2 content of air do occur. The two major influences man has on the cycle are to increase the amount of CO_2 in the air by combustion of fossil fuels, and to alter the plant biosphere significantly, for example removal of large tracts of tropical rain forest, which may disturb the equilibrium established in equations (7.2) and (7.3).

Nitrogen cycle

Nitrogen is a vital element for life being a constituent of amino acids from which proteins are formed. The biosynthesis of amino acids occurs in plants, all the necessary acids being formed. Only some are synthesised by the human metabolism, the remaining essential amino acids supplemented through food

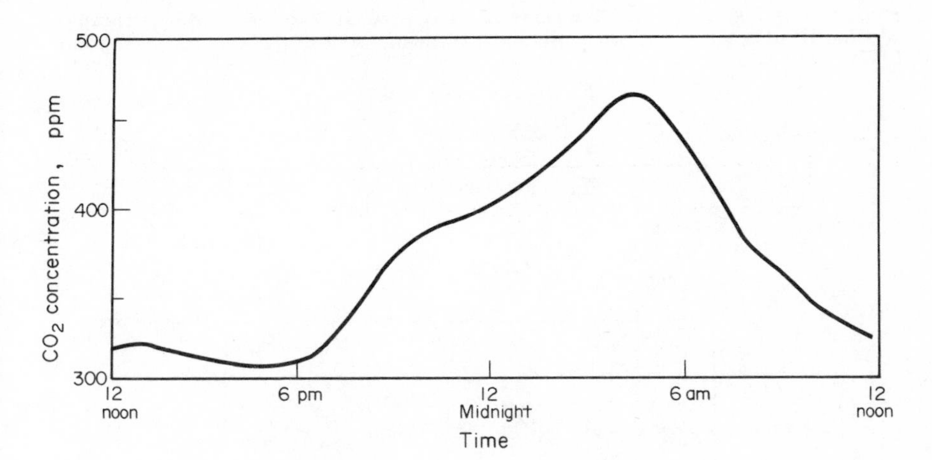

Fig. 7.3. The daily variation in atmospheric CO_2 concentration 1 m above a wheat field during sunny weather

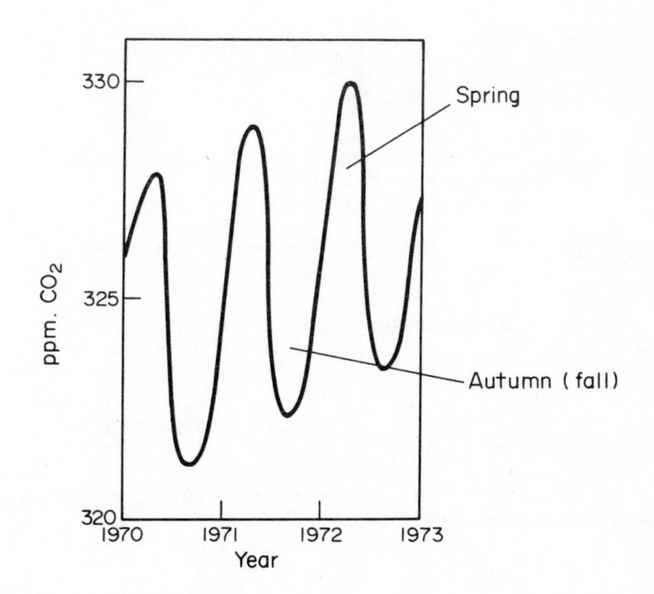

Fig. 7.4. The annual variation of CO_2 concentration in the atmosphere (for the northern hemisphere)

intake. As indicated in Chapter 6.1, and also in Fig. 7.5, the principal source of nitrogen is atmospheric dinitrogen. Once this is fixed, by the

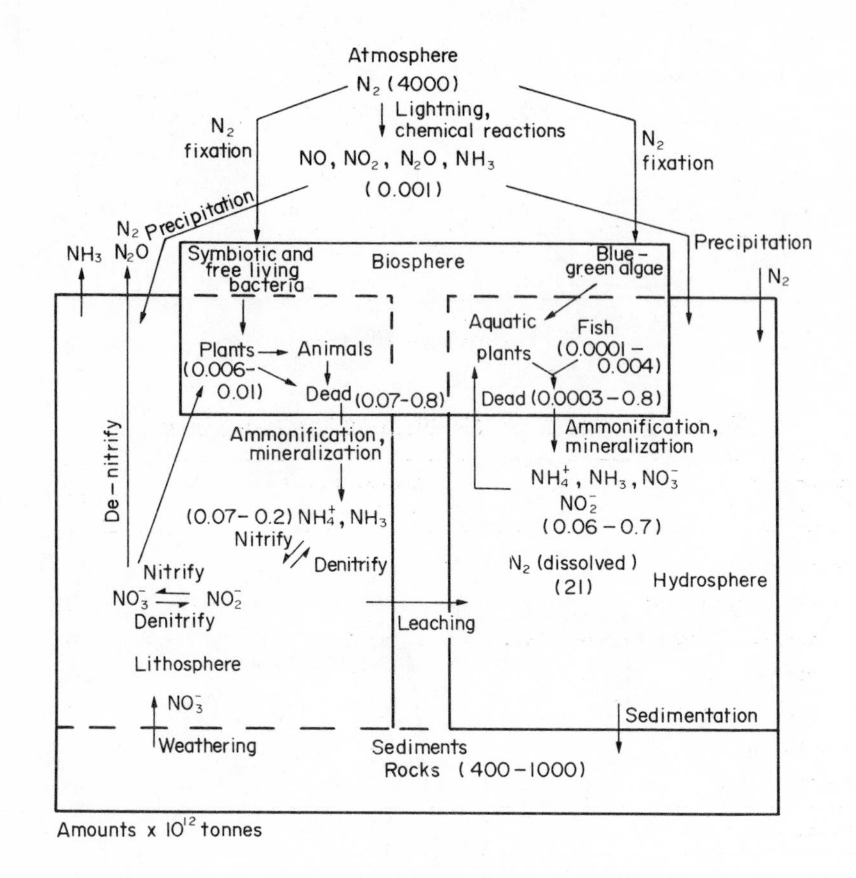

Fig. 7.5. The global nitrogen cycle

methods listed in Table 7.1, the nitrogen is involved in biochemical cycles in the soil-plant system (Fig. 7.6). Within this cycle there are two closed loops. It has been estimated (1974) that around 251 x 10^6 tonnes of nitrogen are fixed each year (Table 7.1) but around 7300 x 10^6 tonnes are utilized each year. This means that the closed loops in the soil-plant cycle are of major importance, as approximately 29 times more nitrogen is involved in them than fixed annually. Therefore any detrimental affect on these cycles could have serious consequences. The various processes in the biochemical nitrogen cycle are catalysed by bacteria, and some of the reactions are listed in Table 7.2. Nitrogen is reduced in the nitrogen fixation and denitrification reactions, while it is oxidised in the nitrification reactions. The two nitrification reactions have favourable free energies;

$$NH_3 + \frac{3}{2}O_2 \rightarrow H^+ + NO_2^- + H_2O, \quad \Delta G = -271 \text{ kJ}, \tag{7.6}$$

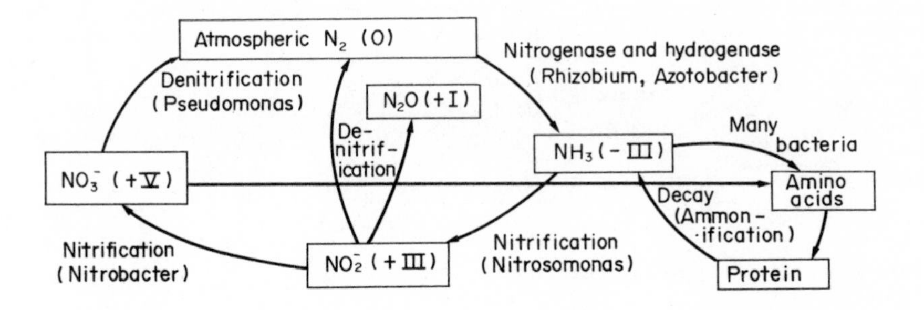

Fig. 7.6. The biological nitrogen cycle

TABLE 7.1 Annual Amounts of Fixed Nitrogen (1974)

Method of Fixation	Amount Fixed x 10^6 tonnes	Percent Fixed
Biological		
Agriculture land	95	38
Forested & uncultivated land	62.5	25
Oceans	~ 1	~ 0.4
Fertilizer production	60	24
Combustion	22.5	9
Lightning	10	4
Total	251	100.4

$$NO_2^- + \tfrac{1}{2}O_2 \rightarrow NO_3^-, \quad \Delta G = -75 \text{ kJ}, \tag{7.7}$$

and the overall process;

$$NH_3 + 2O_2 \rightarrow NO_3^- + H^+ + H_2O, \tag{7.8}$$

is thermodynamically favourable with log K = 7.6. Reactions (7.6) and (7.7) are catalysed by nitrosomonas and nitrobacter respectively. The nitrate ion is the form of nitrogen most readily assimilated by plants.

The influence man has on the natural cycle is an increase in nitrogen by application of fertilizers, and an increase in the removal of NH_4^+, NO_2^- and NO_3^- from soils through irrigation. Both these factors have led to pollution of waterways, and there is concern over a possible increase in N_2O in the atmosphere from denitrification reactions (Table 7.2). Nitrous oxide is

TABLE 7.2 Some Reactions Within the Biological Nitrogen Cycle

Process	Reaction	Organism
Nitrogen Fixation	$N_2 \xrightarrow[\text{reducing agent}]{\text{starch, ATP}} NH_3$	Rhizobium, Azotobacter
Amino Acid Synthesis	$2NH_3 + 2H_2O + 4CO_2 \rightarrow 2CH_2NH_2COOH + 3O_2$ (glycine)	Numerous bacteria
Ammonification (decay)	reverse reaction	
Nitrification	$2NH_4^+ + 3O_2 \rightarrow 2NO_2^- + 4H^+ + 2H_2O$	Nitrosomonas
	$2NO_2^- + O_2 \rightarrow 2NO_3^-$	Nitrobacter
Denitrification	$6NO_3^- + C_6H_{12}O_6 \rightarrow 6CO_2 + 3H_2O + 6OH^- + 3N_2O$ (glucose)	Numerous bacteria
	$4NO_3^- + 2H_2O \rightarrow 2N_2 + 5O_2 + 4OH^-$	Pseudomonas
	$5S + 6KNO_3 + 2CaCO_3 \rightarrow 3K_2SO_4 + 2CO_2 + 3N_2 + 2CaSO_4$	Thiobacillus denitrificans

unreactive in the trophosphere and finds its way into the stratosphere where it may influence the ozone levels.

Oxygen cycle

Because oxygen is so widely distributed in the earth in numerous chemical species, it is difficult to construct a comprehensive diagram of the global oxygen cycle, the one presented in Fig. 7.7 is largely restricted to the species O_2 and CO_2. Oxygen as molecular oxygen or in chemical combination is involved in many processes, including oxidative weathering, the activity of aerobic organisms, respiration, combustion, decay and photosynthesis. Because of the reactivity of dioxygen and the relatively high positive reduction potentials for the reactions;

$$O_2 + 2H_2O + 4e \rightarrow 4OH^-, \quad E^o = +0.04V, \tag{7.9}$$

and

$$O_2 + 4H^+ + 4e \rightarrow 2H_2O, \quad E^o = +1.23V, \tag{7.10}$$

many natural processes consume dioxygen, for example;

$$4FeO + O_2 \rightarrow 2Fe_2O_3, \tag{7.11}$$

$$C + O_2 \rightarrow CO_2, \tag{7.12}$$

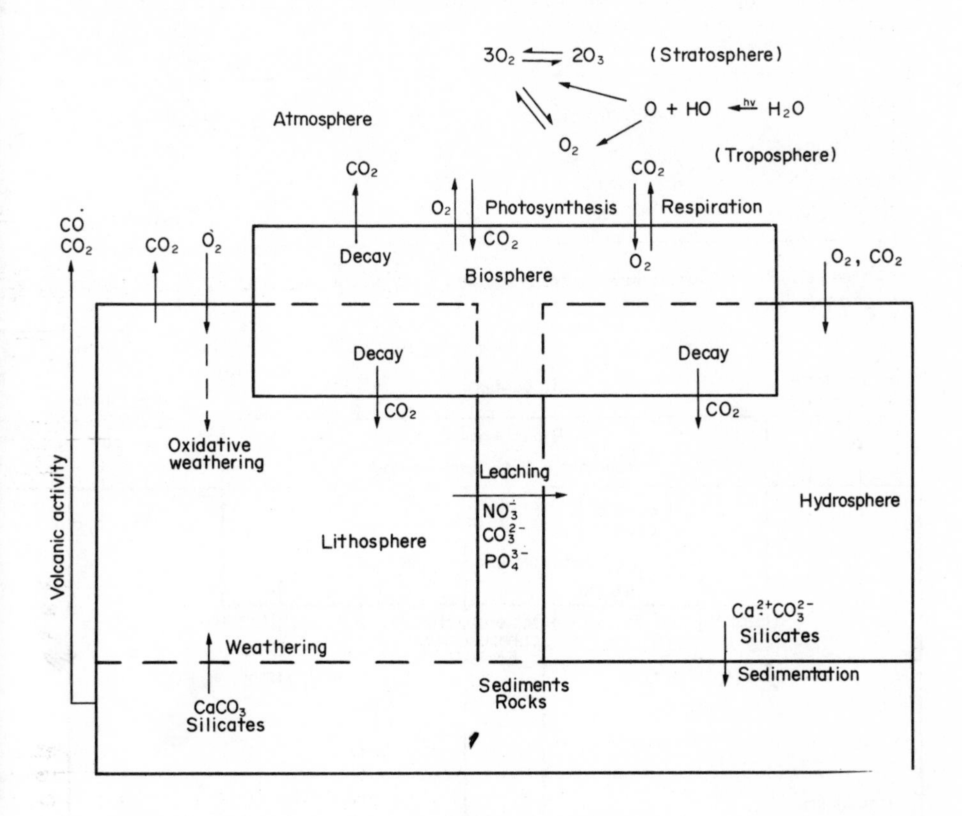

Fig. 7.7. The global oxygen cycle

$$C_6H_{12}O_6 + 6O_2 \rightarrow 6CO_2 + 6H_2O. \quad (7.13)$$

The main exception is photosynthesis, effectively the reverse of reaction (7.13), which highlights the importance of photosynthesis in maintaining the dioxygen balance. The wide occurrence and the important position of oxygen in the world scene is shown by the data presented in Table 7.3. It is clear that oxygen is a significant component of each of the four spheres, and accounts for 48% (by weight) of the four areas.

Phosphorus cycle

Phosphorus is an essential macronutrient for plants and animals, and for plants it is often the limiting nutrient. This is because of the chemical properties of phosphates, and that there is no rapid way for cycling phosphorus through the global cycle (Fig. 7.8). Phosphorus only occurs in the atmosphere as dust particles, therefore cycling depends on transfers within solution, within the solid state and between the two states. These transfers are restricted because of the low solubility of most phosphates especially those (Ca^{2+}, Al^{3+} and Fe^{3+}) that occur, or form in nature. The most soluble

TABLE 7.3 The Distribution of Oxygen in the Earth

Area	Atoms percent	Mass percent	Approximate mass of oxygen 10^{15} tonnes
Lithosphere	62	46.6	11,040*
Hydrosphere	33	89	1,290**
Atmosphere	21	23	1.2
Biosphere	25	52	0.004***

* To depth of about 17 km. ** Includes ice and fresh water. *** Wet weight of living organisms.

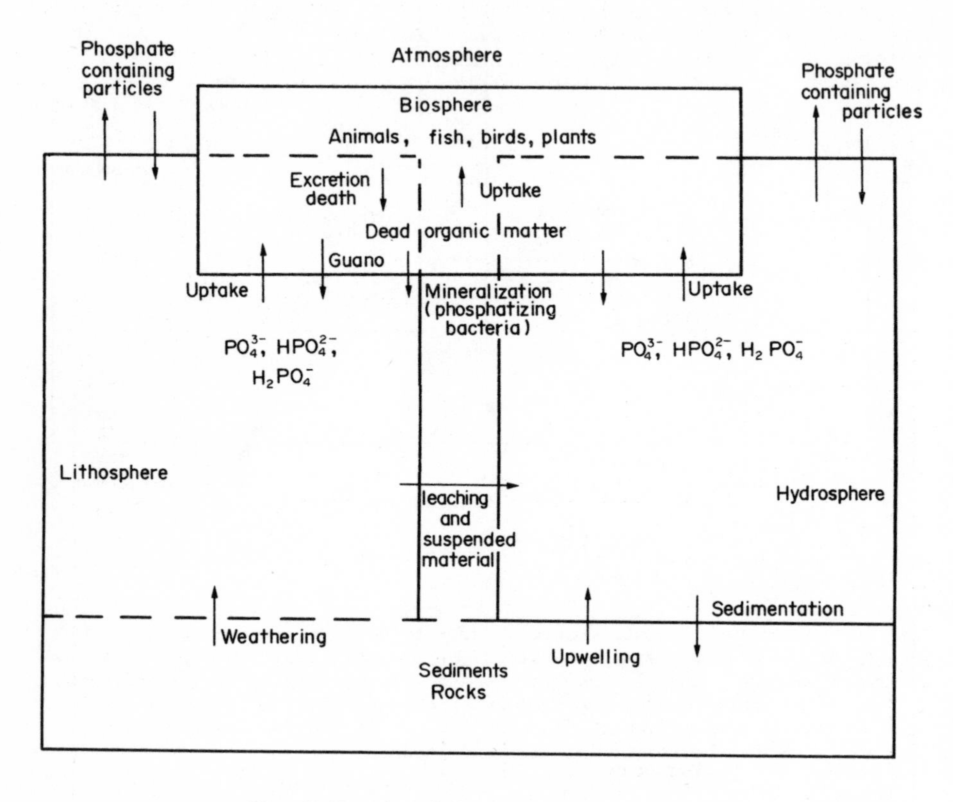

Fig. 7.8. The global phosphorus cycle

form of phosphate that occurs in the natural system is $Ca(H_2PO_4)_2$, which exists over a limited pH range as shown in the distribution diagram (Fig. 6.7).

Since the movement within the phosphorus cycle is slow, the addition of phosphate fertilizers can have a significant influence on plant growth. The efficiency of nitrogen fixing crops, such as clover, depends on an adequate supply of phosphate. In New Zealand for example, where agriculture depends on grasslands, most nitrogen is obtained by biological fixation, however it is necessary to add phosphate fertilizers to ensure sufficient nitrogen is obtained. Unfortunately when a phosphate fertilizer is added to soil it is rapidly fixed and becomes unavailable to the plants, either by absorption on to the clay and minerals within the soil, or by formation of insoluble salts depending on the pH and type of soil. Man can also influence the phosphorus cycle by the use of phosphate detergents, which initially, end up in the hydrosphere and act as a nutrient source for water plants.

Sulphur cycle

Sulphur, an essential nutrient element, is involved in a complex geo-biochemical cycle (Fig. 7.9). The complexity is similar to that for nitrogen and is due to the range of accessible oxidation states of sulphur, and the ease of their interconversions in the environment. The principal forms encountered in the environment, and some of the interconversions between the species are given in Table 7.4. Hydrogen sulphide, the product of decay

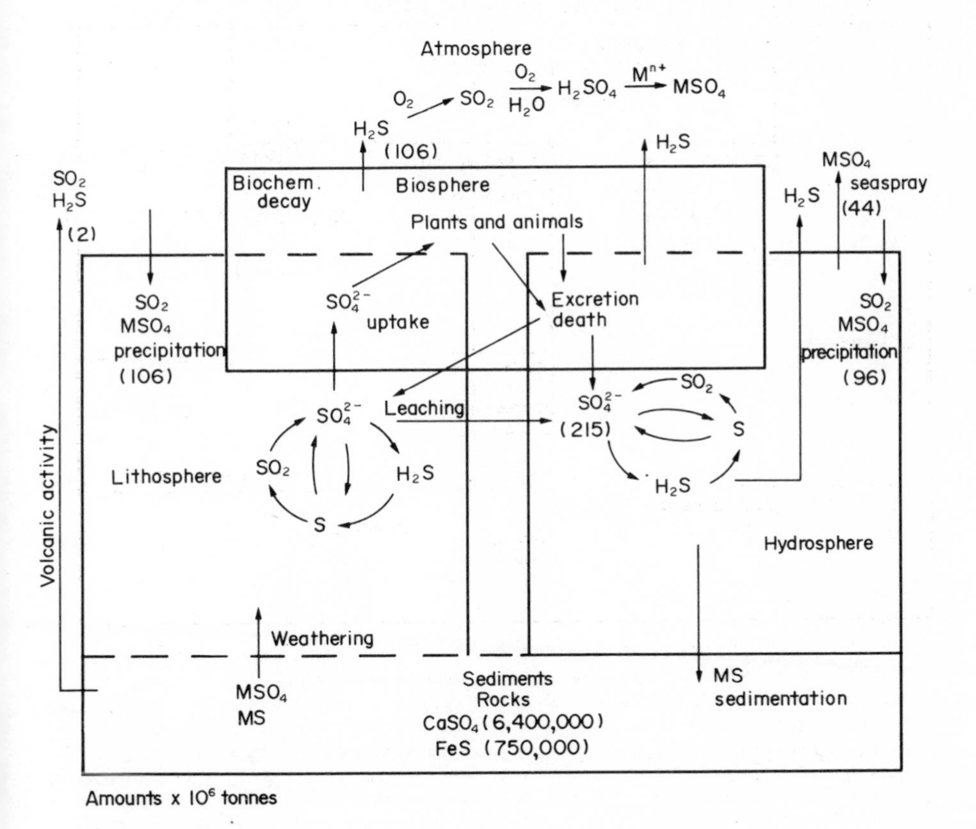

Fig. 7.9. The global sulphur cycle

TABLE 7.4 Sulphur in the Environment

Oxidation	Form found in Environment	Chemical Transformation Processes	Comments on Transformations
-II	H_2S, decay of organic material	$H_2S \rightarrow S \rightarrow SO_4^{2-}$	aerobic bacteria (Thiobacillus)
		$H_2S \rightarrow SO_2$	in atmosphere catalysed by transition metals
	SH^-,RR'S, amino acids	$RR'S \rightarrow S \rightarrow SO_4^{2-}$	variety of bacteria
		$RR'S \rightarrow H_2S$	anerobic bacteria (desulphovibrio)
	S^{2-}, metallic sulphides	$S^{2-} \rightarrow SO_4^{2-}$	oxidative weathering
-I	S_2^{2-} metallic sulphides, in proteins	$S_2^{2-} \rightarrow SO_4^{2-}$	oxidative weathering
O	S_8 free in nature	$H_2S \rightarrow S$	aerobic bacteria
		$RR'S \rightarrow S$	
+IV	SO_2, H_2SO_3 combustion processes	$SO_2 \rightarrow SO_3$	aerial oxidation
		$(H_2SO_3 \rightarrow H_2SO_4)$	catalysed by transition metals
		$SO_2 \rightarrow RR'S$	assimilation in plants
+VI	SO_3, H_2SO_4, SO_4^{2-} oxidative processes	see reactions above	chemical and biochemical oxidations

of organic material is rapidly oxidised to SO_4^{2-} in water provided dioxygen is present;

$$H_2S + 4H_2O - 8e \rightarrow SO_4^{2-} + 10H^+ \quad (7.14)$$

$$O_2 + 4H^+ + 4e \rightarrow 2H_2O, \quad (7.15)$$

$$\text{i.e.} \quad H_2S + 2O_2 \rightarrow SO_4^{2-} + 2H^+. \quad (7.16)$$

The reaction is slower (few hours to days) in the atmosphere, and probably occurs in water aerosol droplets catalysed by metal ions. The two main environmental sulphur compounds in the atmosphere, SO_2 and H_2S are both quite toxic to man. Man's contribution to the atmospheric levels of SO_2 are as high as one third to one half of the natural sources. Anthropogenic sources also tend to be localized around point sources in industrialised urban areas, and therefore reach quite high concentrations. The end products of the reactions of sulphur compounds in the atmosphere are either H_2SO_4 or metal sulphates. The former precipitates as acid rain which is becoming more of a destructive influence in the environment. The large amounts of sulphur precipitated on land, 106 x 10^6 tonnes annually, contribute to plant sulphur requirements, making it unnecessary, in some areas, to add sulphur fertilizers.

The Cycling of Other Elements

It is possible, to a limited extent, to construct environmental global cycles for other chemical elements, and this will be done when required in later chapters. Most elements are involved in cycling processes, but not necessarily in all of the four spheres of the habital earth. For example, there are no natural volatile compounds of iron in the atmosphere, however iron does occur in dust and aerosol particles to some extent. The same applies for the element lead, and also this element does not have a role in the biosphere. When chemicals become involved in areas in which they do not occur naturally this can give rise to problems, such as toxicity in the biosphere. This is especially critical when one section of the biosphere apparently copes with material, such as organomercury in fish, while for another section the material is toxic.

The amounts of chemical species added to the environment through the activities of man, are often much less than from natural sources. But, in general, natural source emissions are spread over the earth's surface, while man's influence is often localized in urban areas. Also some of the added materials are chemically stable in an environment dominated by the three principal chemical change-agents, water, dioxygen and radiation. Such materials have a long residence time and can become involved in chemical reactions which initially were not considered likely, or envisaged.

It has become increasingly clear that in chemical environmental cycles, biochemical processes are of importance. What were once considered to be unreactive forms of some elements, and therefore "safe" are now being found to be incorporated into the geo-bio-chemical processes. An example of this is the biochemical methylation of Hg^{2+}, once thought to be inactive as insoluble HgS in river, lake and ocean sediments. Recent work indicates similar methylation occurs for lead and arsenic, and therefore one is less sure about the non-existence of volatile lead compounds in the atmosphere.

CHAPTER 8

Air Pollution

Human beings receive their nutrient and energy requirements from food, water and air. Of the three, man cannot survive long without air. The intakes for an "average" adult are listed in Table 8.1. The figures are approximate, and

TABLE 8.1 Average Air, Food and Water Needs of an Adult

Material	Average Daily Intake	Average Intake kg day^{-1}
Air	20 m^3 [a]	24 [b]
Food	1.5 kg	1.5
Water	1 litre	1

[a] This varies with the person and physical activity. [b] Taking the density of air as 1.225 kg m^{-3} at sea level (288 K).

vary from individual to individual, but they indicate the significant position of air for man. Much of what is breathed in is exhaled, but the lung tissue does come into contact with a large quantity of air daily. The absorption of material from the lungs into the blood stream is more effective (perhaps by a factor of 2 to 5) than absorption from the alimentary tract. These facts emphasise the need for good quality air for healthy existence, and why the levels of air pollutants are required to be at least an order of magnitude less than in food and water.

Air pollution is not a new phenomenon, and dates back a number of centuries. King Edward I of England (1272-1307), prohibited the burning of coal in London while Parliament was in session, owing to the smoke and odour produced. The penalty for burning coal was rather severe.

> "Be it known to all within the sound of my voice, whosoever shall be found guilty of burning coal shall suffer the loss of his head"
>
> King Edward I (ca 1300)

A few examples of serious air pollution episodes are listed in Table 8.2. In most cases the pollutants are particulates and sulphur dioxide. The term

TABLE 8.2 Some Air Pollution Episodes

Date	Location	Pollutants	Effects
Winters 1873 1880 1892	London, UK	Smoke, SO_2	Cattle deaths, Excess human deaths 600 - 1000
Dec 1-5, 1930	Meuse Valley, Belgium	Particulates SO_2(10-38 ppm)	63 Excess deaths, chest pain, cough, eye and nasal irritation - all ages
Oct 27-31, 1948	Donora, USA	Particulates SO_2(0.5-2 ppm)	20 Excess deaths as above, older people mostly affected.
Nov 24, 1950	Poza Rica, Mexico	H_2S	22 Excess deaths,320 hospitalized, all ages
Dec 5-9, 1952	London, UK	Particulates (4500 $\mu g\ m^{-3}$), SO_2 (1.3 ppm)	4000 Excess deaths.
Dec 5-10, 1962	London, UK	SO_2 (2.0 ppm)	700 Excess deaths
Jan 29,1963- Feb 12, 1963	New York, USA	Particulates (coef. haze=7) SO_2 (0.5 ppm)	200-400 Excess deaths.
Nov 24-30, 1966	New York, USA	Particulates(COH=5) SO_2 (1.0 ppm)	168 Excess deaths.

smog (smoke and fog) probably appeared in the early part of the 20th century. Air pollution is not a new problem, but is one which many more people are aware of. The problem has grown however, as a result of the widespread use of the internal combustion engine, and until recently a cheap fuel source, viz. petroleum.

AIR POLLUTION: GENERAL COMMENTS

We will consider some general aspects of air pollution, such as the amounts of atmospheric emissions, and different ways of viewing air pollution.

Emissions Into the Atmosphere

Emissions into the atmosphere are either gases or particles. Gases fill the volume available, can become very mobile and exert pressure. Particulate material is either liquid or solid and can have a wide range of sizes. Particulate material may also absorb gases on to their surface.

A material emitted into the atmosphere becomes involved in a number of events, such as mixing and dilution with the air, transport both vertically and horizontally, and chemical and physical changes. Emissions are either primary, i.e. they remain unchanged, or secondary, i.e. the material undergoes change, such as change in size or participation in chemical reactions with formation of new chemical species.

Atmospheric Pollutant Sinks

Ultimately the material emitted into the atmosphere ends up in areas called *sinks*, such as the oceans or on the land. Some materials are accumulating in the atmosphere, a good example of this is carbon dioxide. Finally some emissions, in particular those that are unreactive in the troposphere, (e.g. N_2O and halocarbons) escape into the stratosphere where they do participate in chemical reactions. The average time a material remains in the atmosphere, called the residence time, varies from minutes to years depending on the substance and the climatic conditions. In some cases the duration depends on the method of estimating the time. Estimates of the residence time for some atmospheric emissions are given in the last column of Table 8.3.

Quantities of Pollutants Emitted

Numerous attempts have been made to estimate the quantity of material emitted into the atmosphere each year from natural and anthropogenic sources. A list of estimates, made from data collected over the period 1965 - 1975, is given in Table 8.3. The information must be considered as tentative, but indicative of the magnitude of the emissions of different substances. Except for halocarbons and sulphur dioxide, the anthropogenic emissions are quite a bit less than those from natural sources. However, anthropogenic sources are localised in areas of dense population, and emissions can produce high local concentrations. The concentration of a material is the critical factor in affecting health. Unpolluted air contains around 3 to 30 $\mu g\ m^{-3}$ of SO_2 and 1000 $\mu g\ m^{-3}$ of CO, while the levels in industrialised urban areas can reach 50 to 5000 $\mu g\ m^{-3}$ of SO_2 and 5000 to 230,000 $\mu g\ m^{-3}$ of CO. It is estimated that around 50% of the atmospheric pollution in the U.S.A. occurs on 1.5% of the land area.

The major sources of some air pollutants (in the U.S.A.) are listed in Table 8.4. Transport produces approximately 50% of atmospheric pollution, particularly CO, NO_x and hydrocarbons. The next major contributor is combustion (24%) forming principally, SO_2, NO_x and particles. The amount of pollution emitted from a particular source (e.g. coal burning) depends, to some extent, on the mode of utilisation. The emissions, in kilograms per tonne of coal consumed, are listed in Table 8.5 for different uses of coal. From the data is may be suggested that domestic and commercial burners are not as efficient as power plants and industrial burners. Greater emissions of CO and hydrocarbons, occur because of the lower temperatures reached in the domestic and commercial burners. However, because of the lower temperatures less NO_x is produced. In terms of absolute emissions, because more coal is used in power plants more CO and hydrocarbons are emitted. The amount of SO_2 produced does not vary with the way the coal is used because sulphur is a contaminant which is fully released as SO_2 on burning. However, pollution control measures at power plants remove some of the SO_2 from the emitted gases, while domestic and commercial combustion of coal is, generally, uncontrolled.

Air Pollution Categories

Air pollution may be considered from different points of view. Attention can be given to the species that are discharged into the atmosphere. With some

TABLE 8.3 Estimates of Annual Global Emissions into the Atmosphere (1965 - 1975)

Material	Major Pollution Source	Natural Source	Approximate Emissions (Annual) Pollution $x10^6$ tonnes	Natural $x10^6$ tonnes	Pollution as % of total	Estimated residence time in atmosphere
Carbon dioxide	Combustion	Biological decay	13000	10×10^6	0.13	4 years
Carbon monoxide	Transport, Combustion	Forest fires	360	3500	9.3	1-4 months
Hydrocarbons	Transport, Combustion	Biological processes	80	1604	5	16 years
Halocarbons	Aerosol sprays, refrigeration	Nil	10	-	100	>20 years
Sulphur dioxide	Combustion coal and oil	Volcanoes	146	73	67	3-7 days
Hydrogen sulphide	Chemical industry	Volcanoes, Biological processes	3	98	3	2 days
Nitrogen oxides (NO, NO_2, N_2O)	Combustion	Biological processes	53	500 (NO_2) 592 (N_2O)	9.6 (NO_2) 4.6 (total)	4 days (NO, NO_2)
Ammonia	Waste treatment	Biological decay	4	3700	0.1	2 days
Particulates	Combustion	Salt spray, Dust (soil)				
Primary			92	1400	6	minutes to weeks
Total			296	2500	10	

Sources: Various.

TABLE 8.4 Emission of Air Pollutants (USA) 1975 x 10^6 tonnes (% of total)

Source	CO	Hydrocarbons	NO_x	SO_x	Particles	Total
Transport	77.4 (80.4)	11.7 (37.9)	10.7 (44.2)	0.8 (2.4)	1.3 (7.2)	101.9 (50.4)
Combustion (fuels)	1.2 (1.3)	1.4 (4.5)	12.4 (51.3	26.3 (79.8)	6.6 (36.8)	47.9 (23.7)
Industrial Processes	9.4 (9.8)	3.5 (11.3)	0.7 (2.9)	5.7 (17.3)	8.7 (48.3)	28.0 (13.8)
Solid Waste Disposal	3.3 (3.4)	0.9 (2.9)	0.2 (0.8)	<0.1 (0.3)	0.6 (3.3)	5.1 (2.5)
Other, e.g. Agriculture	4.9 (5.1)	13.4 (43.4)	0.2 (0.8)	0.1 (0.3)	0.8 (4.4)	19.4 (9.6)
Total	96.2 (100)	30.9 (100)	24.2 (100)	32.9 (100)	18.0 (100)	202.3 (100)

Source: Cleaning our Environment - A chemical perspective, ACS, 1978.

TABLE 8.5 Gaseous Emissions for Coal Combustion
(kg $tonne^{-1}$ coal burned)

Pollutant	Power Plant	Industrial	Domestic and Commercial
Aldehydes	0.0023	0.0023	0.0023
Carbon monoxide	0.23	1.4	22.7
Hydrocarbons	0.09	0.45	4.5
Oxides of nitrogen	9.1	9.1	3.6
Sulphur dioxide	20S*	20S*	20S*

* S = % sulpur in coal

notable exceptions, this amounts to a discussion of the oxides of the non-metals, carbon, nitrogen and sulphur. Alternatively, air pollution may be discussed in relation to the sources of a particular pollutant (Table 8.4), or the type of pollutants that arise from a particular industry (Table 8.6).

TABLE 8.6 Some Emissions from Selected Industries

Industry	Pollutants
Chloro-alkali	Cl_2, Hg (Hg cell only)
Copper smelters	Particles, SO_2
Pulp and Paper (Kraft)	Particles, SO_2, CO, H_2S, CH_3SH, $(CH_3)_2S$, $(CH_3)_2S_2$
Petroleum refining	SO_2, CO, hydrocarbons, nitrogen oxides, aldehydes, NH_3
Portland cement	Particles, SO_2, nitrogen oxides
Aluminium production	Particles, gaseous and particulate fluorides

Source: Cleaning our Environment, A Chemical Perspective, ACS 1978.

In more detail, discharges from different sections of an industry can be considered, as outlined for the Kraft Pulp and Paper industry in Table 8.7. A third way to approach air pollution is in terms of the chemistry involved, which includes photochemistry, acid-base chemistry, catalysis and the importance of synergism. We will consider air pollution principally from the aspect of the type of pollutant (Chapter 9). Before doing so it is necessary to mention the affect climate has on air pollution.

METEOROLOGY IN THE TROPOSPHERE

The troposphere contains most of the material in the atmosphere, 50% up to 5 km height and 90% up to 12 km height. An understanding of the movement of the air within the troposphere is important for determining the fate of

TABLE 8.7 Emissions (uncontrolled) from Pulp and Paper Industry (Kraft)

(kg per tonne of air dried unbleached pulp)

Source	Particles	SO_2	CO	H_2S	organic sulphur compounds
Digester relief and blow tank	-	-	-	0.05	0.75
Brown stock washers	-	0.005	-	0.01	0.1
Multiple effect evaporators	-	0.005	-	0.05	0.2
Recovery boiler & direct contact evaporator	75	2.5	1-30	6	0.5
Smelt dissolving tank	2.5	0.05	-	0.02	0.2
Lime kilns	22.5	0.15	5	0.25	0.125
Turpentine condenser	-	-	-	0.005	0.25

Source: Cleaning our Environment, A Chemical Perspective, ACS 1978.

atmospheric pollutants. The two dominant air movements are vertical (air currents) and horizontal (winds).

Vertical Movement

Vertical movement of air masses depends on the stability of the atmosphere, in unstable air there is the greatest amount of vertical movement. For a parcel of gas in the atmosphere at some particular height, its temperature will initially be the same as the surrounding air. If the gas rises it will expand as the surrounding air pressure drops. Assuming the expansion is adiabatic, i.e. no heat transfer occurs, the temperature of the gas will fall. If the temperature drop - called the adiabatic lapse rate - is greater than the change in the temperature of the surrounding air, then the atmosphere is called stable and the parcel of air will drop down to its original position. This can be demonstrated with the use of the ideal gas law;

$$PV = nRT, \tag{8.1}$$

$$\text{or} \quad P = \rho RT \tag{8.2}$$

where ρ = gas density.

For the parcel of gas (g);

$$P_g = \rho_g RT_g, \tag{8.3}$$

while for the surrounding air (a),

$$P_a = \rho_a R T_a . \tag{8.4}$$

Since the two pressures P_g and P_a will be the same we can write;

$$\rho_g = \rho_a \left(\frac{T_a}{T_g}\right) \tag{8.5}$$

Initially $T_g = T_a$, but when the gas rises and its temperature fall is greater than for the surrounding air (i.e. $T_g < T_a$), then $\rho_g > \rho_a$ and the gas will descend. If the adiabatic lapse rate for the gas is less than for the surrounding air (i.e. $T_a < T_g$), then $\rho_g < \rho_a$, and the gas will continue to rise. This corresponds to an unstable situation which allows for dispersal of atmospheric emissions. This is illustrated in Fig 8.1 in which the full

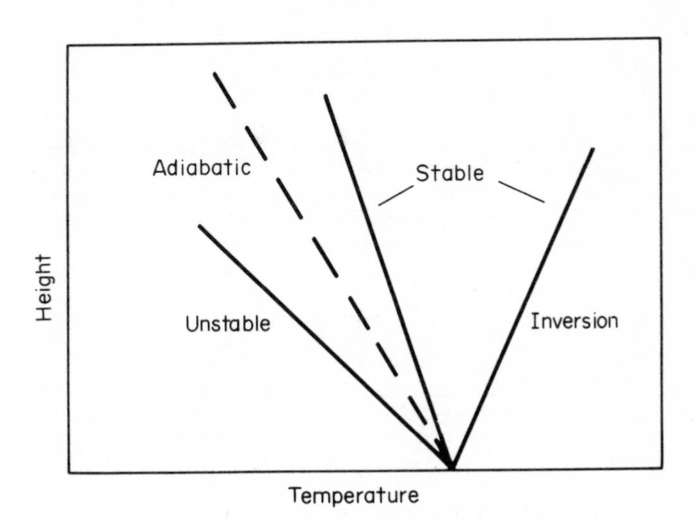

Fig. 8.1. The adiabatic lapse rate divides stable from unstable air.

lines represent the temperature change of the surrounding air and the dotted line the temperature change of the gas. An extreme case occurs when the temperature of the surrounding air increases with height, a temperature inversion.

Atmospheric Inversion

Normally in the troposphere the temperature falls with an increase in altitude. However, circumstances can reverse this situation over a certain height. This is called a temperature inversion, a condition associated with air pollution problems. There are a number of ways that temperature inversion may occur.

Subsidence inversion. A simple model of the atmosphere assumes that most of the heating occurs around the equator and the hot air produced will rise until it reaches the tropopause. As the air rises it will expand, cool and lose its moisture, hence the high rainfall in the equatorial tropical regions. The cooled air moves towards the poles, to around latitude 30°N or S, where it begins to descend, and because of the pressure increase the air temperature increases trapping cooler air below it giving subsidence inversion (Fig. 8.2). The circulation-cells of air shown in the figure are called Hadley cells (after George Hadley, 18th century). The areas of dry,

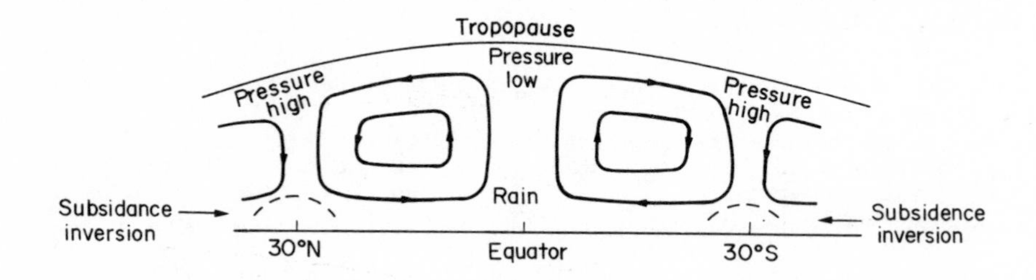

Fig. 8.2. Mechanism for the formation of subsidence inversions at 30°N and 30°S.

high pressure air around 30°N and S are associated with extensive deserts, and subsidence inversion as experienced in California, U.S.A. Seasonal variation means that the system moves north or south with the zone of greatest heating. This model assumes the earth's surface is uniform, ignoring its roughness, however it does explain the origin of the prevailing winds over the world.

Radiative inversion. On clear cold nights the earth's surface cools rapidly, radiating heat into the atmosphere, cooling the air near the ground compared with air at higher altitudes. This produces the inversion illustrated in Fig. 8.3a. The following day the surface air is warmed, producing a limited mixing area above which sits warm air trapping in low level material (Fig. 8.3b). During the day the air warms up and the inversion is likely to disappear (Fig. 8.3c).

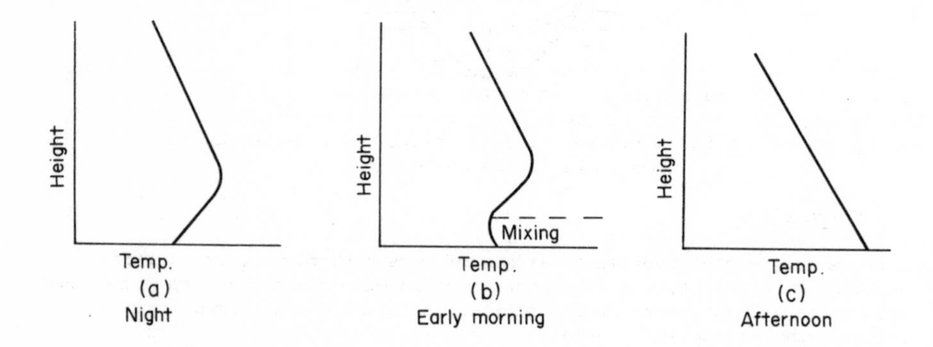

Fig. 8.3. Development of an inversion on (a) clear cold nights, (b) trapping in of surface air in the early morning as the earth warms up and (c) later in the day

Frontal inversion. Cold air, being heavier than warmer air can at times slip under the warmer air producing an inversion. For example, cooler air above the ocean can slip under the warmer air above the land. This often leads to a sea breeze, which helps in dispersal of coastal pollution, but when the air movement is slow inversion can occur on the land close to the sea (Fig. 8.4).

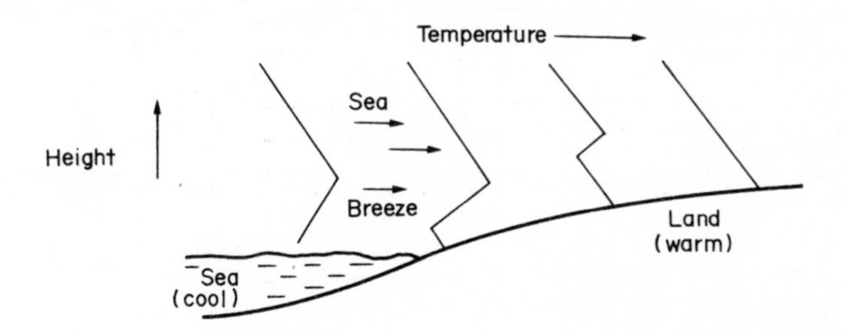

Fig. 8.4. Development of an inversion on land close to the sea, when air movement is slow.

Advective inversion. When the sides of valleys cool at night the adjacent air is also cooled and falls into the valley, while warmer air may flow over the top of the valley trapping the cold air, and any pollution, underneath (Fig. 8.5). The pollution episodes at the Meuse Valley, Belgium and at Donora, Pennsylvania, U.S.A. arose from this type of inversion.

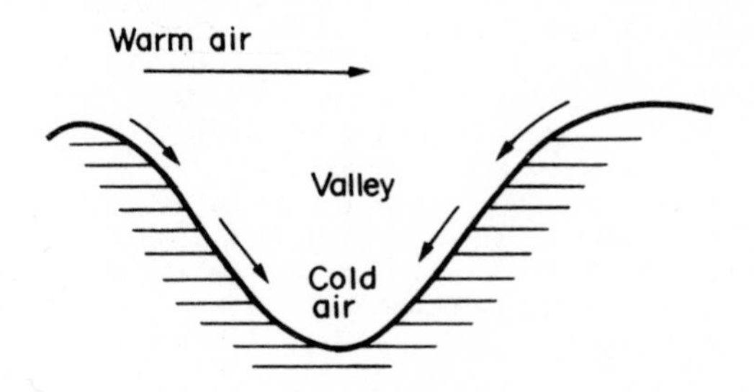

Fig. 8.5. Formation of an inversion in a valley as cool air sinks to the bottom.

Smoke Behaviour

A plume of smoke behaves in a variety of ways depending on the vertical temperature gradient, and the presence or absence of temperature inversions. Some modes of behaviour are illustrated in Fig. 8.6a-f. The dotted lines on the diagrams on the left represent the adiabatic lapse rate and the full lines the vertical temperature gradient of the air. Fastest dispersal of the smoke occurs through *looping* (Fig. 8.6a) followed by *coning* (Fig. 8.6b). The looping may be quite large and at times could reach the ground. In the last four examples (Fig. 8.6c-f) smoke is entrapped by inversions, the most serious regarding low-level pollution is *fumigation*, where the stack top is at a lower height than the mixing height of the inversion. In this case the smoke is kept low and can readily reach ground level. The *fanning* behaviour may turn into fumigation if the mixing height drops as the ground air temperature increases during the day.

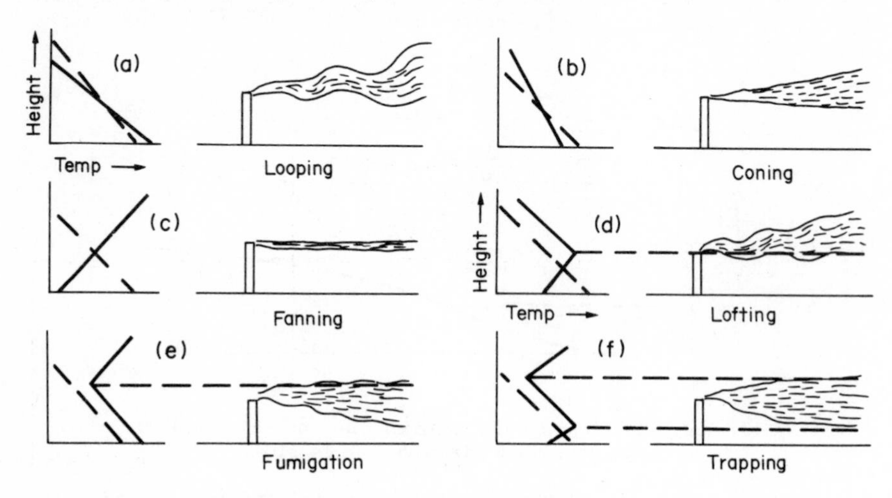

Fig. 8.6. Stack smoke behaviour in relation to air temperature. The broken line is the adiabatic lapse rate and full line the vertical temperature gradient.

Winds

Wind is the horizontal movement of air and in general moderate to strong winds promote the dilution and dispersal of air pollutants. The amount of aerosol collected alongside a busy urban street (in mg m^{-3}) (at two hourly intervals) is shown in Fig. 8.7. Line (a) is for a calm day, the profile being explained by changes in traffic density. Line (b) has a similar profile during the early part of the day, but as the wind strength picked up, and in spite of the homeward bound traffic, the aerosol levels fell due to dispersal by the wind.

The horizontal motion of air arises as the resultant of three influences, the pressure gradient, the coriolis deflection and friction with the earth's surface. The coriolis deflection originates from the effect of the spin of the earth on the surrounding air. As a parcel of air travels north (in the northern hemisphere), it moves with a greater velocity than any point north of it due to the fact that a point on the equator has a greater velocity than other points on the spinning earth's surface. Therefore to an observer, the parcel of air appears to be deflected (to the east in the northern hemisphere). The result is that the wind moves along the isobars (lines of constant pressure) with some deviation because of the effect of friction (Fig. 8.8) rather than across the isobars.

Wind strength is reduced by rough terrain (e.g. by buildings in a city) compared with smooth terrain. Air turbulence is due to eddies, i.e. packets of air moving randomly but with a circular motion. If the cross section of an eddy is comparable with the dimensions of a gas plume, this provides for rapid dispersal.

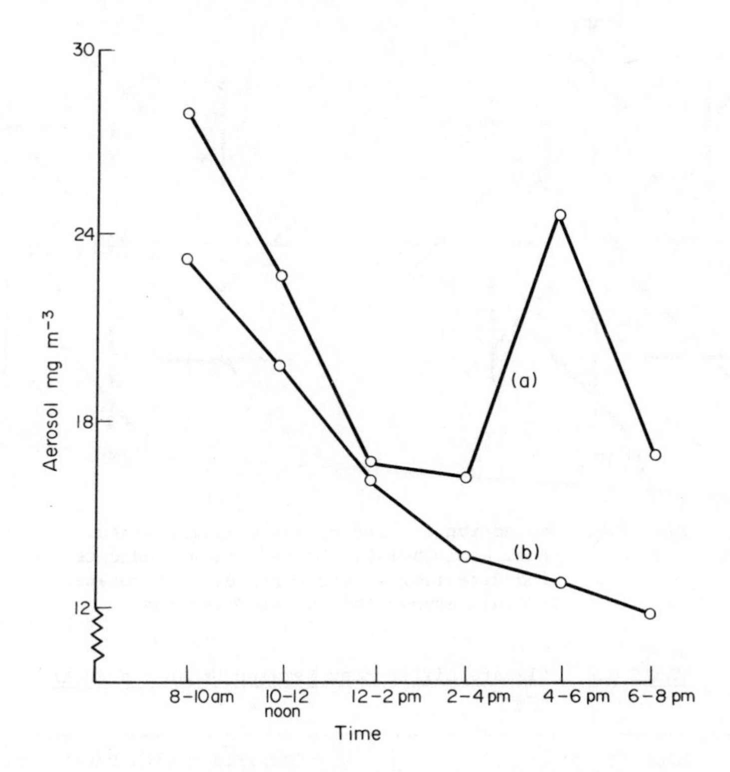

Fig. 8.7. Levels of aerosol along a busy urban road, (a) on a calm day, and (b) a wind springs up in the afternoon.

Conditions For a Pollution Episode

In addition to a pollution source the atmospheric conditions that favour a pollution episode are; several days with a low mixing height (i.e. <1500 m), low wind speed (<4 m s^{-1}), and no significant precipitation. If these conditions last for at least two days a pollution episode could occur. As it is possible to decide if a particular area of land has the potential atmospheric conditions to trap pollutants, this information should be included in industrial development planning.

Climate in Cities

Factors such as; the rough terrain (due to buildings), high energy consumption and subsequent loss to the atmosphere, and strongly reflecting surfaces, modify the climate of cities relative to neighbouring rural areas. The results of a study of this difference are given in Table 8.8. Certain of the factors, including reduced wind speed, loss of heat to the atmosphere at night from surfaces which are good heat conductors (concrete, bitumen roads, metal roofs), assist in providing conditions that trap pollutants.

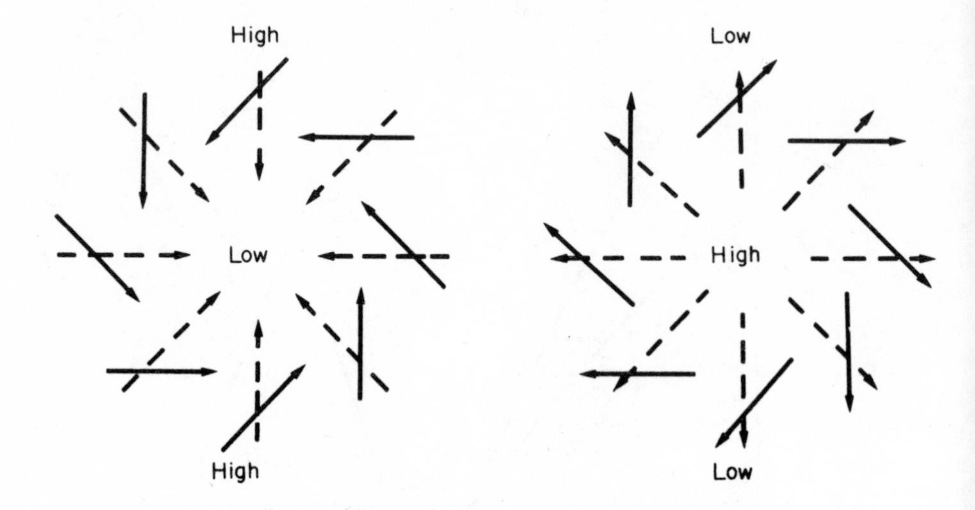

Fig. 8.8. Deflection of wind by the Coriolis effect (Nth. Hemisphere). Dotted arrows indicate wind direction if the earth did not rotate, and full arrows the actual direction.

TABLE 8.8 Climate Differences Between Urban and Rural Areas

Element	Comparison with Rural Areas
Temperature	0.5 - 1.0°C higher
Relative humidity	6% lower
Dust particles	10 times more
Cloudiness	5 - 10% more
Radiation	15 - 20% less
Wind speed	20 - 30% lower
Calms	5 - 20% more
Precipitation	5 - 10% more

Source: Landsberg 1962 in "Air over Cities Symposium", U.S. Dept. Health Education & Welfare.

In addition the heat island produced by city activities and the resulting air circulation (Fig. 8.9) work to keep material within the city. Therefore air pollution episodes, besides being a result of sources and meteorology, are also a result of the micro-climate of urban environments. Estimates of pollution concentrations in a city can be made by considering a city as a box - a rather simplified model. Pollution is generated within the box and

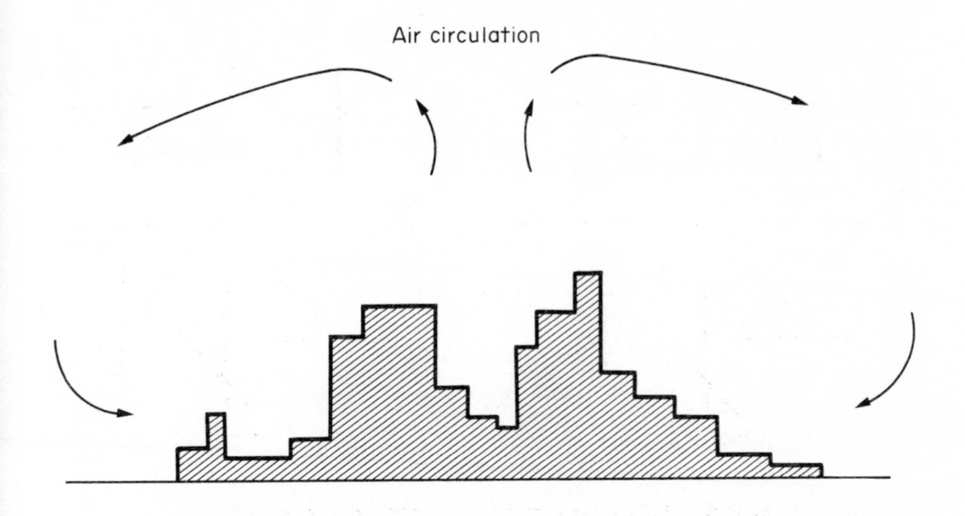

Fig. 8.9. Air circulation associated with a city.

air flows in and out of the box. In a steady state the concentration of the pollutants is given by;

$$C_{wt/vol} = \frac{\text{Pollution emission rate (wt/time)}}{\text{Air flow rate (vol/time)}}, \quad (8.6)$$

assuming rapid and complete mixing of air and pollutants within the box. The air flow rate into a box of width ℓ and mixing height h and with a wind speed v is $v\ell h$.

Hence $$C = \frac{\text{Pollution emission rate (P)}}{v\ell h} \quad (8.7)$$

The concentration of pollutants will increase within the box, when v and h are low. For a city with ℓ = 30 km, h = 200 m and v = 8000 m hr^{-1} and P = 200 x 10^9 mg hr^{-1} of carbon monoxide;

$$C = \frac{200 \times 10^9 (\text{mg hr}^{-1})}{8 \times 10^3 (\text{m hr}^{-1})\ 3 \times 10^4 (\text{m})\ 2 \times 10^2 (\text{m})} \quad (8.8)$$

$$\therefore \quad C = 4.2 \text{ mg m}^{-3}.$$

This concentration is easily reached from car emissions at rush hour in a city. If the wind speed is halved the pollution concentration would double. This demonstrates the importance of wind, especially as of all the variables, it is the one that varies widely and cannot be controlled.

PHOTOCHEMISTRY

Photochemistry is the study of chemical changes and related physical properties that are produced by the interaction of electromagnetic radiation and matter. Photochemical reactions are of vital importance to living materials and play a significant part in atmospheric pollution.

The preliminary step in a photochemical reaction is the absorption of a photon of light energy by atoms, molecules, ions or radicals;

$$A + h\nu \rightarrow A^*, \quad (8.10)$$

where A* stands for species A in an electronically excited state and $h\nu$ is the photon of light energy. The excited species A* can then take part in one of a number of further processes:

$$A^* \rightarrow X + Y + \quad \text{dissociation} \quad (8.11)$$

$$A^* + B \rightarrow M + N + \quad \text{direct reaction} \quad (8.12)$$

$$A^* \rightarrow A^+ + e \quad \text{ionization} \quad (8.13)$$

$$A^* \rightarrow A + h\nu \quad \text{fluorescence} \quad (8.14)$$

$$A^* \rightarrow A + h\nu' \quad \text{phosphorescence} \quad (8.15)$$

$$A^* + M \rightarrow A + M \quad \text{collisional deactivation} \quad (8.16)$$

The reactions (8.11) to (8.13) lead to chemical change, while the last three reactions return A to its ground state.

The Nature of Light

Light is energy emitted or absorbed in distinct quanta (or photons), the propogation of which is described by the wave theory. The energy of a single photon is given by;

$$E = h\nu, \quad (8.17)$$

where h is Planck's constant and ν the frequency. For one mole;

$$E = Nh\nu = Nhc/\lambda, \quad (8.18)$$

where N is Avogadro's number, c the velocity of light and λ the wavelength of the light. The relationship between the energy (kJ mol^{-1}), wavelength (nm) and frequency (cm^{-1}) for the ultra-violet and visible region of the electromagnetic spectrum is given in Fig. 8.10. The energies of typical

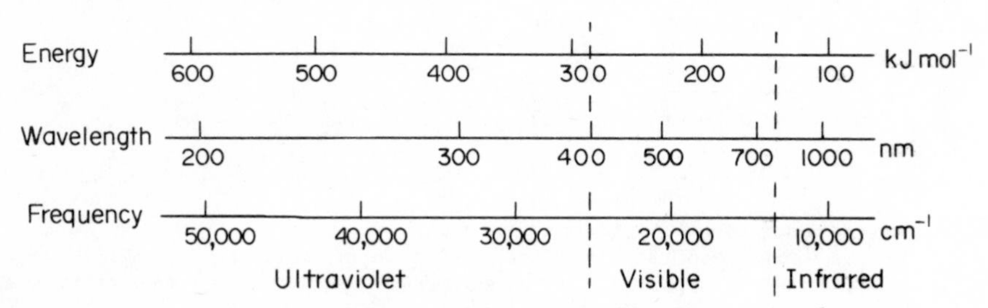

Fig. 8.10. The relationship between energy (kJ mol^{-1}), wavelength (nm) and frequency (cm^{-1}).

chemical bonds are comparable with the energy of visible and ultra-violet radiation, hence absorption of photons of UV-visible radiation can cause bond dissociation. For example, the O-H bond in water has a bond energy of 464 kJ mol^{-1}, hence absorption of a photon of radiation with $\lambda < 258$ nm will give the molecule sufficient energy to dissociate a hydrogen atom from water.

From equation (8.18);

$$\lambda = \frac{Nhc}{E} \tag{8.19}$$

i.e.
$$\lambda = \frac{6.02 \times 10^{23}(\text{mol}^{-1})\ 6.63 \times 10^{-34}(\text{Js})\ 3.0 \times 10^{8}(\text{ms}^{-1})}{464 \times 10^{3}(\text{J mol}^{-1})}$$

$$\lambda = 2.58 \times 10^{-7}\text{m},$$

i.e.
$$\lambda = 258 \text{ nm}$$

The high energy radiation from the sun is filtered out before reaching the earths' atmosphere, and in the upper atmosphere UV radiation with $\lambda < 290$ nm is involved in photochemical reactions. In the troposphere radiation with $\lambda < 400$ nm may, depending on the pollutants, be involved in photochemical processes.

CHAPTER 9

Air Pollutants

In this chapter we will discuss the sources, chemistry and effects of some of the more important air pollutants such as the oxides of carbon, nitrogen and sulphur. The oxides are gaseous and are the products of combustion.

OXIDES OF CARBON

The oxides CO and CO_2 are the two gaseous pollutants produced in largest quantities (Table 8.3) from natural and anthropogenic sources. They originate from oxidation processes, such as decomposition of organic material, oxidation of CH_4, burning of fossil fuels or other materials such as wood. Carbon monoxide has a triple bond C≡O (bond energy 1080 kJ mol^{-1}), while the linear shape of CO_2, O=C=O, is a consequence of π-bonding. The dioxide is acidic, whereas CO is a weak Lewis base forming coordination compounds with the transition metals. Carbon dioxide is the more soluble in water.

Oxidation of carbon is summarized by the equations;

$$2C + O_2 \rightarrow 2CO, \quad \Delta G_{298K} -275 \text{ kJ}, \qquad (9.1)$$

$$2CO + O_2 \rightarrow 2CO_2, \quad \Delta G_{298K} -514 \text{ kJ} \qquad (9.2)$$

Even though the last reaction has a favourable ΔG, the reaction is slow owing to kinetic factors, in particular the large energy of activation, because of the strong C≡O bond, and a low dipole moment.

Sources

A recent and surprising discovery is that much more CO is produced naturally than from anthropogenic sources. Some tentative figures are given in Table 9.1 for both natural and man-made sources of CO. Most natural CO comes from oxidation of CH_4, by anaerobic decomposition of organic material, particularly in swamps and in the humid tropics. The CH_4 produced, approximately 1600 x 10^6 tonnes annually, would account for the concentration of CO in the troposphere. Oxidation of CH_4 is initiated by a hydroxyl radical;

TABLE 9.1 Global Sources of Carbon Monoxide

Source	Amount produced annually x 10^6 tonnes
Natural	
Oxidation of CH_4	3000
Biosynthesis and decay of chlorophyll (in autumn)	100
Oceans and Other	400
Total	3500
Anthropogenic	
Motor vehicles	
Petrol	197
Diesel	2
Aircraft	5
Watercraft	18
Other vehicles	28
Industry	46
Solid wastes disposal	23
Other	41
Total	360
Total	3860

$$CH_4 + \cdot OH \rightarrow \cdot CH_3 + H_2O, \quad (9.3)$$

followed by a number of complex reactions, including the following;

$$\cdot CH_3 + O_2 + M \rightarrow CH_3O_2 + M, \quad (9.4)$$

$$CH_3O_2 + NO_2 \rightarrow HCHO + HNO_3, \quad (9.5)$$

$$HCHO + h\nu \rightarrow H_2 + CO \quad (9.6)$$

The large quantities of CO produced naturally means that other additions do not greatly add to the total. But also, efficient removal processes prevent a build up of CO. This is however, not the situation for CO_2, whose concentration is increasing in the atmosphere. From late in the 19th century to the 1960's the concentration has increased from 290 ppm to 315 ppm (9% increase).

Consumption of transport fuels are the principal source of CO and CO_2 (cf Tables 9.1 and 9.2). Seventy percent of CO pollution comes from all forms of transport, and 79% of this comes from petrol burnt in automobiles, while 31% of man-made CO_2 comes from the burning of petrol. In Los Angeles 3.5×10^6 tonnes of CO are emitted per year, 97% of this comes from the car. The amounts of CO and CO_2 emitted from vehicles depend on the mode of operation (Table 9.3). Most CO is produced during the idling and deceleration stages, because in these stages the combustion mix tends to be fuel rich.

TABLE 9.2 Global Anthropogenic Sources of Carbon Dioxide

Source	Amount produced annually x 10^6 tonnes
Coal	7000
Petrol	4000
Natural gas	1000
Other	1000
Total	13000

TABLE 9.3 Carbon Monoxide and Dioxide Emissions from Automobiles

Car Operation	In Exhaust Gases	
	CO Vol %	CO_2 Vol %
Idle	4-9	9-10
Cruise	1-7	12-13
Acceleration	0-8	10-11
Deceleration	2-9	9-10

As expected these stages are lowest in CO_2 emissions.

Considering petrol as pure 2,2-dimethyl-4-methylpentane (iso-octane) the combustion, using air as the oxidant is;

$$C_8H_{18} + \underbrace{12.5O_2 + 46.6N_2}_{\text{air, 1:3.73 moles}} \rightarrow 8CO_2 + 9H_2O + 46.6N_2, \quad \Delta H = -5062 \text{ kJ.} \quad (9.7)$$

Complete combustion requires an air:fuel mole ratio of 59.1:1, or on a mass basis 15:1, i.e. 1 kg of fuel requires 15 kg of air. However, this assumes ideal conditions. A lower ratio (fuel rich) gives incomplete combustion and more CO produced, an air:fuel ratio of 15 gives around 1% of CO in the exhaust gases, while a ratio of 13 gives around 4%. The effect of the air: fuel ratio on CO emissions is illustrated in Fig. 9.1.

The critical chemical reaction, as regards the relative amounts of CO and CO_2 in automobile emissions is the equilibrium;

$$CO_2 \rightleftharpoons CO + \tfrac{1}{2}O_2. \quad \Delta H = 1825 \text{ kJ.} \quad (9.8)$$

In order to reduce the formation of CO, the temperature needs to be low, pressure high, and a good supply of oxygen, all which favour pushing the equilibrium to the left. The $\log_{10}K$ values: -45 at 298K, -2.9 at 2000K and 0.7 at 4000K, demonstrate the effect of temperature. However, creating optimum conditions to reduce the CO emissions gives other problems such as increased amounts of other pollutants, especially nitrogen oxides (discussed later).

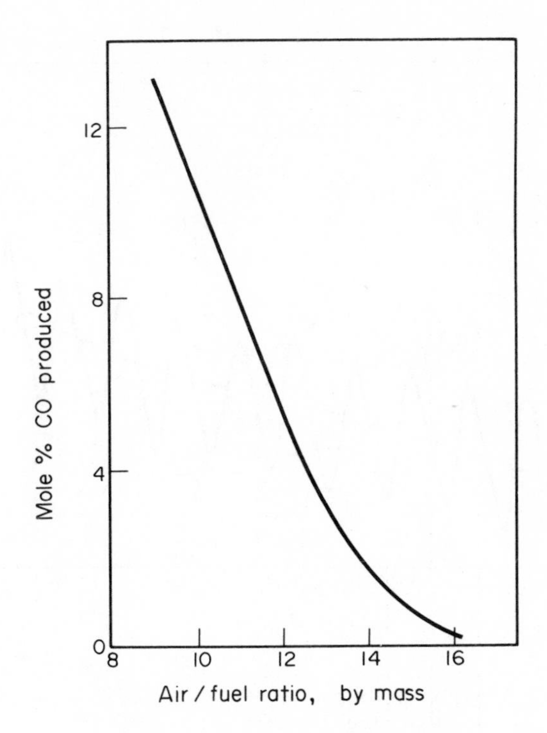

Fig 9.1. Exhaust CO concentration for different air-fuel ratios

Carbon dioxide

The atmospheric levels of CO_2 show a steady increase in concentration from around 1870. Measurements at Mauna Loa, Hawaii, an area away from heavy industrial concentrations, indicate a yearly increase of approximately 0.7 ppm (during 1958-65) and 0.5 ppm from 1965 (Fig. 9.2). An average increase of 0.6 ppm yr^{-1} (0.2% yr^{-1}) approximates to the amount of CO_2 formed from the burning of fossil fuels. Regular seasonal variations in the concentration of CO_2 correlates with photosynthetic activity, and in the Northern Hemisphere, the concentration peaks around April, and is lowest around September-October. It is believed that the major contribution to the seasonal variation is the photosynthetic activity of mid-latitude forests. Forest destruction could have a serious affect on the levels of atmospheric CO_2. Carbon dioxide undergoes photochemical reactions, producing CO at higher altitudes in the atmosphere.

$$CO_2 + \underset{(uv)}{h\nu} \rightarrow CO + O \qquad (9.9)$$

This is a significant source of CO at these levels.

The world's climate is dependent on the global heat balance, a mean temperature change of 2-3K would have a marked effect. Of the 8.4×10^4 J m^{-2} min^{-1} (1400 watt m^{-2}) energy that enters the earth's atmosphere, approximately

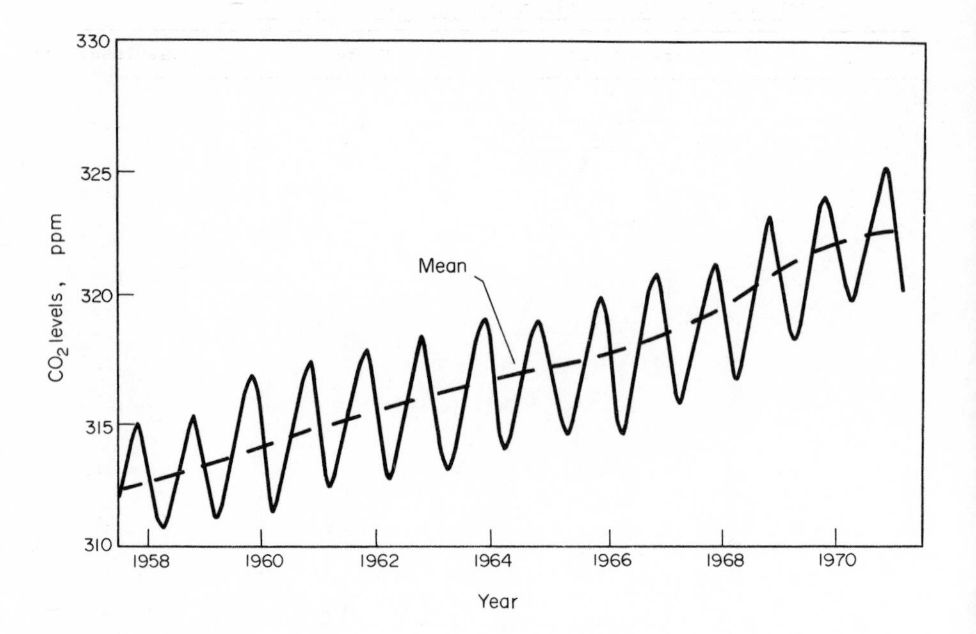

Fig. 9.2. Build up of CO_2 in the atmosphere, measured at Mauna Loa, Hawaii.

47% reaches the earth's surface and adjacent atmosphere, directly or after being scattered. The incoming ultraviolet, visible and infrared energy has a max intensity, at the earth's surface around 483 nm, while energy re-emitted from the earth is only in the infrared region (2000-40,000 nm) with a maximum intensity around 10,000 nm. If this energy was lost to the atmosphere the temperature of the earth's surface would be around -20° to $-40^{o}C$. However, some of the infrared is absorbed by water vapour and carbon dioxide in the air and re-emitted in all directions, some of which returns to the earth producing an average temperature of around $14^{o}C$.

Water has three fundamental vibration modes, ν_1, ν_2 and ν_3 (Table 9.4), while CO_2 has four modes, ν_1, ν_{2a}, ν_{2b} and ν_3 (modes ν_{2a} and ν_{2b} are degenerate). The strong infrared absorptions due to water occur below 8000 nm (and >20,000 nm) while the ν_2 absorption of CO_2 is the most intense and absorbs over the region 13,000-18,000 nm. This leaves an infrared exit window between 8000-13,000 nm.

The earth's mean temperature rose $0.4^{o}C$ from 1880 to 1940, and from 1945 has dropped around $0.1^{o}C$. The initial rise has been correlated with an increase in atmospheric concentration of CO_2 (Fig. 9.3). The effect has been called the "greenhouse effect", an increase in the CO_2 concentration means more infrared radiation is trapped and re-emitted back to the earth (Fig. 9.4), producing a build up of infrared radiation in the atmosphere and a mean global temperature rise. Estimates of the rise in temperature with increasing CO_2 concentration range from 0.1 to $4.9C^{o}$ with a mean around $2C^{o}$ for doubling the CO_2 concentration to 600 ppm. As there is a logarithmic

TABLE 9.4 Vibrational Modes of H_2O Vapour and CO_2

	Normal Modes	Symbol	Wavelength nm
H_2O	↑O, H H	ν_1	2730
	↑O, H↘ ↙H	ν_2	6300
	O→, ↙H ↖H	ν_3	2660
CO_2	← O — C — O →	ν_1	7400
	O↓ — C↑ — O↓	ν_{2a}	15,000
	O — C — O	ν_{2b}	
	← O — C →← O	ν_3	4300

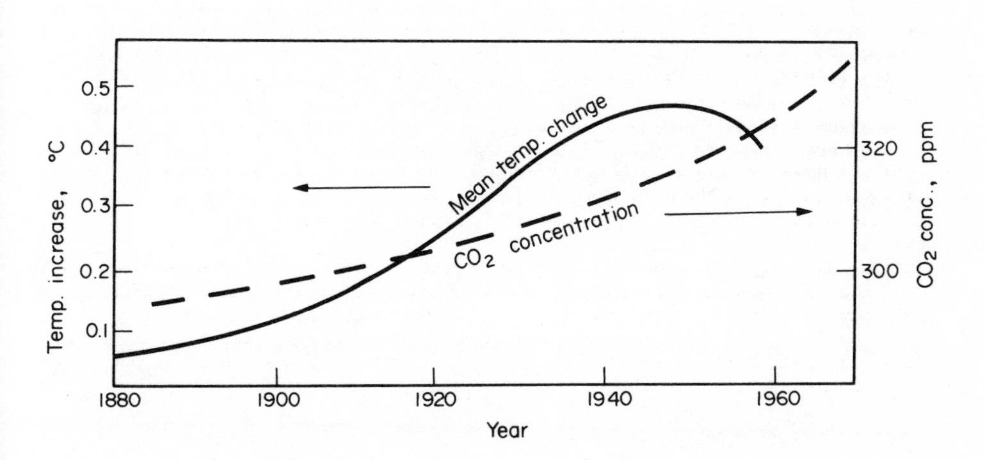

Fig. 9.3. Changes in mean world temperature and CO_2 levels in the atmosphere.

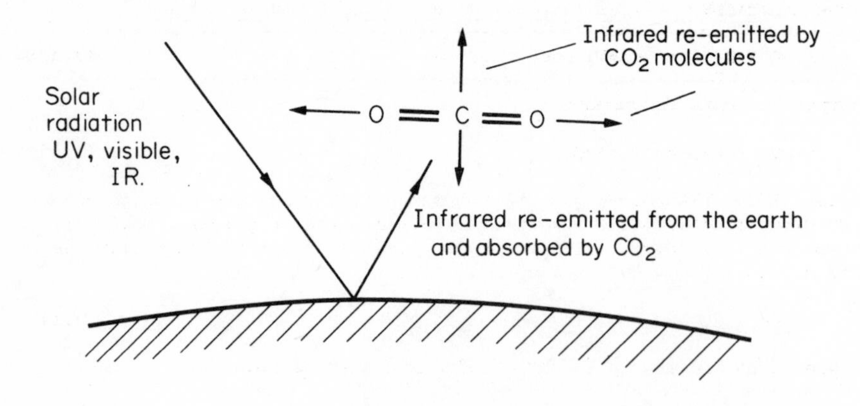

Fig. 9.4. A schematic drawing illustrating the greenhouse effect. Infrared radiation from the earth absorbed by CO_2 molecules is re-emitted in all directions returning some back to the earth's surface.

relationship between surface temperature and CO_2 concentration it is likely that a temperature will be reached, beyond which further increases of CO_2 levels have little effect. A maximum increase of 2.5K has been suggested. Since 1945 the mean global temperature has dropped suggesting that either the greenhouse effect is not a satisfactory explanation, or some other effect is beginning to dominate. Aerosols in the size range 0.1-5 μm, could reduce incoming radiation by 10% through back scattering and any increase could lead to cooling of the atmosphere. The rate of increase of atmospheric turbidity has been greater than the increase in CO_2 levels, and it could be that this is the more important factor today. It is not yet possible to adequately explain the change in the temperature over the last century, but at least an increase in the CO_2 and aerosol content of the atmosphere has the potential to alter the world's climate. Pollutants such as N_2O, CH_4 and CCl_3F, which have strong infrared absorptions, could also influence the mean global temperature.

Carbon Monoxide

Even though carbon monoxide is discharged into the atmosphere in large quantities the ambient, non-urban concentrations remain relatively constant, with a mean around 0.1-0.2 ppm (0.11-0.23 mg m^{-3}). This, and an atmospheric life-time of 1 to 4 months, suggests efficient sinks exist for the removal of CO, up to five methods are known. In the troposphere and upper stratosphere the rates of input and removal of CO are similar, while in the lower stratosphere rapid removal of CO occurs. The sharp fall off in CO concentration just above the tropopause is due to the reaction;

$$\cdot OH + CO \rightarrow CO_2 + \cdot H. \qquad (9.10)$$

The ·OH radicals come from photolysis of water;

$$H_2O + h\nu \rightarrow \cdot OH + \cdot H \qquad (9.11)$$

and the reaction;

$$\cdot O + H_2O \rightarrow 2\cdot OH. \quad (9.12)$$

The oxygen atom is produced by;

$$O_3 + h\nu \rightarrow O_2 + \cdot O \quad (9.13)$$

Reaction (9.10) also occurs in the troposphere, and may remove up to 50% of the CO discharged into the atmosphere. Perhaps the principal removal of tropospheric CO is bacterial and fungal processes occurring in soil. The reaction;

$$CO + \tfrac{1}{2}O_2 \rightarrow CO_2 \quad (9.14)$$

is achieved by *bacillus oligocarbophilus* and the reaction;

$$4CO + 2H_2O \rightarrow CH_4 + 3CO_2 \quad (9.15)$$

by *methanosarcina backerii*. The reaction;

$$CO + 3H_2 \rightarrow CH_4 + H_2O \quad (9.16)$$

can also occur. Sterilization of the soil inhibits the removal of CO. The uptake of CO is rapid, and varies from 7 to 115 mg CO hr^{-1} m^{-2} depending on the soil, tropical soils being the most active and desert soils the least. For an uptake rate of 19 mg CO hr^{-1} m^{-2}, and an area equal to that of the U.S.A., it is estimated that 1.3×10^9 tonnes of CO would be consumed per year. While an approximate figure, it is 3 to 4 times the annual global anthropogenic input of CO, therefore it would appear that soil bacteria may cope with CO emissions for a long time to come.

Two other chemical sinks for CO are;

$$CO + \cdot O + M \rightarrow CO_2 + M \quad (9.17)$$

and conversion of CO to CO_2 or amino acids at plant leaves.

$$CO + \text{plant leaves} \xrightarrow{\text{in daylight}} \text{amino acids} \quad (9.18)$$

$$CO + \text{plant leaves} \xrightarrow{\text{at night}} CO_2 \quad (9.19)$$

The number of efficient sinks suggest that in the long term CO pollution is not a problem. However, in the short term, levels can reach 100-300 ppm (115-344 mg m^{-3}) in congested streets which is a problem, because of the toxicity of CO to human beings.

The reaction (9.10) may also be important in photochemical smog formation, as the H atoms react further as follows;

$$\cdot H + O_2 + M \xrightarrow{\text{fast}} HO_2\cdot + M, \quad (9.20)$$

followed by;

$$HO_2\cdot + NO \rightarrow NO_2 + \cdot OH. \quad (9.21)$$

Hence CO may accelerate the oxidation of NO to NO_2, and thereby conserve ozone. However, the reaction is only of significance in the absence of hydrocarbons, or when the competing reaction;

$$\text{Hydrocarbon} + \cdot OH \rightarrow \text{products}, \qquad (9.22)$$

is slow, and even then the concentration of CO needs to be >100 ppm in order to have a noticeable effect on the rate of formation of photochemical smog.

The levels of CO vary between the hemispheres, between urban and rural areas and during the day in urban areas. The average level of CO in the northern hemisphere is around 0.14 ppm (maximum at $50^{o}N$) dropping to 0.04 ppm in the southern hemisphere (minimum at $50^{o}S$). The difference probably reflects greater inputs of CO in the north, and since the atmospheric life time of CO (1 to 4 months) is less than the mixing time between the hemispheres (one year), the CO is removed before it is transported from north to south. Levels of CO in urban areas can reach quite high values in congested traffic reaching 300 ppm at bad times. During a day in a busy street, the concentration of CO peaks twice, the early morning and late afternoon, corresponding to rush hour traffic (Fig. 9.5). Mean concentrations of 10 to 18 ppm over an eight hour day are common.

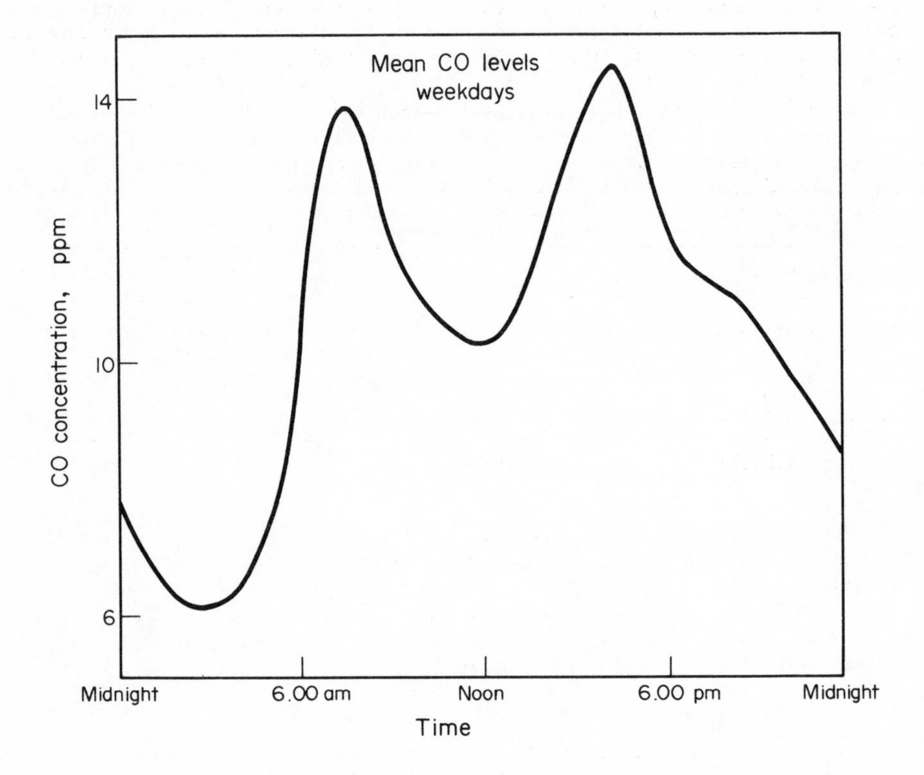

Fig. 9.5. Diurnal variation of mean CO concentrations in urban areas.

OXIDES OF SULPHUR

Gaseous sulphur dioxide and sulphur trioxide are serious pollutants, SO_2 is the primary pollutant and SO_3 a secondary product. Sulphur dioxide has a bent structure, with a lone pair of electrons on the sulphur and double bonding between the S and O. In the trioxide the lone pair is involved in bonding a third oxygen producing a planar molecule. The oxides are acidic, and with water give sulphurous and sulphuric acids.

$$SO_2 + H_2O \rightarrow H_2SO_3 \rightleftharpoons 2H^+ + SO_3^{2-} \quad (9.23)$$

$$SO_3 + H_2O \rightarrow H_2SO_4 \rightleftharpoons 2H^+ + SO_4^{2-} \quad (9.24)$$

Sulphur has a range of oxidation states which helps explain its varied environmental chemistry.

Sources

Around 146 x 10^6 tonnes annually, or 67% of the total SO_2 discharged into the atmosphere has anthropogenic origins. Around 70% of this amount comes from the combustion of coal and oil, mainly for producing electricity (Table 9.5). Over 90% of SO_2 pollution occurs in the northern hemisphere. Sulphur dioxide emissions have increased from 5 x 10^6 tonnes annually in 1860,to around 146 x 10^6 tonnes today, the greatest increase occurring since 1950. Some SO_3 is also produced, but only 1 to 2% of the total sulphur oxide emissions.

TABLE 9.5 Principal Sources of Sulphur Dioxide

Source	Percentage of total*	
Coal Combustion	70%	(2/3 in electrical production)
Oil Combustion and Refining	20%	(3/4 of residual oil in electric power production)
Cu, Zn, Pb smelting and other	10%	

* Approximately 1 to 2% of emissions are SO_3

The amount of sulphur in coal varies from 0 to 6% (by weight), and depends on the type of coal, bituminous coal has the highest sulphur content. Low sulphur coal has 0-1% S and high sulphur coal has >3% S. The sulphur is converted to SO_2 on combustion, according to the reactions;

$$\geqslant C\text{-}SH + O_2 \rightarrow SO_2 + CO_2/CO + H_2O \quad (9.25)$$

and

$$4FeS_2 + 11O_2 \rightarrow 2Fe_2O_3 + 8SO_2 \quad (9.26)$$

Less sulphur occurs in oil, but it is concentrated in the residual oil (fuel oil) on refining. The sulphur content varies from 0.75 to 3%, while in petrol the sulphur content is around 0.05%. Some emission factors for SO_2 from different industries are given in Table 9.6.

TABLE 9.6 Sulphur Dioxide Emission Factors

Source	Factor
Coal combustion	20% S kg per tonne coal*
Fuel oil combustion	16% S kg per 1000 litre oil*
Waste combustion	0.5 - 1 kg per tonne refuse
H_2SO_4 manufacture	9-32 kg per tonne acid (100%)
Cu smelting	635 kg per tonne conc. ore
Pb smelting	300 kg per tonne conc. ore
Zn smelting	495 kg per tonne conc. ore
Kraft mill (recovery furnace)	1 - 6 kg per tonne air-dried pulp
Sulphite mill (recovery furnace)	18 kg per tonne air-dried pulp.

* % S = % of sulphur in coal or oil.

Concentration in the Air

The ambient concentration of sulphur dioxide in the air, well removed from pollution sources, is 0.1 to 0.3 ppb (0.26 to 0.78 $\mu g\ m^{-3}$) in the northern hemisphere, and a mean of 0.1 ppb in the southern hemisphere. Figures as high as 1 ppb have been reported at places as widely separated as Panama and the Antartica. In heavily industrialized areas the concentration is orders of magnitude greater, for example in Chicago (1967) the levels of SO_2 were: hourly 1.11 ppm, daily 0.65 ppm, and annually 0.12 ppm. Near point sources the levels can be even higher. Sulphur dioxide levels in cities display seasonal variation, maximum levels in the winter and minimum in the summer. This correlates with coal and oil consumption. The life-time of SO_2 in the atmosphere is short, ranging from 3 to 7 days, depending on weather conditions.

Oxidation of Sulphur Dioxide

The foremost chemical reaction of SO_2 in the atmosphere is oxidation to SO_3, which with water gives sulphuric acid. The acid or sulphates, in particular $(NH_4)_2SO_4$ and NH_4HSO_4, occur as aerosols. The process can be represented by;

$$SO_2 \xrightarrow{\text{oxidation}} SO_3 \xrightarrow{H_2O} \underbrace{H_2SO_4, (NH_4)_2SO_4, NH_4HSO_4}_{\text{aerosols}} \qquad (9.27)$$

The oxidation step is the least well understood, but appears to proceed in one or more of three ways; catalytic oxidation or photooxidation or oxidation by radicals. In order for reactions (9.27) to occur a number of environmental factors need to be favourable, including the humidity, temperature, light intensity, aerosol levels and material transport within the atmosphere.

Catalysis. The reaction;

$$2SO_2 + 2H_2O + O_2 \rightarrow 2H_2SO_4, \quad \Delta H = -600\ kJ, \qquad (9.28)$$

is slow in clean air, but is catalysed by aerosols containing metal ions. Catalysts that increase the rate 10 to 100 times are salts of Mn(II),

Fe(III) and Cu(II), and oxides of Cr(III), Al(III), Pb(II) and Ca(II). Surfaces, such as on buildings, may also act as catalytic centres. The reaction rate is greatest in humidity >32%, and especially at high humidities >70%. The SO_2 is oxidised in water droplets, in the presence of heterogenous catalysts in the droplet. Increase in acidity inhibits the reaction, which proceeds best in neutral or alkaline solutions. Therefore since the reaction produces H_2SO_4 it slows down as acid accumulates. The formation of HSO_3^- and SO_3^{2-} (from SO_2 in water) are suppressed by increasing acidity driving reaction (9.28) to the left, reducing the solubility of SO_2 in the droplet, and therefore the reaction rate. The solubility of SO_2 in water is a function of pH and can be described by the equilibria;

$$SO_{2(g)} + H_2O \rightleftharpoons SO_{2(aq)}, \tag{9.29}$$

$$SO_{2(aq)} + H_2O \rightleftharpoons H_2SO_{3(aq)}, \tag{9.30}$$

$$H_2SO_3 + H_2O \rightleftharpoons H_3O^+ + HSO_3^-, \tag{9.31}$$

$$HSO_3^- + H_2O \rightleftharpoons H_3O^+ + SO_3^{2-}, \tag{9.32}$$

$$2HSO_3^- \rightleftharpoons S_2O_5^{2-} + H_2O. \tag{9.33}$$

The sulphur species in water are either covalent ($SO_{2(aq)}$ and H_2SO_3), or ionic (HSO_3^-, SO_3^{2-} and $S_2O_5^{2-}$), the latter being more soluble because of the polarity of water. In the equilibria (9.31) and (9.32) the H_3O^+ occurs on the right and therefore its removal favours the ionic species. This agrees with the fact that below pH 4, SO_2 solubility drops markedly, while it rises above pH 4. Addition of ammonia promotes the reaction rate by reducing the acidity and forming $(NH_4)_2SO_4$ and NH_4HSO_4, the main sulphate aerosols.

The reaction mechanism for equation (9.28) probably involves oxidation of sulphite species rather than SO_2;

$$HSO_3^- + H_2O - 2e \rightarrow HSO_4^- + 2H^+ \tag{9.34}$$

$$O_2 + 4H^+ + 4e \rightarrow 2H_2O \tag{9.35}$$

i.e.

$$2HSO_3^- + O_2 \rightarrow 2HSO_4^- \tag{9.36}$$

The metal catalyst may act as an oxidising agent, for example Cu^{2+} would be reduced to Cu^+ which is reoxidised by dioxygen. Alternatively a coordination catalyst may be involved. For example the Mn^{2+} ion could effect the oxidation;

$$Mn^{2+} + SO_2 \rightarrow [Mn - SO_2]^{2+}, \tag{9.37}$$

$$2[Mn - SO_2]^{2+} + O_2 \rightarrow [Mn - SO_2]_2^{2+}O_2 \rightarrow 2[Mn - SO_3]^{2+} \tag{9.38}$$

$$[Mn - SO_3]^{2+} + H_2O \rightarrow Mn^{2+} + HSO_4^- + H^+, \tag{9.39}$$

Photochemical oxidation. Sulphur dioxide absorbs radiation producing excited states;

$$SO_2(^1A_1) + h\nu \rightarrow SO_2(^1A_2, ^1B_1) \quad (9.40)$$

240 - 330 nm (intense absorption) ↓ intercrossing

$$SO_2(^1A_1) + h\nu \rightarrow SO_2(^3B_1) \quad (9.41)$$

340-400nm (weak absorption)

The triplet state has a relatively long lifetime, and may react by a variety of routes to give SO_3, for example;

$$SO_3(^3B_1) + O_2 \rightarrow SO_3 + O(^3P) \quad (9.42)$$

$$SO_2(^3B_1) + SO_2 \rightarrow SO_3 + SO \quad (9.43)$$

and

$$SO + SO_2 \rightarrow SO_3 + S \quad (9.44)$$

For levels of SO_2 of 5-30 ppm, reaction (9.42) achieves around 0.1 - 0.2% conversion to SO_3 per hour.

Oxidation by radicals. Hydrocarbons and nitrogen dioxide both increase the rate of oxidation of SO_2, as do radicals and peroxy compounds. These materials are either components of automobile emissions or secondary products. The following reactions are feasible routes to the oxidation of SO_2.

$$SO_2 + O_3 \rightarrow SO_3 + O_2 \quad (9.45)$$

This reaction is slow in the gas phase, but rapid in solution (water droplet). The reaction;

$$SO_2 + NO_2 \rightarrow SO_3 + NO \quad (9.46)$$

may have a similar mechanism to the reactions producing SO_3 by the Lead Chamber Process (Chapter 5). The reverse reaction is also possible in the atmosphere.

Other feasible reactions are;

$$SO_2 + HO_2\cdot \rightarrow SO_3 + HO\cdot, \quad (9.47)$$

$$SO_2 + RO_2\cdot \rightarrow SO_3 + RO\cdot, \ (R{=}CH_3C{=}O), \quad (9.48)$$

$$HO\cdot + SO_2 \rightarrow HOSO_2\cdot, \quad (9.49)$$

$$HOSO_2\cdot + O_2 \rightarrow HOSO_2O_2\cdot, \quad (9.50)$$

$$HOSO_2O_2\cdot + NO \rightarrow HOSO_3 + NO_2 \quad (9.51)$$

$$HOSO_3 \downarrow H_2SO_4$$

$$RH + SO_2(^3B_1) \rightarrow RSO_2H \rightarrow H_2SO_4 + RH, \quad (9.52)$$

sulphinic acid

A number of reactions are available for the oxidation of SO_2 to SO_3, which occurs extensively in the atmosphere as indicated by the presence of H_2SO_4 aerosol.

Acid Rain

Acid rain is produced by the dissolving of SO_2 in water to give H_2SO_3 and subsequently H_2SO_4 as outlined above; as well as by the two reactions;

$$H_2SO_4 + 2NaCl \rightarrow Na_2SO_4 + 2H^+ + 2Cl^- \tag{9.53}$$

and

$$2NH_4^+ + SO_4^{2-} + 2H_2O \rightarrow 2NH_3 + 2H_3O^+ + SO_4^{2-} \tag{9.54}$$

Relatively pure rain water has a pH around 5.5-5.7, but owing to SO_2 emissions the pH of rain can drop as low as 2. This increases the acidity of waterways, in particular lakes. Some Scandanavian lakes receiving rain, in which the H_2SO_4 aerosol originated in Europe and the U.K. have a pH of 5.5 to 4. Three quarters of the sulphur precipitated in Scandanavia originates from Europe and the U.K. In addition to increasing acidity of lakes and affecting aquatic life, acid rain accelerates leaching of nutrients from soils, affects the metabolism of soil organisms, increases metallic corrosion and destruction of basic building materials such as limestone and marble.

The acidity of natural water systems also increases by the dissolution of SO_2. Approximately 4 to 8 x 10^7 tonnes of SO_2 dissolve in the oceans annually.

Smog

In the majority of the episodes where smog has caused serious harm (Table 8.2), SO_2 has been involved. However, water and particulate material are also necessary, the three factors together cause a more serious situation than individually, this is called the synergistic effect. The particles act as centres for, nucleation of water droplets, and catalytic oxidation of SO_2 to SO_3, (which can be quite rapid on the large surface area of the small particles). In the small water droplets H_2SO_4 is formed, which can penetrate the respiratory system producing severe distress and even death.

The Effects of Sulphur Dioxide and Sulphuric Acid Aerosol

Sulphur dioxide in association with particulate material and at high humidities can have serious effect on human beings, plants and materials. Some of these effects, and the levels of SO_2 producing them are summarized in Table 9.7.

Human beings. Controlled, short-term inhalation experiments have shown that a slight effect on the respiratory system occurs at 0.75 ppm (2.1 mg m^{-3}) SO_2, but no effect at 0.37 ppm. For H_2SO_4 aerosol (depending on particle size) levels as low as 0.13 ppm have respiratory effects. The SO_2 is taken into the respiratory tract by inhalation, either as H_2SO_4 aerosol, or as SO_2 absorbed onto particulates, which is then converted to H_2SO_4 in the lungs. Either way, it appears that H_2SO_4 is the principal damaging agent. Health effects are; aggravation of respiratory diseases including asthma, chronic bronchitis and emphysema, reduced lung function, irritation to eyes and

TABLE 9.7 Effect of SO_2/SO_3 on Humans, Plants and Animals

Subject	Effect	Single Concentration	Average Concentration	Exposure	Comments
Humans	Odour	0.5-0.7ppm		1 sec	
	Taste	0.3-0.1ppm		few secs	
	Epidemio-logical significance	0.2ppm	0.015ppm	24 hrs (annual average)	
	Pulmonary Function	1.6ppm		10 min.	SO_2/SO_3 + particulates aggravate effects
	Discomfort	5ppm		10 min sensitive subjects	
	Severe distress	5-10ppm			
Plants	Plant leaf symptons		0.28ppm	24hrs	Bleached spots on edge & area between leaf veins
	Plant Chlorosis	>0.25ppm	0.03ppm	Annual average	
	Plant growth altered		0.05-0.2ppm	24 hrs for growing season	Growth suppression, loss in yield
Animals	Lethal (Rabbit)		50ppm	30 days 6 hr day	
	Central Nervous System	0.2ppm		10 sec several times	

Source: Bond, R.G. and Straub, C.P., Handbook Envir. Control, I, Air Pollution, (1972) CRC Press, Cleveland.

respiratory tract and death. The relationship between health effects, exposure time and concentration of SO_2 are illustrated in Fig. 9.6. Sulphur dioxide stimulates contraction of airways at 2-5 ppm, which becomes quite severe at levels >5 ppm, making it very difficult to breath. The air pollution episode in London, U.K., December 5-8, 1952, produced 3500-4000 excess deaths. The relationship between smoke, SO_2 levels and deaths is clearly shown in Fig. 9.7, as is the levels of pollutants and atmospheric temperature. The temperature inversion lifted on December 8th to 9th, corresponding to a rise in the atmospheric temperature.

Plants. Sulphur dioxide destroys leaf tissue (Table 9.7), and generally, long term low-level exposure is more serious than short-term high exposures. Yields were reduced in controlled experiments, 50% loss occurred for long exposures (3 days per week) of SO_2 at 0.15 ppm, but affects were negligible for 3 hours per week exposure at 1.20 ppm. Alfalfa, barley, cotton and wheat are all sensitive to SO_2. The oxide may inhibit photorespiration as glyoxalate hydrogensulphite, $HC(OH)(COO^-)(SO_3^-)$ has been isolated from rice

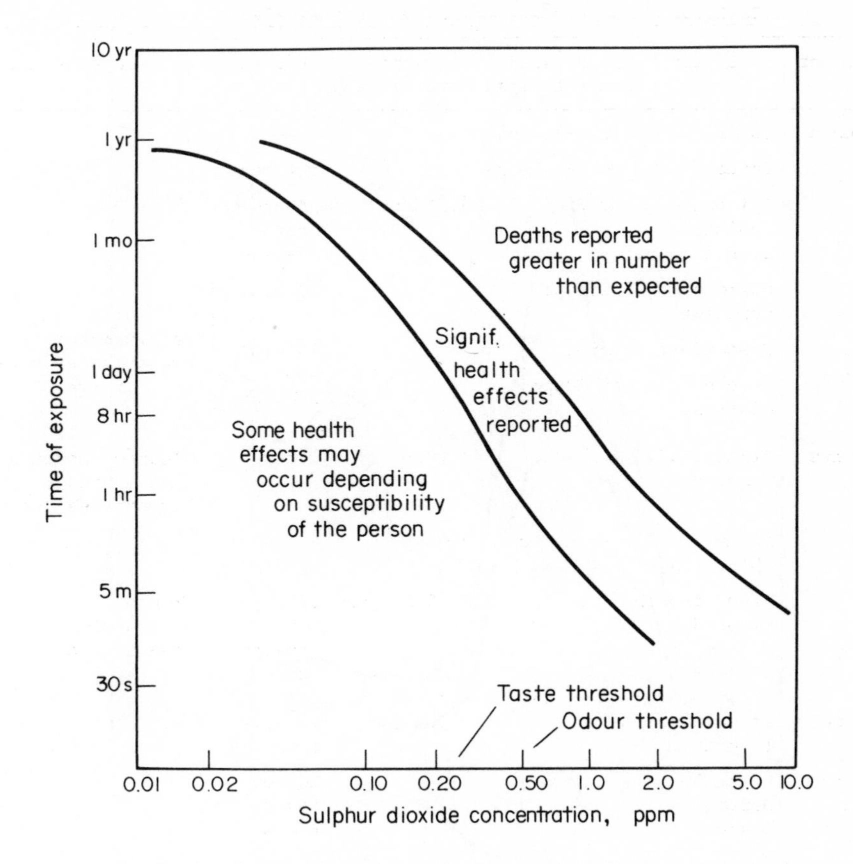

Fig. 9.6. A summary of health effects of SO_2 in relation to concentration and time (NB: curves for individuals will differ according to their susceptibility to SO_2 poisoning (see Fig. 13.2).

exposed to SO_2. The hydrogensulphite probably inhibits the enzyme *glycolic oxidase* involved in photorespiration reactions.

Materials. Both SO_2 and H_2SO_4 speed up the deterioration of building materials such as carbonate stones (limestone, marble)

$$CaCO_3 + H_2SO_4 \rightarrow CaSO_4 + H_2O + CO_2 \tag{9.55}$$

$$CaCO_3 + SO_2 + 1/2O_2 \rightarrow CaSO_4 + CO_2 \tag{9.56}$$

The $CaSO_4$ is more soluble than $CaCO_3$ and eventually washes out. But it can dissolve in moisture in crevices in the stone work and then recrystallise when the water evaporates. Since the molar volume of $CaSO_4$ is greater than for $CaCO_3$ stress is produced in the stone causing it to break. The rate of corrosion of iron is increased by as much as 50% in the presence of 0.1ppm of SO_2 and particulate matter. Ferrous sulphate is produced;

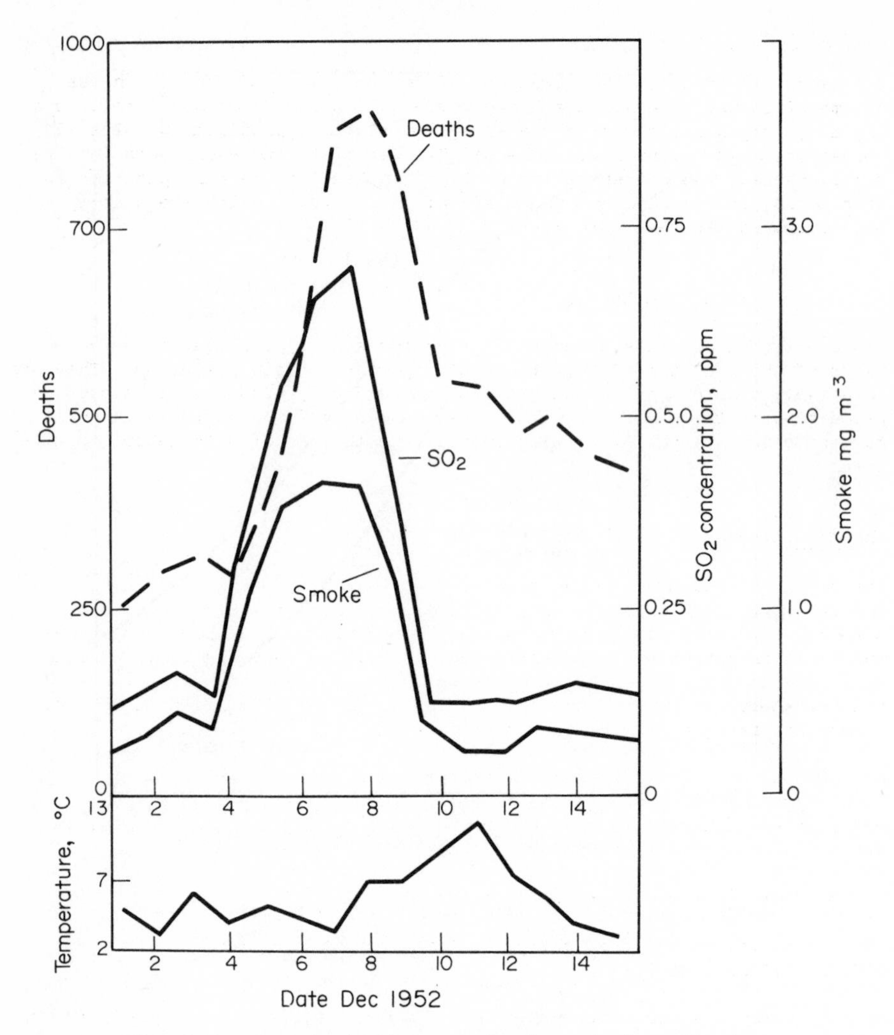

Fig. 9.7. Correlation between deaths and air pollution in London, U.K., December 1952.

$$Fe^{2+} + O_2 + SO_2 \rightarrow FeSO_4, \qquad (9.57)$$

which provides ionic conduction in the electrochemical corrosion process. For other metals, such as Zn, Cu and Al, sulphuric acid dissolves the various protective corrosion layers and the sulphates produced are dissolved away. Sulphur dioxide destroys paper and leather, making the materials brittle, in the case of paper, acid hydrolysis of cellulose is a factor in the breakdown.

Sulphates

Sulphates are added to the atmosphere from sea spray and dust which has become airborne. The average troposphere concentration of SO_4^{2-} is around 2μg m^{-3}, sulphate added from sea-spray amounts to approximately 132 x 10^6 tonnes (as SO_4^{2-}) annually. Some of this material, and H_2SO_4 gets into the stratosphere. The concentration of SO_4^{2-} is high in the precipitation in arid areas, presumably from the soils of these areas, which are high in SO_4^{2-}, and have become airborne.

OXIDES OF NITROGEN

Since nitrogen exhibits a range of oxidation states, a number of oxides exist of which NO and NO_2 are the most important in air pollution. They are relatively stable and are the primary and secondary products of combustion processes. A third oxide N_2O, produced by denitrifying bacteria in soils may increase as a pollutant with increase in the use of nitrogenous fertilizers.

The Chemistry of Nitrogen Oxides

The two major features of the chemistry of nitrogen in the environment are the stability and low reactivity of dinitrogen, and the wide range of nitrogen oxidation states. The oxidation states, typical compounds, some of which are important in the environment, are listed in Table 9.8. Gaseous nitrogen compounds occur under both oxidising and reducing conditions. The reduction potentials for some nitrogen oxy-compounds are given in Fig. 9.8, from which it is clear that the stability of N_2 is a dominant factor.

Even though N_2 is thermodynamically stable, and has low reactivity it will, with oxygen under the right conditions, produce NO and eventually NO_2. The temperatures achieved in the internal combustion engine lead to reasonable amounts of NO. The free energy of the equilibrium reaction;

$$N_{2(g)} + O_{2(g)} \rightarrow 2NO_{(g)} \quad (9.58)$$

is +173 kJ (twice the free energy of formation of NO). From log K = $-\Delta G/2.303RT$ one obtains;

$$\log K = -30.3 \text{ at } 298K, \quad (9.59)$$

indicating the equilibrium lies largely to the left. If air is the reaction mixture the amount of NO produced can be calculated as follows;

$$\log K = \frac{P^2_{NO}}{P_{O_2} P_{N_2}}, \quad (9.60)$$

$$\text{therefore } P^2_{NO} = 10^{-30.3} \times 20.3 \times 81, \quad (9.61)$$

where 20.3 and 81 are the pressures in kPa of O_2 and N_2 in the air respectively.

$$\text{Therefore } P_{NO} = 2.9 \times 10^{-14} \text{ kPa} \quad (9.62)$$

TABLE 9.8 Oxidation States of Nitrogen

Oxidation State	Example	Importance in the Environment
+V	NO_3^-, HNO_3	Important fertilizer Produced in atmosphere from NO_2
+IV	$NO_2 \rightleftharpoons N_2O_4$	Serious air pollutant in urban areas
+III	NO_2^-, HNO_2	Intermediate in nitrogen cycle in soil, toxic to infants.
+II	NO	Serious air pollutant in urban areas
+I	N_2O	Product of denitrifying bacteria, gets into statosphere.
O	N_2	Major source of nitrogen, main component of air
-I	NH_2OH	Intermediates in nitrogen fixation?
-II	N_2H_4	
-III	NH_3, NH_4^+	Important ferilizer, product of nitrogen fixation.

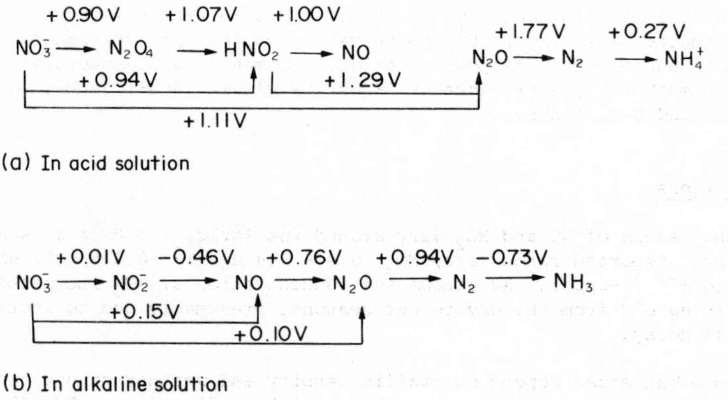

$$NO_3^- \xrightarrow{+0.90\,V} N_2O_4 \xrightarrow{+1.07\,V} HNO_2 \xrightarrow{+1.00\,V} NO \qquad N_2O \xrightarrow{+1.77\,V} N_2 \xrightarrow{+0.27\,V} NH_4^+$$

$NO_3^- \rightarrow HNO_2$: +0.94 V; $NO \rightarrow N_2O$: +1.29 V; $NO_3^- \rightarrow N_2O$: +1.11 V

(a) In acid solution

$$NO_3^- \xrightarrow{+0.01\,V} NO_2^- \xrightarrow{-0.46\,V} NO \xrightarrow{+0.76\,V} N_2O \xrightarrow{+0.94\,V} N_2 \xrightarrow{-0.73\,V} NH_3$$

$NO_3^- \rightarrow NO$: +0.15 V; $NO_3^- \rightarrow N_2O$: +0.10 V

(b) In alkaline solution

Fig. 9.8. Reduction potentials for nitrogen-oxygen species

However, at 2000 K, log K = -4.5 hence;

$$P^2_{NO} = 10^{-4.5} \times 20.3 \times 81 \qquad (9.63)$$

i.e. $$P_{NO} = 0.23 \text{ kPa}. \qquad (9.64)$$

At the high temperature considerably more NO is produced, and with rapid cooling of the exhaust gases the NO can be trapped before decomposition.

The two species NO and NO_2 are odd electron compounds which, explains in part, their reactivity, especially of NO_2. The dioxide dimerises;

$$2NO_2 \rightleftharpoons N_2O_4, \qquad (9.65)$$

N_2O_4 predominating at low temperatures. Significant amounts exist at ambient temperatures, but in what follows we will use the formulation NO_2.

Sources

Estimates of the annual emissions of nitrogen oxides from both natural and man-made sources are listed in Table 8.3. Natural sources are mainly from lightning, volcanic and bacterial activity. A breakdown of the global emissions is given in Table 9.9, amounts from anthropogenic sources are much less than from natural, but are concentrated in urban areas where high levels can be attained. Approximately 95% of the global emissions are in the northern hemisphere.

The principal source from combustion is coal, producing 51% of NO_x (approximately 90-95% as NO), and combustion of petrol accounts for around 14%, depending on the way the fuel is used. The production of NO_x from fuels increases in the order; natural gas < petroleum < coal, this correlates with an increase in the temperature necessary for combustion.

In the U.S.A. (1970), nitrogen oxides accounted for 43% of the total gaseous emissions, 51% of this came from transport sources, approximately 11.7×10^6 tonnes annually, while 10×10^6 tonnes were produced from stationary combustion sources, such as power stations. Domestic and commercial sources contribute much less, approximately 2% of the global figure, and 5 to 6% of the U.S.A. and U.K. figures.

Concentrations

Background levels of NO and NO_2 vary around the world, and have seasonal variations. Reported ranges are; NO_2, 0.4 - 9.4 $\mu g\ m^{-3}$ (0.2-5ppb), NO, 0 - 7.4 $\mu g\ m^{-3}$ (0-6ppb). At Panama the concentration of NO_2 increased from 1.7 to 618 $\mu g\ m^{-3}$ from the dry to wet seasons, presumably due to increased biological decay.

Levels in urban areas depend on traffic density and weather conditions, and for large cities the annual mean can be as high as 50 - 95 $\mu g\ m^{-3}$ (40-77ppb). With increased use of the motor-car, levels have risen over the years, at Frankfurt-on-Main a mean annual concentration of 19 $\mu g\ m^{-3}$ in 1962 had increased to 80 $\mu g\ m^{-3}$ in 1971. Urban levels may be expressed for different time intervals, which is necessary to give a complete picture, for example during 1975 in Tokyo; annual mean, 86; maximum hourly, 840; maximum 24 hour

TABLE 9.9 Estimated Global Emissions of Nitrogen Oxides (1970)

Oxide	Source	Emissions X 10^6 tonnes (annually)
NO_x ($NO + NO_2$) estimated as NO_2	Coal combustion*	26.9
	Petroleum refining	0.7
	Petrol combustion	7.5
	Oil combustion	14.1
	Natural gas combustion	2.1
	Other combustion	1.6
	Total	52.9
NO	Biological activity	500
N_2O	Biological activity	592

*Power generation 12.2X10^6, Industrial uses 13.7X10^6, Domestic and commercial 1.0X10^6 tonnes/year. Source: Robinson, E. and Robbins, R.C., J. Air Pollut. Control Ass.,20,303 (1970).

mean, 426; and maximum monthly mean, 105 $\mu g(NO_x)m^{-3}$.

The nitrogen oxide levels can also be expressed as the number of days over a period, above a certain level. During July - September, 1975, in Los Angeles, more than 50% of the days had NO_x levels between 200 - 4000 $\mu g\ m^{-3}$.

Urban levels also follow seasonal variations, and in Los Angeles and Tokyo the concentration of NO_x is 2 to 3 times higher in the winter. Daily variations on the other hand, are principally influenced by traffic density with early morning and late afternoon peaks. The concentration peaks for NO_2 lag behind that of NO since NO_2 is formed from NO.

High levels (such as 2000 $\mu g\ m^{-3}$, or 1.1 ppm) of nitrogen oxides have been reported inside houses, due to the use of space heaters and gas cookers. Cigarette smoking is also a source of NO_x and between 160 - 500 μg of NO can be produced per cigarette, while levels in the smoke can reach 80 - 110 ppm of NO, and 80 - 120 ppm of NO_2.

Photochemical Smog

An area where nitrogen oxides are of major significance, is in the formation of photochemical smog. The smog is restricted mainly to the zone between 30° and 60° N. The products, which appear as a white-yellow haze, are respiratory irritants with long term health effects. Evidence of the smog was first found from plant damage in 1944, in Los Angeles, U.S.A.

Smog Conditions. Four conditions for photochemical smog are, a stable temperature inversion, adequate sunlight, an enclosed topography and certain chemical inputs. These conditions commonly occur together in Los Angeles. Los Angeles has a stable subsidence temperature inversion, between 200 - 1000 m high, for approximately 80% of the summer, and the mid-summer sun's zenith angle is 10°, so there is minimum loss of energy from passage through the atmosphere. The topography of Los Angeles is such that the sea-breeze is light and the surrounding hills provide a basin which is like a reaction flask, into which around 4 million cars push over 10,000 tonnes of exhaust gases per day. The inputs from automobiles and other sources are given in Table 9.10, 68% of the nitrogen oxide emissions arise from the motor-car.

TABLE 9.10 Daily Air Pollution Inputs into Los Angeles (tonnes day^{-1})

	NO_x	SO_2	CO	Hydro-carbons	Particles	Total
From automobiles	645	30	9470	1730	45	11920
From other sources	305[a]	195	220	820[b]	65	1605
% from automobiles	68	13	98	68	41	88

[a] From other fuel sources. [b] Other sources include petroleum and solvents.

Observations. Some of the primary and secondary chemical pollutants in photochemical smog are NO, NO_2, hydrocarbons, O_3 and aldehydes. The variation in the concentration of each of these materials over a day is presented in Fig. 9.9. The concentration of the primary pollutants NO and hydrocarbons rise during the early morning rush hour, then the concentration of NO_2 builds up as NO is oxidised. The NO_2 concentration falls as the sun-light intensity increases, due to NO_2 photolysis. This corresponds to significant increases in ozone and aldehydes, and a decrease in hydrocarbons which are involved in the formation of aldehydes and other materials. An increase in NO, NO_2 and hydrocarbon in the late afternoon does not eventuate, because they are removed as produced by the ozone in the atmosphere, leading to a drop in its concentration. A suggested mechanism must account for these observations.

Chemistry. Some possible chemical reactions involved in photochemical smog are listed in Table 9.11. A number of these have been studied in detail, but as the exact stoichiometry of photochemical smog is not known many of the reactions are illustrative of the chemistry involved. Smog formation may be broken up into; production of NO, production of NO_2, formation of O_3 and chemical reactions involving hydrocarbons.

The formation of NO from N_2 and O_2 in the car engine has been mentioned above. Formation of hydroxyl (.OH) radicals, reaction (5) (Table 9.11) and subsequent reactions (6 to 8) are also important in producing NO, and in atmospheric chemistry.

The oxidation of NO to NO_2 is termolecular, and slow at atmospheric concentrations. At a NO concentration of 1000 ppm reaction (9) occurs in seconds, but for 1 ppm of NO the half-life is 100 hours, and is 1000 hours for 0.1 ppm NO. Therefore the rapid loss of NO (Fig. 9.9) suggests other methods of

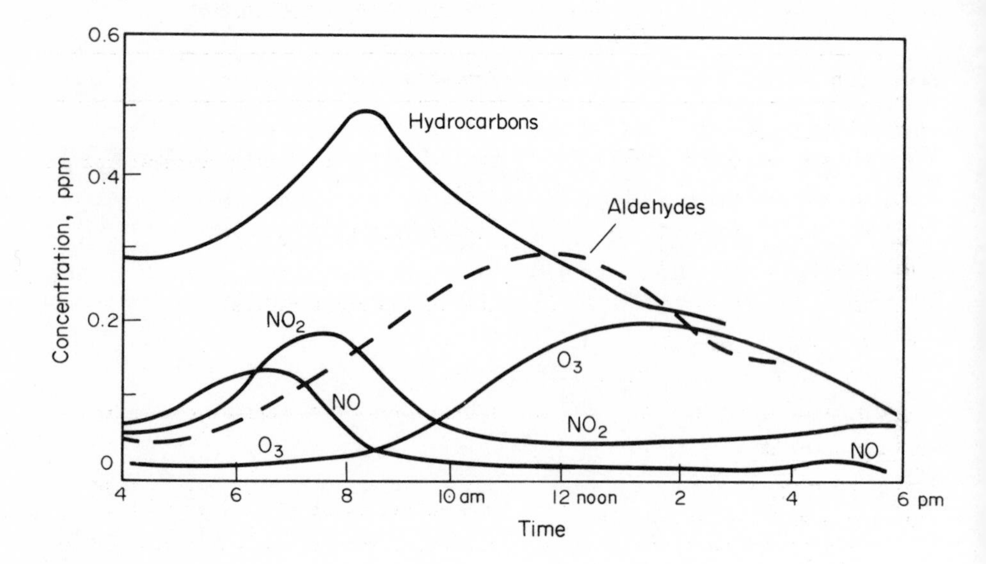

Fig. 9.9. Variation in some atmospheric constituents during days of photochemical smog in Los Angeles, U.S.A.

oxidation including reactions (10, 13, 14, 26, 33 and 38) in Table 9.11. For example, 0.1 ppm NO in the presence of O_3 (0.1 ppm) is oxidised in 18 seconds. Many of the oxidants are free radicals, which in low concentrations have half-lives of minutes to hours.

The photochemical step in the smog formation is reaction (11), light of wavelength < 395 nm is sufficient to break a N-O bond (300 kJ). The next step, formation of ozone, reaction (12), is a consequence of the production of oxygen atoms in reaction (11). However, consideration of reactions (11), (12) and (13) suggests that the concentrations of NO, NO_2 and O_3 should reach a steady state, but the concentration of ozone increases at the expense of NO_2, i.e. O_3 is conserved in some manner. This would occur by the other oxidations of NO listed above. While background levels of O_3 are around 0.005 - 0.05 ppm (hourly), in dense urban areas the highest hourly range can reach 0.15 - 0.4 ppm.

The production of other oxidants, which affect the respiratory system and are eye irritants, are initiated by reactions of O, O_3 and NO_2 with hydrocarbons emitted into the air. The hydrocarbon composition of Los Angeles atmosphere (1967) was 53% paraffin, 16% olefin, 20% alkyl benzenes and 11% acetylene. Of these materials, the olefins, followed by substituted benzenes, are the most reactive. Reactions from (23) onwards are based on olefins. The formation of free radicals lead to peroxy-compounds, aldehydes, alcohols and ketones (Table 9.11). Peroxy-compounds, such as peroxymethylnitrate, CH_3-CO-O-O-NO_2, and peroxybenzylnitrate, C_6H_5-CO-O-O-NO_2, are produced which have serious effects on plants, health and materials.

TABLE 9.11 Some Chemical Processes in Photochemical Smog

Reaction	Comments
Production of NO	
1. $O_2 + M \rightleftharpoons O + O + M$	Reactions occur at high temperatures in flames
2. $N_2 + M \rightleftharpoons N + N + M$	
3. $O + N_2 \rightleftharpoons NO + N$	Effectively $N_2 + O_2 \rightleftharpoons 2NO$
4. $N + O_2 \rightleftharpoons NO + O$	
5. $RH + O \rightarrow R. + .OH$	.OH an important radical
6. $N + .OH \rightarrow NO + H.$	
7. $CO + .OH \rightarrow CO_2 + H.$	CO sink
8. $H. + O_2 \rightarrow .OH + O$	Main source of O atoms in combustion
Oxidation of NO	
9. $2NO + O_2 \rightleftharpoons 2NO_2$	Termolecular, only feasible at high concentration of NO (car exhaust?).
10. $NO + HO_2. \rightarrow NO_2 + .OH$ plus other reactions given below	Important reaction for oxidation of NO
Photolysis of NO_2	
11. $NO_2 + h\nu \rightarrow NO(^2\pi) + O(^3P_2)$	$\lambda < 395$ nm, breaks NO bond of 300 kJ
Production of O_3 and other Inorganic Species	
12. $O(^3P_2) + O_2 + M \rightarrow O_3 + M$	Fast reaction
13. $NO + O_3 \rightarrow NO_2 + O_2$	Fast reaction
14. $O + NO + M \rightarrow NO_2 + M$	
15. $O + NO_2 \rightarrow NO + O_2$	
16. $O + NO_2 + M \rightarrow NO_3 + M$	
17. $O_3 + NO_2 \rightarrow NO_3 + O_2$	
18. $NO + NO_3 \rightarrow 2NO_2$	
19. $NO_2 + NO_3 \rightarrow N_2O_5$	Very temperature dependant
20. $N_2O_5 + H_2O \rightarrow 2HNO_3$	Assisted by aerosols and surfaces
21. $NO + NO_2 + H_2O \rightarrow 2HNO_2$	
22. $HNO_2 + h\nu \rightarrow HO. + NO$; $\rightarrow H. + NO_2$	$\lambda < 360$ nm may initiate smog formation
Reactions of Hydrocarbons (olefins) with O	
23. $R\text{-}CH{=}CH_2 + O \rightarrow RCH\text{-}CH_2O.$	
24. $RCH\text{-}CH_2O. \rightarrow RCH_2. + H\dot{C}{=}O$	
25. $RCH_2. + O_2 \rightarrow RCH_2O_2.$	
26. $RCH_2O_2. + NO \rightarrow RCH_2O. + NO_2$	Method of oxidizing NO and conserving O_3
27. $RCH_2O. + O_2 \rightarrow RCHO + HO_2.$	Aldehyde formation

TABLE 9.11 (continued)

No.	Reaction	Comment
28.	$RCH_2O\cdot + NO_2 \rightarrow RCH_2ONO_2$	Chain terminating
29.	$RC(=O)-H + HO\cdot + O_2 \rightarrow RC(=O)-OO\cdot + H_2O$	
Reactions of Hydrocarbons (Olefins) with O_3		
30.	$RCH{=}CH_2 + O_3 \xrightarrow{1} RCHO + H\overset{+}{C}HO\overset{-}{O}$	(R)(H)C–O–C(H)(H) ozonide ring, with –O–O– bridge, bonds labelled 2 and 1
31.	$\xrightarrow{2} R\overset{+}{C}HOO^- + HCHO$	Ozonide intermediate
32.	$R\overset{+}{C}HO\overset{-}{O} + O_2 \rightarrow \cdot OH + RC(=O)-OO\cdot$	Oxidant
33.	$RC(=O)-OO\cdot + NO \rightarrow NO_2 + RC(=O)O$	R mainly H or CH_3
34.	$RC(=O)-OO\cdot + NO_2 \rightarrow RC(=O)OONO_2$	Peroxyacylnitrate (PAN)
35.	$RC(=O)-OO\cdot + HO_2\cdot \rightarrow R-C(=O)-OOH + O_2$	
36.	$RC(=O)-O\cdot \rightarrow R\cdot + CO_2$	
37.	$R\cdot + O_2 \rightarrow RO_2\cdot$	
38.	$RO_2\cdot + NO \rightarrow RO\cdot + NO_2$	
39.	$RO\cdot + NO \rightarrow RONO$	
40.	$RO\cdot + NO_2 \rightarrow RONO_2$	Alkyl nitrate
Other Reactions		
41.	$\cdot OH + CH_3-CH{=}CH_2 \rightarrow CH_3-\dot{C}H-CH_2OH$	
42.	$CH_3-\dot{C}H-CH_2OH + O_2 \rightarrow CH_3-CH(\cdot O_2)-CH_2OH$	
43.	$O + CH_3CHO \rightarrow \cdot OH + CH_3C(=O)\cdot$	
44.	$CH_3C(=O)\cdot + O_2 \rightarrow CH_3C(=O)-OO\cdot$	Most important reaction of $RC(=O)\cdot$
45.	$HC(=O)\cdot + O_2 \rightarrow HO_2\cdot + CO$	
46.	$CH_3CHO + h\nu \rightarrow CH_3\cdot + HC(=O)\cdot$	Photochemical smog initiating (46–47)
47.	$CH_3ONO + h\nu \rightarrow CH_3O\cdot + NO$	

Other reactions may occur involving inorganic species (reactions 16-22) and producing HNO_2 and HNO_3 which are removed by precipitation, or aerosol formation, i.e. NH_4NO_3 and $NaNO_3$.

It has been proposed to liberate free radical scavengers into the atmosphere to prevent the formation of photochemical smog, for example diethylhydrolamine, $(C_2H_5)_2NOH$. However, such an approach must be considered with extreme caution because of other reactions that may occur.

A generalized scheme for the formation of photochemical smog is given in Fig. 9.10.

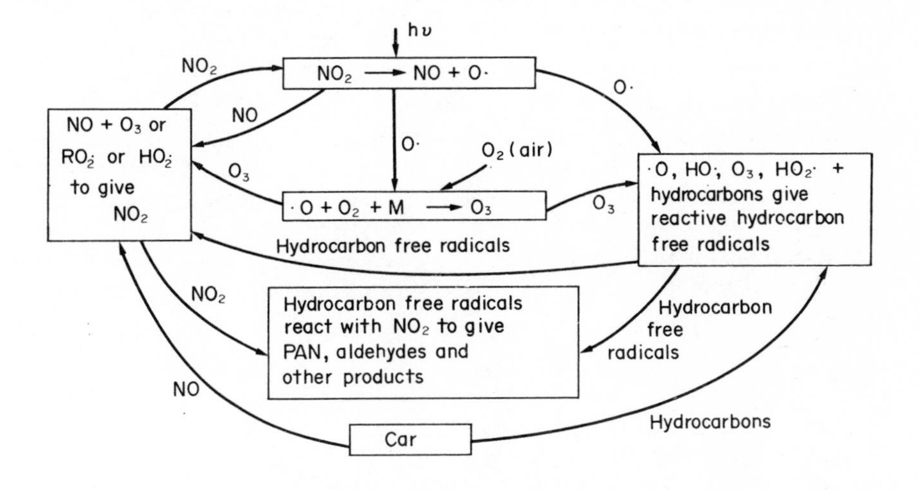

Fig. 9.10. A generalised scheme for the production of photochemical smog.

<u>Health effects</u>. The detrimental effects of nitrogen oxides and some products of photochemical smog are summarised in Table 9.12. As some products are powerful oxidising agents, quite low levels are dangerous. While no deaths can be directly related to photochemical smog, long term health effects have certainly occurred. Plants are particularly sensitive to the oxidising materials. Ozone produces cracking in rubber, due to the attack of O_3 on the double bonds giving an ozonide and then bond cleavage.

$$\left[-CH_2 - \overset{H}{\overset{|}{C}} = \overset{CH_3}{\overset{|}{C}} - CH_2- \right]_n + nO_3 \longrightarrow \left[-CH_2 - \overset{H}{\overset{|}{C}} \overset{O - O}{\underset{O}{\diagup\!\diagdown}} \overset{CH_3}{\overset{|}{C}} - CH_2- \right]_n \qquad (9.66)$$

break along here

<u>The automobile</u>

The automobile is the principle source of materials involved in photochemical smog. Most of the material comes from the exhaust, 100% of the CO, NO_x and lead and 65% of the hydrocarbons. Factors which influence the emissions are: the air-fuel ratio, ignition timing, compression ratio, combustion chamber geometry, engine speed and type of fuel. Twenty percent of the hydrocarbon emission comes from the crankcase blowby, especially during the compression stage and power stroke, where the hydrocarbon concentration in the gas can reach 6000 - 15,000 ppm. The remaining 15% comes from the fuel tank and carburettor, and in the latter case mainly just after the engine has been shut

TABLE 9.12 The Effects of NO_2 and Photochemical Smog Products on Humans, Animals and Plants

Pollutant	Effect	Single Concentration	Mean Concentration	Exposure	Comments
Humans					
NO_2	Odour threshold	1-3ppm		Immediate	
	Lethal	500ppm	500ppm	48 hrs	
O_3	Odour	<0.02-0.05ppm		Immediate	
	Respiratory		0.4ppm	120 min	Increase air-way resistance
	Eye,nose irritation		0.1ppm, 0.3-1ppm	15-60 min	
	Severe distress		1.5-2ppm	120 min	Impaired lung function, chest pains, coughing
PAN	Pulmonary function	>0.30ppm		5 min	Increase in O_2 uptake
	Eye irritation	>0.50ppm		12 min	
HCHO	Odour	1ppm			
	Severe distress	10-20ppm			
Animals					
NO_2	Discomfort	10-20ppm			Mice,rabbits, cats
	Severe distress	20-100ppm			
PAN	Lethal		105ppm	120 min	Mouse
HCHO	Lethal		800ppm	30 min	Rat
Plants					
NO_2	Leaf symptons	3ppm	2.5ppm	4 hrs	White, brown lesions
	Inhibit photosynthesis		>0.6ppm		
O_3	Leaf lesions, inhibit photosynthesis		0.1ppm	2-4 hrs	
PAN	Collapse of young cells		0.01 ppm	6 hrs	
HCHO	Leaf symptons		>0.2ppm	2 days	

Source: Bond, R.G. and Straub, C.P., Handbook of Environmental Control, I, Air Pollution (1972) Chem. Rubber Co., Cleveland.

off and the carburettor is still hot.

The air-fuel ratio is a significant factor in determining emissions and at the stoichiometric ratio (approximately 15:1, air:fuel) the CO and hydrocarbon emissions are low but NO_x is high (Fig. 9.11). To reduce the NO_x and

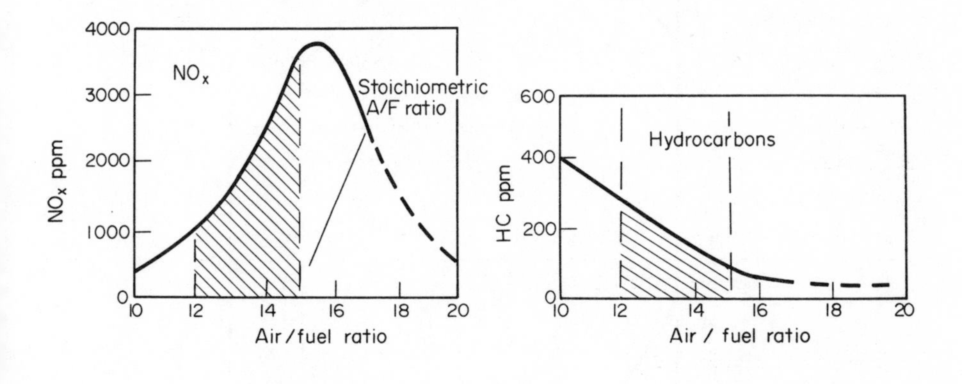

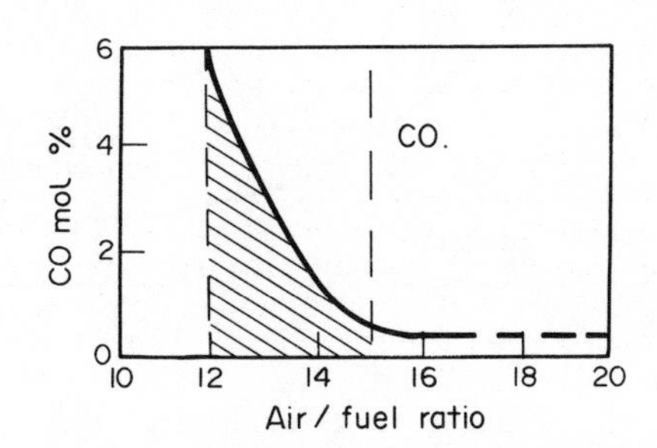

Fig. 9.11. The concentrations of hydrocarbons, CO and NO_x in automobile exhaust gases as a function of the air-fuel ratio. The most common ratio used is shaded.

keep the CO and hydrocarbon emissions low a lean (high air-fuel ratio) mixture needs to be used, but if too lean misfiring will occur. Reducing the NO_x emissions by using a richer mix, increases CO and hydrocarbon pollution. The relationship between the air-fuel ratio and the power obtained from the combustion is given in Fig. 9.12, maximum power produces high CO and hydrocarbon emissions and low NO_x, while for economic running the reverse holds. Hence, optimum running of a car is a compromise between power, economy and emissions. A common air-fuel setting for a car is 10-15% on the rich side.

Exhaust emissions also depend on the operation of the car as indicated in Table 9.13. When emissions of CO and hydrocarbons are down, those of NO_x are up and vice versa. Controlling emissions involves changes to the engine, and the use of catalytic convertors in the exhaust system. It is interesting

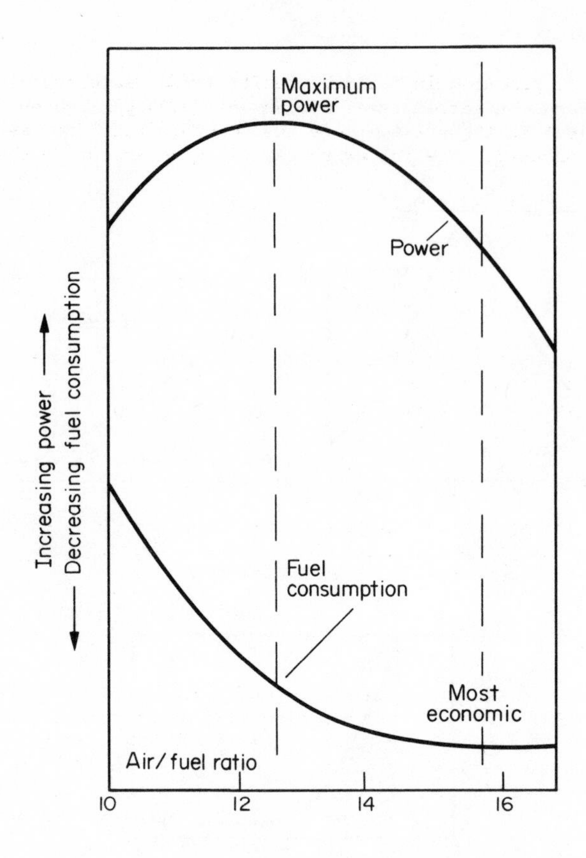

Fig. 9.12. The relationship between fuel consumption and power with respect to the air-fuel ratio.

TABLE 9.13 Car Emissions and Engine Speed

Mode of operation	Materials in Exhaust Gases		
	Unburnt hydrocarbons (ppm)	NO_x (ppm)	CO %
Cruise	200-800	1000-3000	1-7
Idle	500-1000	10-15	4-9
Accelerate	50-800	1000-4000	0-8
Decelerate	3000-12,000	5-50	2-9

to note that since standards have been introduced to help control emissions, the effect has been to reduce CO and hydrocarbons but actually increase the NO_x emissions.

Nitrous Oxide

Nitrous oxide is produced in soils by denitrifying bacteria. The oxide is stable and its concentration remains constant (0.25 ppm) throughout the troposphere. But in the stratosphere the concentration decreases suggesting the oxide is consumed. The following reactions may occur in the stratosphere;

$$N_2O + O \rightarrow N_2 + O_2, \quad (9.67)$$

$$N_2O + O \rightarrow 2NO, \quad (9.68)$$

$$N_2O + h\nu \rightarrow N_2 + O, \quad (\lambda < 337 \text{ nm}) \quad (9.69)$$

$$N_2O + h\nu \rightarrow NO + N, \quad (\lambda < 250 \text{ nm}) \quad (9.70)$$

Since NO is a product, the ozone levels may be reduced. Therefore it is possible that an increase in N_2O from greater use of nitrogen-fertilizers, may influence the ozone layer.

OTHER GASES

Other gaseous materials get into the atmosphere, but in general, are less of a problem than the gases discussed above, except when they may occur in local high concentration as a result of an accident.

Ammonia

Most atmospheric ammonia originates from natural bacterial activity, approximately 3700 x 10^6 tonnes yr^{-1} compared with 4 x 10^6 tonnes from anthropogenic sources. In an acid environment, salts, such as $(NH_4)_2SO_4$, NH_4NO_3 and NH_4Cl, soon form as aerosols, and are important in smog formation. The total atmospheric concentration of NH_3 (i.e. NH_3 + NH_4^+) is around 5 μg m^{-3}, and has a residence time approximately 7 days. Ozone in the air oxidises NH_3 to a number of materials, such as N_2O, N_2, NH_4NO_3.

Hydrogen Sulphide

Nearly all the H_2S in the atmosphere is produced naturally (98 x 10^6 tonnes yr^{-1}) by anaerobic decay of organic material or by bacterial reduction of SO_4^{2-}, for example;

$$SO_4^{2-} + 2\{CH_2O\} + 2H^+ \xrightarrow[\text{desulphovibrio}]{} H_2S + 2CO_2 + 2H_2O \quad (9.71)$$

The actual amount may be much greater because of errors in the measurement of low concentrations of H_2S and because some H_2S is oxidised to SO_2. Approximately one third comes from the oceans, and two thirds comes from the land. Anthropogenic sources accounts for approximately 3 x 10^6 tonnes yr^{-1}, the principal being, the Kraft pulp and paper mills, rayon production, coal gasification and oil refining. The mean global concentration of H_2S is around 0.2 ppb, but in cities it can range from 2 - 20 μg m^{-3} and reach 140 μg m^{-3} at times. Hydrogen sulphide is very toxic, and is a hazard encountered when drilling for natural gas, as pockets of H_2S can occur in deep wells. Hydrogen sulphide blackens lead based paints by forming PbS, but oxidation to $PbSO_4$ may occur later, with disappearance of the black.

Oxidation of H_2S to SO_2 (by O, O_2 and O_3) occurs rapidly in the air suggesting a catalytic process, as normally the oxidation is slow in the gas phase.

$$H_2S + O_3 \rightarrow H_2O + SO_2 \quad (9.72)$$

$$H_2S + O \rightarrow OH + SH \rightarrow SO_2 \text{ (finally)}. \quad (9.73)$$

A mixture of 1 ppb H_2S and 0.5 ppb O_3 in the presence of 15,000 particles cm^{-3} react completely in two hours. Since both O_3 and H_2S are soluble in water, oxidation may also occur in water droplets. Oxidation is also achieved by bacteria (e.g. *thiobacillus thiooxidans*).

$$2H_2S + O_2 \rightarrow 2S + 2H_2O \quad (9.74)$$

Hydrocarbons

Natural global amounts of CH_4, are estimated at 1600 x 10^6 tonnes yr^{-1}, and 170 x 10^6 tonnes yr^{-1} of terpenes (plant volatiles). Man-made sources of around 80 x 10^6 tonnes yr^{-1} comprise approximately 55% from petroleum, 28% from incineration and refuse fires, 11% from solvent evaporation, and the rest (6%) from coal and wood combustion. Ninety-five percent of the pollution is in the Northern hemisphere. More than 60 hydrocarbons have been identified in the Los Angeles atmosphere, this number could increase as analytical detection limits are improved. The major component is methane being 90% of the alkanes which themselves comprise 94% of the total hydrocarbons.

Methane is produced from anaerobic degradation of organic material;

$$CO_2 + 8H^+ + 8e \xrightarrow{\text{bacteria}} CH_4 + 2H_2O, \quad (9.75)$$

$$2\{CH_2O\} + 2H_2O - 8e \rightarrow 2CO_2 + 8H^+, \quad (9.76)$$

i.e. $$2\{CH_2O\} \xrightarrow{\text{bacteria}} CO_2 + CH_4. \quad (9.77)$$

Some of the bacteria concerned are *methano bacterium*, *methano bacillus*, *methano coccus*, and *methano sarcina*.

The Halogens

Gaseous halogen pollutants include the dihalogens, hydrogen halides and halogenated hydrocarbons, such as the freons and some pesticides and herbicides. In addition halide aerosols occur.

Fluorides. Aerosol fluoride comes from sea spray, oxidation products of F_2 with O_3 or formation of HF. Anthropogenic fluoride sources are associated with aluminium, superphosphate (using fluoro-apatite which contains approximately 3% F), steel, glass and ceramics production. The principal emissions are F^-, HF, F_2, SiF_4, H_2SiF_6 and fluorocarbons. The reactivity of F_2 and rapid hydrolysis of SiF_4 and H_2SiF_6 means that in the atmosphere they quickly react to HF. One end product of fluoride emissions is CaF_2;

$$2HF + CaCO_3 \rightarrow CaF_2 + H_2O + CO_2 \quad (9.78)$$

Chlorides. Dichlorine, like difluorine, is very toxic and it has been used as a poison gas in warfare. The global concentration of chloride (as Cl^-,

Cl_2 and HCl) is in the range 0.5 - 5.0 $\mu g\ m^{-3}$. The major natural source of Cl^- is sea-spray, from which HCl forms as a secondary product;

$$2NaCl + H_2SO_4 \rightarrow 2HCl + Na_2SO_4 \tag{9.79}$$

Man-produced chlorides include $PbCl_2$ and PbBrCl aerosols, emitted from automobiles using petrol containing lead tetraalkyls, and $C_2H_4Cl_2$, $C_2H_4Br_2$ scavengers for lead, HCl from the incineration of PVC, and Cl^- from the combustion of coal (which may contain up to 0.5% Cl^-). Dichlorine forms HOCl and HCl in water droplets

$$Cl_2 + H_2O \rightarrow HOCl + HCl \tag{9.80}$$

Bromides. Aerosol bromides are emitted in automobile exhaust gases. Hydrolysis will produce HBr.

Freons. Concern exists over the use of fluoro-chlorocarbons as propellants in aerosol spray cans, because of likely reaction with ozone in the stratosphere. Production of CCl_3F (fluorocarbon -11) rose from 23 x 10^3 tonnes yr^{-1} to 313 x 10^3 tonnes yr^{-1} from 1958 to 1973, and production of CCl_2F_2 (fluorocarbon -12) rose from 59 x 10^3 to 469 x 10^3 tonnes over the same period. Up to 50% of freons are used in aerosol cans, 28% as refrigerant, 10% in plastics and 5% as solvents. However, the proportions are altering as less freons are used in spray cans.

The use of freons in spray cans appeared to be a good idea, because of their lack of reactivity and apparent absence of biological effects. They do not react with O_2 or $O(^3P_2)$, but $O(^1D_2)$ reacts to give COF_2. Stability of the freons, such as resistance to hydrolysis, may be a serious disadvantage as eventually they get into the stratosphere. Photochemical decomposition can then occur and lead to reactions with ozone.

$$CCl_2F_2 + h\nu \rightarrow Cl. + .CClF_2, \quad (\lambda = 175\text{-}220 \text{ nm}) \tag{9.81}$$

$$Cl. + O_3 \rightarrow ClO + O_2, \tag{9.82}$$

also,

$$ClO + O(^3P_2) \rightarrow Cl. + O_2, \tag{9.83}$$

$$Cl. + O_3 \rightarrow ClO + O_2. \tag{9.84}$$

Reactions (9.83) and (9.84) give;

$$O_3 + O(^3P_2) \rightarrow 2O_2, \tag{9.85}$$

i.e. loss of O_3 without absorption of radiation.

The reactions;

$$ClO + NO \rightarrow \cdot Cl + NO_2, \tag{9.86}$$

$$\cdot Cl + O_3 \rightarrow ClO + O_2, \tag{9.87}$$

also remove O_3 without utilising radiation. Termination reactions include;

$$.Cl + CH_4 \rightarrow HCl + .CH_3, \tag{9.88}$$

$$.Cl + H_2 \rightarrow HCl + .H, \tag{9.89}$$

The HCl may eventually diffuse back into the troposphere.

Estimates of the depletion of stratospheric ozone, range from 2 to 20% (mean approximately 7%). Since it takes a number of years for material to get into the stratosphere it could be that the maximum effect of freons is yet to occur.

PARTICULATE MATTER

Atmosphere particles range from molecular species to material that settles out rapidly, i.e. from 10^{-9} to 10^{-4}m in diameter. Aerosols form in two ways, by dispersion into, or condensation from, the atmosphere, of molecular (vapor or gaseous) systems; or from chemical reactions in the atmosphere producing low volatile substances. Different aerosols are classified according to their type and properties. Fine liquid particles are called *mists*, solid particles are called *dusts* or *smoke*, but the latter term is mainly restricted to solid products of combustion. *Fogs* have a high concentration of water droplets, while smoke and fog together are called *smog*. The term *haze* relates to reduced visibility, which can arise from different aerosols.

Naturally produced aerosols are either continental or oceanic. Continental sources include dust storms, volcanic eruptions, photochemical gas reactions and viable particles such as viruses, bacteria, fungi, spores and pollen. Oceanic sources include sea spray and evaporation residues of sea spray. The principal anthropogenic sources are; combustion, secondary products of photochemical and atmospheric reactions (example, oxidation of SO_2 to SO_4^{2-}) and industrial emissions (non-combustion). An estimate of global emissions listed in Table 9.14 indicate that coal combustion, atmospheric oxidation of SO_2 to SO_4^{2-} and NO_x to NO_3^- and organic materials are the major man-made aerosols.

Estimates of the concentration of aerosols depend on the location of the measurement, weather conditions and the mode of measurement. A range of 30-200 $\mu g\ m^{-3}$ is not uncommon for urban areas, while for rural areas a range of 10-40 $\mu g\ m^{-3}$ has been reported. Levels of 2000 $\mu g\ m^{-3}$ can occur for high pollution. Over the oceans the number of particles range from 800 to 5000 x 10^6 particles m^{-3}, while in city air a range from 50,000 to 380,000 x 10^6 particles m^{-3} can occur. The lower troposphere residence time depends on size, and the weather conditions; but 6-14 days is a typical time. Particles with a diameter $>$ 10 μm deposit by virtue of their size and mass, under the influence of gravity. Residence times of 1-5 years occur in the stratosphere.

The Effects of Aerosols

Aerosols influence the weather, chemical reactions in the atmosphere and have adverse health effects. Aerosols account for about 10% of the backward scattering and absorption of solar radiation, and may, if levels increase, reduce the earth's mean temperature. The role of aerosols as condensation nuclei, and a source of catalysts for chemical reactions has been discussed, in relation to smog formation. Aerosols are efficient absorption surfaces for gaseous materials and organic compounds, such as polycyclic aromatic hydrocarbons. Such materials are more readily able to penetrate, and be retained, by the lungs. The relationship between particle size and depth of

TABLE 9.14 Estimated Global Particulate Emissions (1968)

Source	Emissions (Tg x 10)	
	Natural	Anthropogenic
Primary Particle Production:		
Coal fly ash		36
Iron and steel industry		9
Nonfossil fuels		8
Petroleum combustion		2
Incineration		4
Agricultural emission		10
Cement manufacturing		7
Other		16
Sea salt	1000	
Soil dust	200	
Volcanic particles	4	
Forest fires	200	
	1404	92
Gas-to-particle:		
Sulfate from H_2S	202	
Sulfate from SO_2		147
Nitrate from NO_x	430	30
Ammonium from NH_3	269	
Organic aerosol from terpenes, hydrocarbons, etc.	198	27
	1099	204
TOTAL	2503	296

Source: Seinfeld, J.H. Air Pollution (1975) McGraw-Hill, N.Y.

penetration into the respiratory system is given in Fig. 9.13

Particle Size and Shape

The shapes of aerosol particles range from spherical to quite irregular, and the relationship between size (diameters), radiation wavelength and the size range of various aerosols is listed in Table 9.15. Particles in the range 0.38 - 0.76 μm are comparable to the wavelength of visible radiation, and will affect the transmission of light, producing haze. The majority of particles lie within 0.01 - 100 μm, and the mean diameter of particles contributing to smoke is around 0.075 μm.

The size of a particle can be expressed in terms of the Stokes' radius, which is the radius of a sphere having the same falling velocity and density as the particle. The Stokes' radius approximates to the true radius of fume particles, but not for the coagulated particles.

The size distribution of particles normally fit a log-normal frequency curve (Fig. 9.14). Over land the maximum frequency is approximately 0.03 μm, so most particles are Aitken particles (diameter <0.1 μm). However, in terms of mass the Aitken particles are <28% of the total mass. Over the oceans the maximum in the distribution curve is 0.1 μm, but the size increases with humidity, by as much as a factor of 10 from low to high humidity.

TABLE 9.15 Particle Size

Size (μm) (NB log. scale)	0.001	0.001 (1nm)	0.01	0.1	1	10	100	1000 (1mm)	10,000 (1cm)
Electromagnetic spectrum		X-rays		UV	visible Near IR		Far IR		microwaves
Classification: Solid			Fume				Dust		
Liquid					Mist		Spray		
Solid/Liquid				Smog					
Typical Particles	gas molecules			Oil smokes		Fly ash			
				Tobacco smoke		Coal dust			
					Metallurgical Fumes and Dust				
			Aitken nuclei			Cement Dust	Beach sand		
			Viruses			Bacteria			
						Plant Spores	Pollen		
				Atmospheric Dust					
				Sea salt nuclei					
			Combustion nuclei		Lung damaging dust (0.3 - 4 μm)				

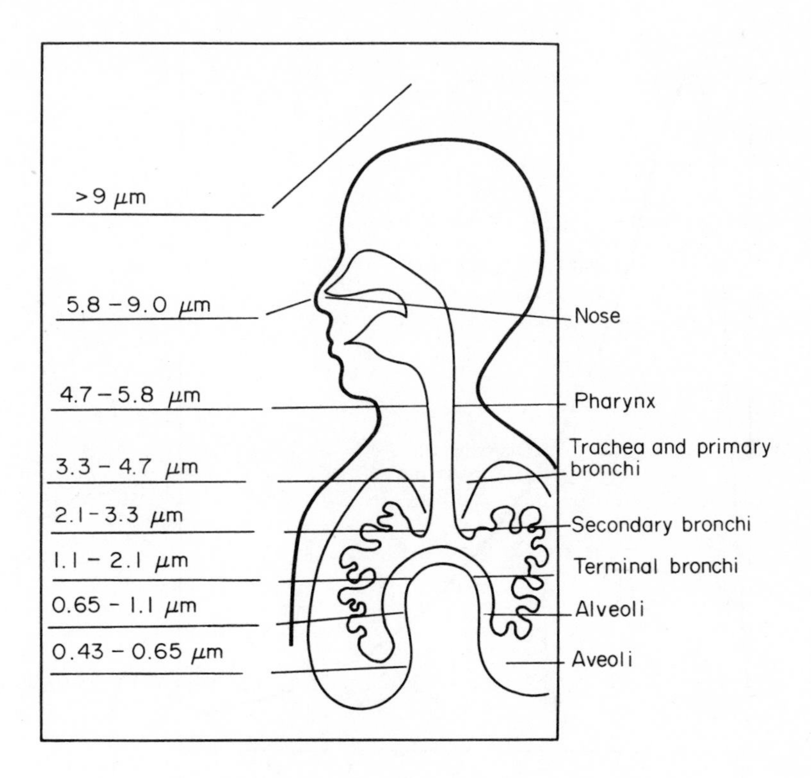

Fig. 9.13. Particle size and penetration into the respiratory system.

Sedimentation and Coagulation

Sedimentation is the natural fall of particles, the terminal velocity of which is given by the Stokes' flow for particles in free fall;

$$V = \frac{2[\rho_p - \rho_m]R^2 g}{9\mu} \qquad (9.90)$$

where ρ_p and ρ_m are the densities of the particles and air respectively, R the radius of the particle, g the gravitational acceleration and μ the viscosity of the medium. At atmospheric pressure and 20^oC, and with ρ_p = 1 g cm^{-3} the terminal velocity is 8.7×10^{-5} cm s^{-1} for R = 0.05 μm, and 1.3×10^{-1} cm s^{-1} for R = 10 μm. The equation does not apply to very small particles which are subject to Brownian movement, that is random movement as a consequence of collisions of the particles with molecular species, e.g. N_2 and O_2.

Coagulation is when particles clump together to form larger particles, contact being controlled primarily by Brownian motion. Assuming each collision produces coagulation, then reduction in the concentration of the particles is given by;

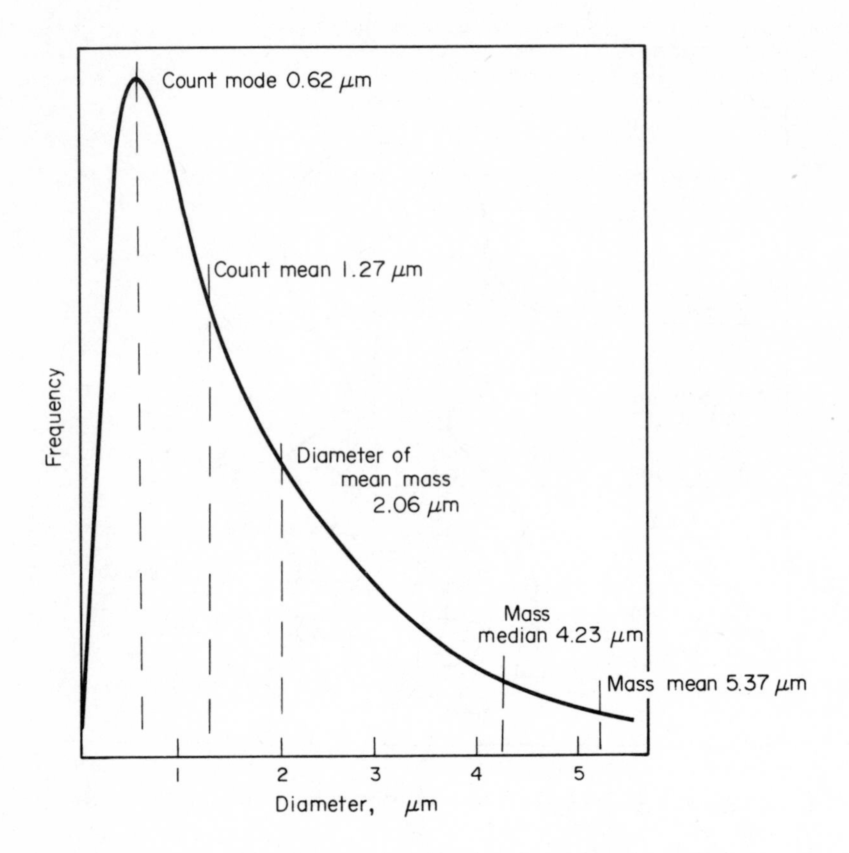

Fig. 9.14. Log-normal distribution for aerosol particle size.

$$n = \frac{n_o}{1 + kn_o t} \tag{9.91}$$

where n_o is the initial concentration and n the final concentration, t the time and k a constant of magnitude 10^{-10} to 10^{-8} cm^3 s^{-1}. Decrease in the number of particles <0.1 μm is mainly due to coagulation. Particles with radius 0.01 μm, disappear at a rate of around 50% per hour, while for 0.05 μm particles the rate is 50% per day.

Aerosol Chemistry and Composition

Aerosols formed over the oceans tend to have a composition similar to that of the dissolved salts in the oceans. Over land the situation is more complex. One of the principal compounds is $(NH_4^+)_2SO_4^{2-}$, but near sea water Na^+Cl^- is a major component. For industrial areas, the ions NH_4^+, SO_4^{2-} and Cl^- are prominent, while in cities with high traffic SO_4^{2-}, Pb^{2+}, Fe^{2+} and NO_3^- are common. The elements V and Mn, found in aerosols, are probably entirely man-produced, therefore measurement of the amounts can be a guide to the influence of man on the environment. Some of the major inorganic species and elements found in aerosols are given in Table 9.16, the species with an asterisk are the principal components.

TABLE 9.16 Some of the More Common Components of Aerosols

Source	Species or Element
From atmospheric reactions, or condensation	H_2O*, NO_3^-*, NH_4^+*, SO_4^{2-}*, SO_3^{2-}
Introduced by man	Ca*, Mg*, Ba, Pb*, F, Cl*, Br, Ti, V, Mn, Fe*, Cu*. Zn.
From natural sources	Na*, K*, Ca*, Mg*, Al*, Si*, Cl*, Br, I, Ti, Fe*

* Major components.

Aerosols in relation to smog formation have been discussed. The following reactions are further examples of aerosol chemistry (* signifies an aerosol);

$$\underset{\text{(in coal)}}{3FeS_2} + 8O_2 \rightarrow Fe_3O_4^* + 6SO_2 \qquad (9.92)$$

$$\underset{\text{(in oil)}}{2V} + 5/2\,O_2 \rightarrow V_2O_5^* \qquad (9.93)$$

$$\underset{\text{(in coal)}}{CaCO_3} \rightarrow CaO^* + CO_2 \qquad (9.94)$$

$$CaO^* + \underset{\text{(droplets)}}{H_2SO_4^*} \rightarrow CaSO_4^* + H_2O \qquad (9.95)$$

$$Pb(C_2H_5)_4 + O_2 + C_2H_4Cl_2/C_2H_4Br_2 \rightarrow CO_2 + H_2O + PbCl_2^* + PbBr_2^* + PbClBr^* + NH_4Cl.PbBrCl^* \qquad (9.96)$$

$$PbX_2^* + H_2SO_4^* \rightarrow PbSO_4^* + 2HCl \qquad (9.97)$$

The composition of airborne particles emitted from the internal combustion engine are listed in Table 9.17. Lead and carbon make significant contributions to urban aerosols.

CONTROLS AND STANDARDS

If air pollution is to be reduced, or avoided, it is necessary to institute emission regulations, and control techniques. Either the offending elements can be removed before the source materials are used, or from the emissions before getting into the atmosphere. A major problem in the latter process, is the removal of what may be trace amounts of material from a large volume of fast moving gas.

TABLE 9.17 The Composition of Airborne Particles from Car Emissions

Element	% Mass	Element or species	% Mass
Pb	24.5	H	5.8
Fe	0.9	NO_3^-	7.3
Cl	8.6	NH_3	5.4
Br	4.0	alkaline earths	2.6
C	28.0		

Sulphur Dioxide

Iron disulphide in coal can be partially removed by crushing to about 100μm and separating the FeS_2 (Sp.G.=5) from the coal (Sp.G.=1.3) by gravitational techniques. Chemical leaching agents, such as 10% NaOH, can also be employed. At best a low-sulphur coal is produced. High-sulphur oil is commonly blended with low-sulphur oil, or the S removed chemically.

$$\underset{\text{in oil}}{S} + CaO \rightarrow CaS \xrightarrow[O_2]{\text{heat}} CaO + \underset{(\text{used to make } H_2SO_4)}{SO_2}, \tag{9.98}$$

$$\underset{\text{in oil}}{S} + H_2 \rightarrow H_2S \rightarrow \text{remove by scrubbing.} \tag{9.99}$$

The gaseous emissions from a 1000 MW power station, using coal with 2.5-3.0% S, is 800-1000 m^3s^{-1}. This has a SO_2 concentration of 0.2-0.3% by volume, ie 2000-3000 ppm. Methods of removing the SO_2 from flue gases, summarised in Table 9.18, consist of reaction with bases such as MgO, CaO, NH_3, catalytic oxidation, or a regenerative procedure. Ground level control can be achieved by using high smoke stacks.

Hydrogen Sulphide

Hydrogen sulphide, produced during the gasification of coal (0.3-3.0% by volume), is very toxic and has to be removed. Some will be removed as $(NH_4)_2S$ formed with ammonia, also in the coal gas, but scrubbing is the most common procedure;

$$H_2S + \underset{\text{thioarsenate}}{Na_4As_2S_5O_2} \rightarrow Na_4As_2S_6O + H_2O, \tag{9.100}$$

$$\downarrow \tfrac{1}{2}O_2$$

$$Na_4As_2S_5O_2 + S.$$

Other scrubbing chemicals are amines, hot K_2CO_3 solution, hot ferric oxide, or dolomite (Ca/Mg CO_3).

TABLE 9.18 Removing SO_2 from Flue-Gases

Process	Reagent	End Product	Reactions and Comments
1 Limestone or lime scrubbing	$CaCO_3$ or CaO	$CaSO_3/CaSO_4$ sludge	$CaCO_3 \xrightarrow{\Delta} CaO + CO_2$ $CaO + SO_2 + \frac{1}{2}O_2 \rightarrow CaSO_4$ SO_2 absorbed in lime slurry giving insoluble sludge
2 Sodium sulphite scrubbing	Na_2SO_3 (regenerated)	H_2SO_4 or S	$Na_2SO_3 + SO_2 + H_2O \rightleftharpoons 2NaHSO_3$ Collect conc. SO_2 and $H_2O + SO_2 + \frac{1}{2}O_2 \rightarrow H_2SO_4$ or $2H_2S + SO_2 \rightarrow 3S + 2H_2O$ $(4S+CH_4+2H_2O \xrightarrow{Al_2O_3} 4H_2S+CO_2)$
3 Magnesium oxide scrubbing	MgO (regenerated)	H_2SO_4	$MgO + SO_2 \rightleftharpoons MgSO_3$ ↓ conc. SO_2 to make H_2SO_4
4 Citrate scrubbing	NaH_2 citrate (regenerated)	Sulphur	$NaH_2Cit+SO_2+H_2O \rightarrow NaH[HSO_3.H_2Cit]$ ↓ H_2S $3S + NaH_2Cit + 3H_2O$ (H_2S prepared as above)
5 Catalytic oxidation	V_2O_5 as catalyst	H_2SO_4	$SO_2 + \frac{1}{2}O_2 \underset{}{\overset{V_2O_5}{\rightleftharpoons}} SO_3$ ↓ H_2O H_2SO_4 (80%)
6 Ammonia scrubbing	NH_3	$(NH_4)_2SO_4$, S	$NH_3+SO_2+H_2O \rightarrow NH_4HSO_3$ $2NH_4HSO_3+(NH_4)_2S_2O_3 \rightarrow$ $2(NH_4)_2SO_4+2S+H_2O$

Hydrogen Fluoride

Since HF is soluble in water it may be washed out by scrubbing. However, water or 2% NaOH cannot be used if F_2 is present. In stronger NaOH the reactions are;

$$F_2 + 2NaOH \rightarrow \tfrac{1}{2}O_2 + H_2O + 2NaF \quad (9.101)$$

$$HF + NaOH \rightarrow NaF + H_2O \quad (9.102)$$

$$2NaF + CaO + H_2O \rightarrow CaF_2 + \underset{(regenerated)}{2NaOH} \quad (9.103)$$

At aluminium smelters, gaseous and particulate fluorides are passed through a bed of fluidized Al_2O_3 (60-165°C), and the product is used to supplement cryolite losses. This process is 99% efficient for gaseous fluorides, and 91% efficient for particulates.

Nitrogen Oxides

As a large proportion of nitrogen oxides are produced by the internal combustion engine, emission controls are concerned with engine performance. Aspects of this have already been discussed. As yet no suitable low temperature decomposition catalyst is available. Catalytic reduction is possible at high temperatures;

$$6NO + 4NH_3 \xrightarrow{\text{catalyst}} 5N_2 + 6H_2O \tag{9.104}$$

and

$$2CO + 2NO \rightarrow N_2 + 2CO_2 \tag{9.105}$$

Absorbents are not efficient, and in water a 1:1 mixture of NO and NO_2 is necessary, i.e. NO_2 has to be added to the gases;

$$NO + NO_2 + H_2O \rightarrow 2HNO_2 \tag{9.106}$$

Catalytic convertors, as part of an exhaust system, are being increasingly used. Two opposing processes need to be carried out, viz. oxidation of CO and hydrocarbons, and reduction of NO. If both are attempted, reduction is carried out first, followed by air intake and then oxidation. Numerous reactions can occur as outlined in Fig. 9.15. Unfortunately some reduction processes achieved in the first chamber are nullified in the oxidation chamber. Catalysts are Pd or Pt supported on Al_2O_3 or oxides such as CoO, CuO, Cr_2O_3 and MnO_2 supported on Al_2O_3, or a mixture of these. The catalysts are readily poisoned by lead, the Pt catalysts become ineffective after 100 hours operation with leaded petrol. The scheme given in Fig. 9.15 is a dual-type catalytic convertor. Attempts to find a single catalyst (called a

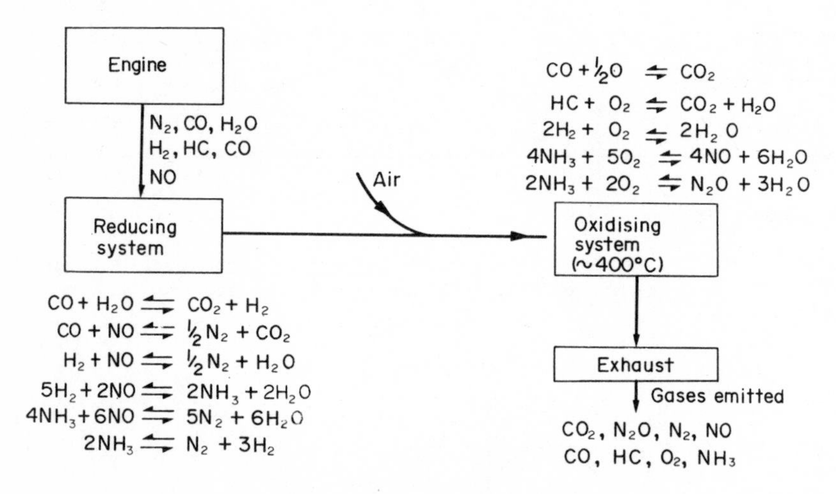

Fig. 9.15. Reactions associated with catalytic convertors used on automobiles.

three-way catalyst) are in progress, which promotes oxidation and reduction between the three materials CO, hydrocarbons and NO. A Pt/Rh catalyst has been used and ruthenium is also a possibility. The best result occurs when the CO, hydrocarbon and NO are in the correct stoichiometry. This does not normally happen unless an electronic fuel injection system is used coupled to an oxygen-sensor feed-back which controls the amount of fuel injected. An exhaust thermal reactor could be employed, where CO and hydrocarbons are oxidised, and if the mix is rich, NO_x emissions are kept to low levels. However, this is poor on fuel economy.

Another approach is to look for alternative engines to the Otto Cycle engine. The Wankel Rotary engine produces less NO, but CO and hydrocarbons emissions are greater. The Stratified-Charge engine (a modification of the Otto Cycle) uses a lean mix, but ignites the mixture with a small amount of rich mix in the vicinity of the spark plug. The Diesel engine produces less CO and hydrocarbon, but more NO, and also particulate material if not maintained. The Diesel engine does not require leaded fuel. Finally electric cars may become more popular, especially if light batteries become readily available. This will shift car generated air pollution from the cities, to the vicinity of power stations.

Particulates

A number of methods are used to reduce particulate material in a gas stream. It is often easier to handle gases in this way, by converting them to solid materials, for example convert SO_2 to $CaSO_4$ and $CaSO_3$ and then collect the solids. In Table 9.19 some of the collecting methods are listed, and

TABLE 9.19 Methods for Removal of Particulate Material

Method	Comment	Diagram in Fig. 9.16
Gravitational Settling Chambers	For large particles diam. >50 μm.	a
Centrifugal separators (cyclone)	Heavy particles thrown to the side and fall into collector diam. 5-40 μm and certainly >1 μm.	b
Wet scrubbers	Use spray towers and cyclone scrubbers. diam 0.2-10 μm. (Can combine with cyclone method)	c
Filters (bag-house)	High efficiency for even small particles at low cost, a disadvantage is the clogging of the filters. diam. >0.01 μm.	d
Electrostatic Precipitator	Handles smallest particles, diam. >0.005 μm. At high temperatures (>1100K) get thermal ionization and hard to maintain discharge voltage.	e

schematic diagrams are given in Fig. 9.16. The choice of method depends on: particle size, cost, efficiency, flow rate of particles, pressure drop, ease

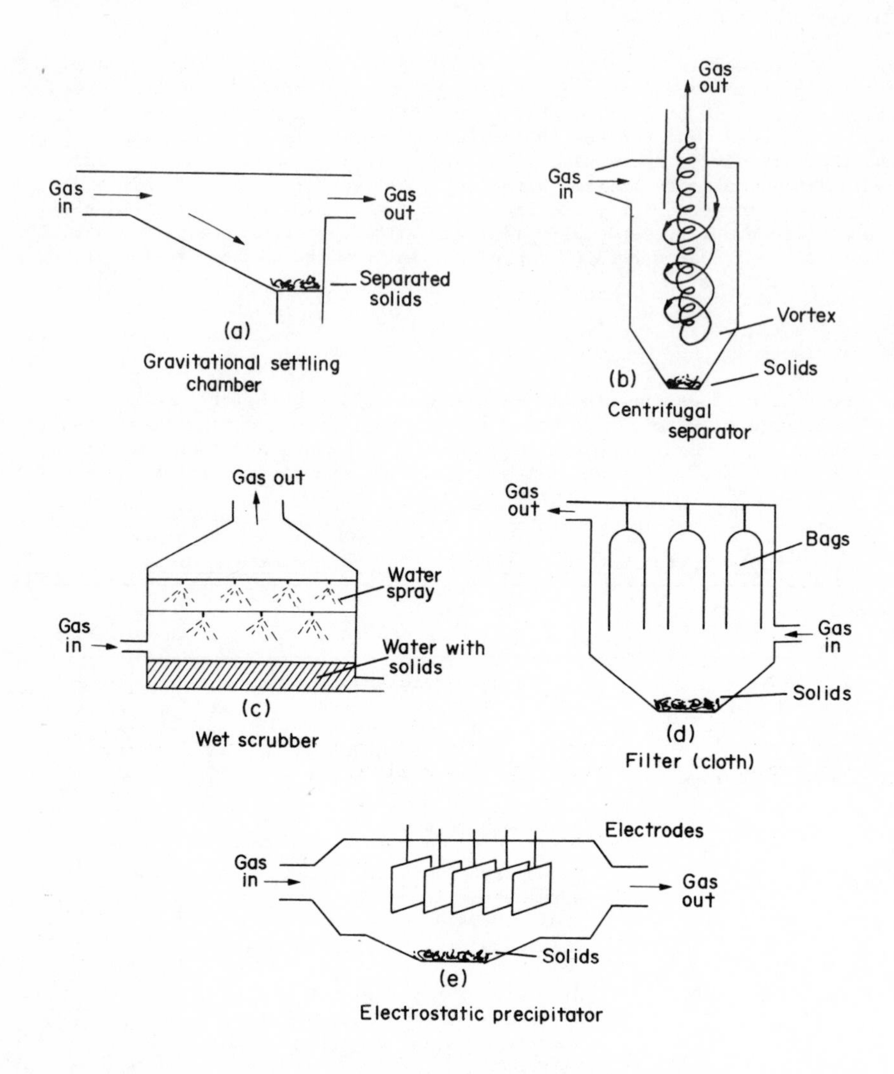

Fig. 9.16. Methods of collecting particulate material.

of disposal of collected material, properties of the carrier gas, ease of maintaining the equipment and its reliability. The electrostatic precipitator has a discharge electrode at 40-60 kV where gas molecules are ionized, which absorb onto the particles moving past so they become charged. The charged particles are then attracted to a collector plate where neutralization occurs, allowing the particles to drop off.

An unfortunate extension of industrial air pollution, particularly as regards aerosols, is the transfer of the material to the homes of the workers via their clothing. Asbestos and lead are two documented examples, where family members have suffered from exposure.

In establishing standards for air pollution it is necessary to consider health and biological effects, the effect on the physical environment (e.g. weather), and costs. The ultimate goal would be to reach background (natural) levels, but in actual fact attainable goals are set at higher levels. These vary from country to country, and from year to year as more information becomes available. In Table 9.20 the ambient air quality standards for the U.S.A. (1970) are given as a sample of levels aimed for. The levels specified are also a function of exposure time, i.e. higher levels are tolerated for short intervals, as long as the average over a longer period (e.g. one day) is not excessive.

TABLE 9.20 Ambient Air Quality Standards (USA 1970)

Material	Duration	Standard		Warning levels	
		ppm	$\mu g\ m^{-3}$	ppm	$\mu g\ m^{-3}$
Particles*	Ann. geom. mean		75		
	24 hr. av.		260		625
SO_2*	Yearly arithm. mean	0.03	80		
	24 hr. av.	0.14	365	0.6	1600
CO	8 hr. av.	9	10	30	34
	1 hr. av.	35	40		
Photochemical oxidants	1 hr. av.	0.08	160	0.4	800
Hydrocarbons (as CH_4)	3 hr. av.	0.24	160		
NO_x	Yearly arithm. mean	0.05	100		
	1 hr. av.			1.2	2260
Aerosol lead (California)	30 day av.		1.5		

* Warning level for particles + SO_2 is $261 \times 10^3\ \mu g\ m^{-3}$ for 24 hr average.

INDOOR AIR POLLUTION

Generally, less thought is given to air pollution inside buildings, coming from penetration from the outside, and generation inside. The latter source can be considerable and reach serious levels. The building walls and the materials on them tend to reduce penetration from the outside due to filtering and absorption on the large surface areas. For indoor generated pollution the walls provide an enclosed space which may allow levels to become quite high in a short time.

Air pollution that penetrates indoors in relation to levels outside, is given by an indoor concentration/outdoor concentration ratio (Table 9.21). Gases such as CO, which are unreactive and do not absorb on the walls strongly, have a ratio close to one, whereas for SO_2, which is reactive, and absorbed onto surfaces and into water the ratio is less than one. The average size of the particles that penetrate a building is less than for the average outside, as the building acts as a filter for the larger particles. Particulate material soluble in benzene, is enriched indoors relative to outdoors, presumably due to combustion process in the house.

TABLE 9.21 Indoors-Outdoors Air Pollution Ratio

Material	Ratio$\left(\frac{\text{Indoors conc.}}{\text{Outdoors conc.}}\right)$	Half-life for disappearance indoors
CO	1.0	-
NO_x	1.0	-
O_3	0.6	6 - 11 min
PAN	1.0	-
SO_2	0.2 - 0.5	40 - 60 min
Particulates	0.4 - 0.9	145 - 300 min

Source: Spedding, D.J. Environmental Chemistry (1977) (Bockris, J.O'M., Ed.) Plenum, N.Y.

Indoor generated air pollution has a number of sources, including cooking, heating (fires, space heaters), use of aerosol spray cans, painting and cigarette smoking. The movement of people tends to keep suspended (or resuspends) particulates in the air longer. Both CO and NO_x can reach high levels owing to the number of combustion processes carried out indoors. Carbon dioxide levels can be 1 to 10 times higher indoors than outdoors due to enclosed combustion in homes, and also from the respiration of people. Carpets, curtains, and bedding act as repositories for particles which can be readily resuspended in the air by normal activity in a home.

CHAPTER 10

Water and Its Pollution

Before discussing water pollution we will consider some of the more important properties of water. Environmentally, water is of considerable importance, and water pollution is a serious problem, with a long history of study.

THE STRUCTURE OF AND BONDING IN WATER

The two hydrogen atoms are covalently bonded to oxygen, the bonding electrons being polarised towards oxygen. The bent shape (bond angle 104.5°) is a consequence of the four valence-shell electron pairs around the oxygen atom, two bonding and two non-bonding, arranged in approximate tetrahedral positions (Fig. 10.1a). Lone pair and lone pair-bond pair repulsions reduce the bond angle to less than 109.5°.

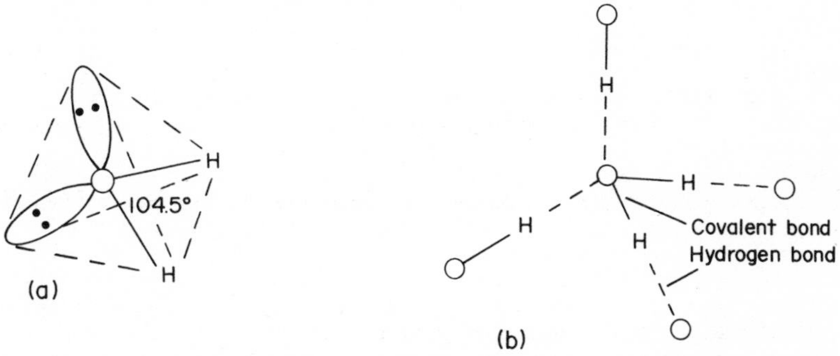

Fig. 10.1. (a) The structure of the water molecule, (b) hydrogen bonding in water and ice.

THE STATES OF WATER

The three physical states of water exist in the environment. The wide liquid range 273-373K (0-100°C) is due to the relatively strong intermolecular forces between water molecules, called hydrogen-bonds. In fully developed hydrogen-bonding, as in ice (Fig. 2.7), each oxygen atom has four neighbouring hydrogen atoms (Fig. 10.1b), a structural feature which persists, to some extent, even close to 100°C. The influence of hydrogen-bonding leads to an elevated boiling point of water (Fig. 10.2) relative to other p-block element hydrides.

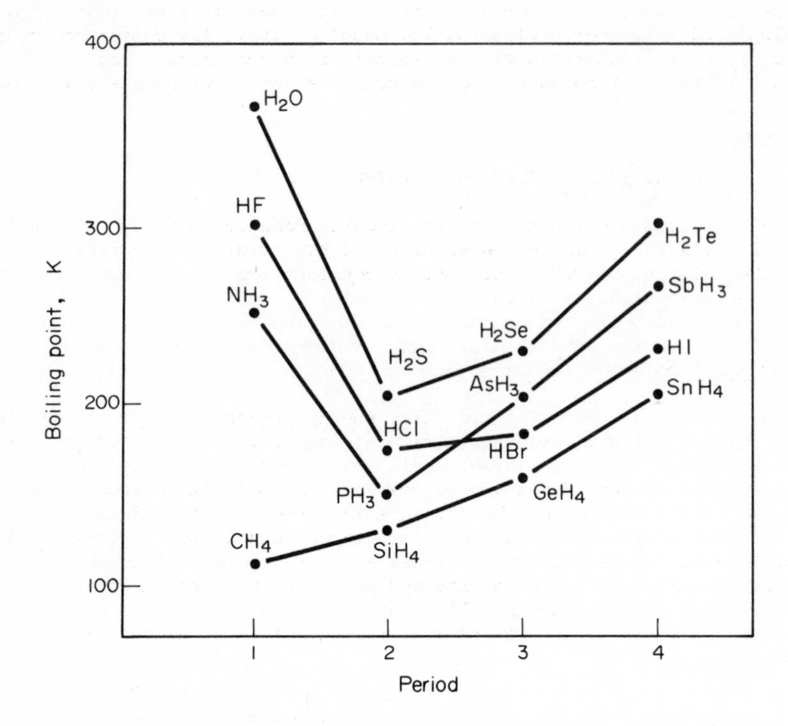

Fig. 10.2. The boiling points of the covalent hydrides of the p-block elements.

The hydrogen-bond is an electrostatic attraction between lone-pairs of electrons on the oxygen atoms and the partially positively charged hydrogen atoms of adjacent H_2O molecules. The range of hydrogen bonds and some bond properties are given in Table 10.1.

TABLE 10.1 The Hydrogen-Bond

Bond	X...Y distance (pm)	Sum of Van der Waals' Radii X and Y	Hydrogen Bond energy (kJ mol^{-1})
F - H - F (symmetrical)	226 - 232	270	113
F - H....F (unsymmetrical)	255	270	28
O - H....O	248 - 276	280	19 - 30
N - H....H	294 - 300	300	25
N - H....F	263	290	21
N - H....O	291	290	10

There are nine structural modifications of ice, each containing fully developed hydrogen-bonding. The resulting tetrahedral arrangement around each oxygen, produces the open structure, with a density approximately half the calculated value for a close-packed model of ice. The common structure (Fig. 2.7) contains channels with hexagonal cross-sections, a feature reflected in the flat hexagonal, or columnar hexagonal prism shape of snow flakes.

SOME PHYSICAL PROPERTIES OF WATER

The polar O-H bond and the bent shape of water make it polar with a dipole moment of 6.1×10^{-30} Cm. Hence water has a high dielectric constant of 78.5. The polarity, and high dielectric constant are reasons why water is a good solvent especially towards ionic materials.

Water as a Solvent

When a solid dissolves in a liquid three energy processes are involved: the solvent-solvent interaction, the lattice energy of the solute and the solute-solvent interactions (solvation energy). Because of the hydrogen-bonding in water, it could be considered a poor solvent compared with solvents such as C_2H_5OH and $CHCl_3$, because of the strong solvent-solvent interaction. This is certainly the case for weakly or non-polar solutes. For ionic solutes, the solvent-solute interactions in water are ion-dipole coulomb attraction, and increase with a decrease in size and increase in charge of the ion. Data on the hydration of the alkali metal cations (Table 10.2) bear this out. The high charge density (charge/radius) for the Li^+ ion leads to extensive hydration.

TABLE 10.2 Hydration of the Alkali Metals

Metal	Li^+	Na^+	K^+	Rb^+	Cs^+
Ionic radius (pm)	60	95	133	148	169
Charge density (charge/radius) (pm^{-1})	0.0167	0.0105	0.0075	0.0068	0.0059
Hydrated radius (pm)	340	276	232	228	228
Ionic mobility (infinite dilution)	33.5	43.5	64.6	67.5	68
Hydration number	25.3	16.6	10.5	10	9.9
Hydration energy kJ mol^{-1}	-498	-393	-310	-284	-251

The force of attraction between oppositely charged ions (q_+ and q_-) is given by:

$$F = \frac{q_+ q_-}{4\pi\varepsilon r^2} . \qquad (10.1)$$

The dielectric constant (ε) of the medium between the charges, is inversely proportional to F. Since for water ε is large, the attractive forces between dissolved positive and negative ions are considerably reduced by the water molecules interposed between them.

The thermochemistry of the dissolution of a solid in water may be represented by the cycle (solvent-solvent interactions are omitted being a constant energy term for different solutes);

$$\begin{array}{ccc} AX_{(s)} & \xrightarrow{\Delta H_{soln}} & A^{+}_{(aq)} + X^{-}_{(aq)} \\ -(\Delta H_{latt}) \searrow & & \nearrow \Delta H_{hyd} \\ & A^{+}_{(g)} + X^{-}_{(g)} & \end{array} \quad (10.2)$$

and the heat of solution is given by;

$$\Delta H_{soln} = -(\Delta H_{latt}) + \Delta H_{hyd} \quad (10.3)$$

For ionic substances both these enthalpies are large, and often the difference, ΔH_{soln}, is small (Table 10.3). A negative value of ΔH_{soln} indicates a substance becomes less soluble as the temperature rises (and vice versa).

TABLE 10.3 Enthalpies of Solution (kJ mol^{-1}) and Solubilities

Compound	ΔH_{latt}	ΔH_{hyd}	ΔH_{soln} [a]	Solubility g/100g
NaF	-918	-853	+65	4.22
NaCl	-788	-778	+10	35.1
NaI	-670	-698	-28	184
KI	-615	-615	0	127
LiI	-729	-803	-74	165
AgCl	-912	-845	+67	2.8×10^{-4}

[a] $\Delta H_{soln} = -(\Delta H_{latt} - \Delta H_{hyd})$

CHEMICAL PROPERTIES OF WATER

Water has a standard enthalpy of formation of -286 kJ mol^{-1}, and a standard free energy of formation of -237 kJ mol^{-1}. The OH bond energy is 464 kJ mol^{-1}, and is the second strongest bond of the p-block hydrides (second to HF 565 kJ mol^{-1}). The strength of this bond is an important factor in the stability of water.

As outlined in Chapter 2, water can be oxidised and reduced, being pH dependent. Water has a wide stability range, an important feature for the existence of species in an aqueous environment. Water has a small, but important, self-ionization;

$$2H_2O_{(\ell)} \rightleftharpoons H_3O^{+}_{(aq)} + OH^{-}_{(aq)}, \quad (10.4)$$

where $$K_w = [H_3O^+][OH^-] \text{ mol}^2 \text{ l}^{-2} \quad (10.5)$$
$$= 10^{-14} \text{ at } 298K.$$

By definition pure water is neutral, and since $[H_3O^+] = [OH^-] = 10^{-7}$ mol l^{-1} its pH = pOH = 7. Water is a good medium for acid-base reactions, strong acids are completely dissociated (0.1 mol l^{-1} HCl contains 0.1 mol l^{-1} of H_3O^+, i.e. a pH of 1), while weak acids undergo partial dissociation for example:

$$CH_3COOH + H_2O \rightleftharpoons CH_3COO^- + H_3O^+ \quad (10.6)$$

where $$K = \frac{[H_3O^+][CH_3COO^-]}{[CH_3COOH]} \quad (10.7)$$
$$= 1.85 \times 10^{-5}$$

The low value of K indicates weak dissociation, i.e. little H_3O^+ and CH_3COO^- exists in solution. The pH of a 0.1 mol l^{-1} solution of acetic acid is 2.9.

As a number of substances are dissolved in both fresh and salt water, their pH's will not be 7. For example surface sea-water has a pH near 8.1 due to dissolved CO_2, which exists principally as the HCO_3^- ion (see Chapter 2).

Hydrolysis

The importance of hydrolysis in the environment has already been mentioned (Chapters 1 and 2). Hydrolysis is a reaction of water where one, or both, of the O-H bonds are broken. Hydrolysis of a salt in water may proceed according to one of the four processes given in Table 10.4. Metallic cations are, in general, solvated first, either through ion-dipole interaction (e.g. $[Mg(H_2O)_6]^{2+}$), or by bond formation (e.g. $[Cu(H_2O)_4]^{2+}$). For the alkali metals, and most alkaline earth metals this is as far as the process proceeds, but for many other cations hydrolysis produces acid solutions;

$$M^{n+} + xH_2O \underset{\text{solvation}}{\rightleftharpoons} [M(H_2O)_x]^{n+} \underset{\text{hydrolysis}}{\rightleftharpoons} [M(H_2O)_{x-1}OH]^{(n-1)^+} + H^+_{aq} \quad (10.8)$$

$$[M(H_2O)_{x-1}OH]^{(n-1)^+} \rightleftharpoons [M(H_2O)_{x-2}(OH)_2]^{(n-2)^+} + H^+_{aq} \quad (10.9)$$

TABLE 10.4 Hydrolysis of Salts

Salt Type (derived from)	Example	Nature of hydrolysis solution	Ion hydrolysed
Strong base - weak acid	$Na(CH_3COO)$	alkaline	anion
Weak base - strong acid	NH_4Cl	acid	cation
Strong base - strong acid	NaCl	neutral	none
Weak base - weak acid	$NH_4(CH_3COO)$	~neutral	both

The process may stop at forming $[M(OH)_x]^{(n-x)+}$ or an intermediate product. The end product is often an insoluble material, especially when n = x, the metal hydroxide.

The extent of hydrolysis is affected by the charge on the cation, its radius and the pH. The process may be represented as follows;

$$M^{n+} + H_2O \xrightarrow{\text{solvation}} M^{n+} \cdots \overset{\delta-}{O}\left[\begin{matrix} H \\ H \end{matrix}\right]^{\delta+} \text{ (hydrate)} \searrow \left[M - O \begin{matrix} H \\ H \end{matrix}\right]^{n+} \text{ (intermediate)} \xrightarrow{\text{hydrolysis}} [M-OH]^{(n-1)+} + H^+_{aq} \tag{10.10}$$

The ease with which the reaction proceeds will depend on the polarising power of the cation, i.e. on its charge density. For example the Be^{2+} ion (radius 31 pm) undergoes extensive hydrolysis. Metal atoms that exhibit variable valency hydrolyse more readily in the higher oxidation states, i.e. where the cation charge is greatest. This is apparent by the existence of oxyanions for the high oxidation states of some transition metals. Acid suppresses cation hydrolysis, and Fe^{2+} ions which are stable in acid hydrolyse to $Fe(OH)_2$ as the pH rises.

For a hydroxide $M(OH)_n$ the solubility product is given by;

$$K_{sp} = [M^{n+}][OH^-]^n. \tag{10.11}$$

Combining this with K_w and eliminating $[OH^-]^n$ gives;

$$K_{sp} = \frac{[M^{n+}]K_w^{\ n}}{[H_3O^+]^n}, \tag{10.12}$$

which can be written as an expression for pH;

$$pH = \frac{1}{n}\log K_{sp} - \frac{1}{n}\log[M^{n+}] - \log K_w. \tag{10.13}$$

In general the solubility of hydroxides decreases as the value n increases. Also if M^{n+} is kept constant, the pH at which precipitation occurs will drop, the higher the charge n on the cation (Table 10.5). Certain cations therefore cannot exist free in natural waters at a pH around 8.

Covalent species may also undergo hydrolysis provided there is a site where water may attack the compound. For example CCl_4 does not hydrolyse as the small carbon is shielded by the four Cl-atoms, and the carbon does not have available orbitals to, initially, accommodate the oxygen lone pair of electrons. The silicon of $SiCl_4$, on the other hand, is less shielded (larger atom) and it can make use of its empty 3d orbitals. Hence;

$$\overset{+\delta}{Si}Cl_4 \leftarrow \overset{-\delta}{O}H_2 \longrightarrow SiCl_3OH + HCl \xrightarrow{H_2O} Si(OH)_4 + 4HCl \tag{10.14}$$

TABLE 10.5 Solubility Products and pH of Precipitation as hydroxides (solution concentration 0.01 mol l^{-1})

Ion	K_{sp} $mol^3\ l^{-3}$	K_{sp} $mol^4\ l^{-4}$	K_{sp} $mol^5\ l^{-5}$	pH of precipitation
Mg^{2+}	1.3×10^{-10}			10.0
Ca^{2+}	1×10^{-17}			6.5
Fe^{2+}	1×10^{-14}			8.0
Cu^{2+}	5×10^{-19}			5.5
Al^{3+}		3.7×10^{-15}		9.8
Cr^{3+}		6.7×10^{-31}		4.1
Fe^{3+}		1×10^{-38}		2.1
Ti^{4+}			7.9×10^{-54}	1.2

Most covalent halides hydrolyse in this way (except NCl_3).

Most oxides, when soluble, react with water giving either acidic or basis solutions. Basic oxides (Groups I and II and low oxidation states of the transition metals) react as follows;

$$CaO + H_2O \rightarrow Ca(OH)_2 \rightleftharpoons Ca^{2+} + 2OH^- \qquad (10.15)$$

$$Na_2O + H_2O \rightarrow 2Na^+ + 2OH^- \qquad (10.16)$$

This is a direct result of the reaction of the oxide ion with water;

$$O^{2-} + H_2O \rightleftharpoons 2OH^-,\ K > 10^{22} \qquad (10.17)$$

Acidic oxides react as follows;

$$SO_3 + H_2O \rightarrow \underset{H_2SO_4}{(HO)_2SO_2} \rightleftharpoons HSO^-_{4(aq)} + H^+_{(aq)} \rightleftharpoons SO^{2-}_{4(aq)} + 2H^+_{(aq)} \qquad (10.18)$$

$$CO_2 + H_2O \rightarrow \underset{H_2CO_3}{(HO)_2CO} \rightleftharpoons HCO^-_{3(aq)} + H^+_{(aq)} \rightleftharpoons CO^{2-}_{3(aq)} + 2H^+_{(aq)} \qquad (10.19)$$

Some oxides function as acids and bases (amphoteric) depending on the pH of the solution. A classification of oxides according to their reactions with water, or acids and bases, is given in Table 10.6.

TYPES OF WATER POLLUTION

Water is essential to life, and as some water pollution can have serious health consequences, particularly if the pollution leads to pathogenic materials, considerable effort is put into water treatment. The two main areas are, provision of potable water by chemical treatment, and the handling and treatment of sewage, domestic and industrial. Patchy effort is made to control the direct pollution of waterways.

TABLE 10.6 Oxides: Acid-Base Properties

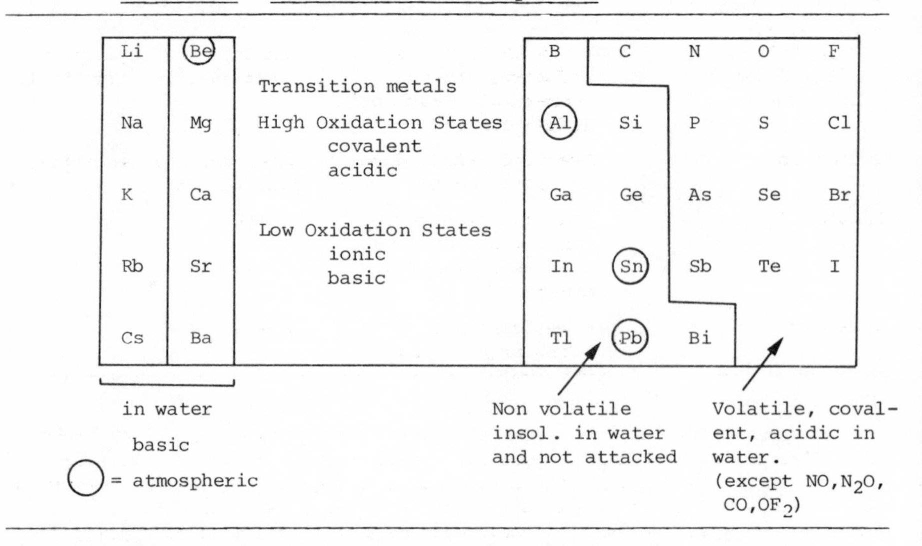

A significant amount of water pollution problems are with systems that are not compatible with water. Two prominent examples are oil spills, and the persistence of organochlorine compounds which do not hydrolyse. It is not easy to classify water pollutants and their sources. However, one attempt is given in Table 10.7. The first four groupings are the major ones.

TABLE 10.7 Types of Water Pollutants

Type of Pollutant	Sources	Effects
Oxygen demanding sewage	Urban environment, domestic, industrial food processing, commercial.	Oxygen levels in water, water quality, eutrophication.
Disease causing		
Chemical carcinogens	Urban environment, Uncontrolled wastes, sewage	Health effects
Pathogens		
Organic chemicals		
Trace amounts	Urban environment, rural, agriculture, oil spills, sewage, industrial, commercial, cleaning oil tankers & use of sea-water as ballast	Health, environmental destruction, eutrophication, wild life esthetics, aquatic life
Pesticides		
Polychlorinated biphenyls		
Petroleum wastes		
Detergents		

Inorganic chemicals and minerals Trace elements metal-organic acid-bases	Urban environment, industry, mining, commercial, leaching refuse dumps	Health, metal transport, water quality, aquatic life
Sediments	Leaching, wind, dumping solid waste	Water quality, wildlife, aquatic life
Radionuclides	Nuclear power stations, waste disposal, nuclear explosions, mining	Health
Thermal	Power stations, urban environment	Aquatic life

Oxygen demand chemicals are materials that directly (e.g. organic wastes) or indirectly (phosphates) reduce the oxygen content of water. Disease causing chemicals are directly concerned with the health of human beings, an area where special attention must continue to be given. Organic and inorganic chemicals are two wide classifications of pollutants that overlap significantly with the first two areas.

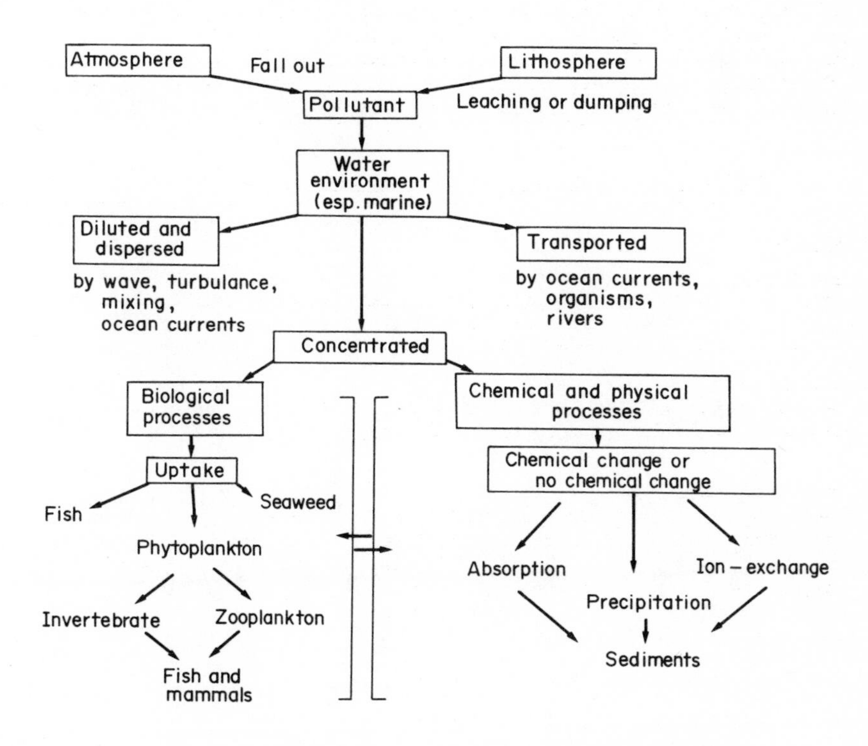

Fig. 10.3. Movement of pollutants in the hydrosphere.

MOVEMENT WITHIN THE HYDROSPHERE

Water is in constant movement, and materials added at one place are transported to other areas. For example, if a material, such as milk whey, is dumped into a river, its movement can be followed as the whey has a significant oxygen demand, and O_2 depletion can be measured downstream. Another main process that can happen in water is dilution of the chemical and its dispersion. The milk whey eventually becomes so diluted that it becomes less easy to trace. But some materials are in fact concentrated, either by biological processes and through feeding chains, or by chemical processes such as absorption and precipitation. The movement of materials in water systems is summarised diagrammatically in Fig. 10.3. The figure shows that the two concentrating processes are not independent, but material can move from one to the other.

CHAPTER 11

Water Pollution

Various aspects of water pollution will now be discussed, with emphasis on the chemistry of individual metal ions and anionic species in water, and water treatment.

SPECIES IN WATER - METALS

The addition of metal ions to natural water systems is an ubiquitous process, involving all avenues of human endeavour, ranging from large industrial enterprises to activities in the home. The philosophy has often been that when "washed away" the material is no further trouble. It is now widely realized that this is not the case, and a variety of complex reactions may take place in the aquatic environment, which can alter a "harmless material" in to one which is toxic. In Table 11.1 a few metals are listed, and some of the industrial processes that contribute to their appearance in water-ways. In addition to industry, metals get into water-ways from weathering of rocks and soils, atmospheric fallout, agricultural activity, domestic sewage, runoff from streets and highways and land clearing. A variety of interrelated factors are important in determining the concentration of a metal in water systems, its availability, transport and its toxicity. Some of these factors will now be mentioned.

Perhaps the most important feature is the chemical form in which the metal exists in solution; whether it is a hydrated ion, or a hydroxy or oxy species, an ion-pair, or whether it is associated with a coordinating agent, and therefore the nature of such reagents is also relevant. The chemical form can also influence the toxicity, for example Ni_2S_3 is considered to be carcinogenic. The chemical species is related to the oxidation state of the metal, which is variable for a number of metals. For example, chromium (VI) is more water soluble and toxic than chromium (III).

The solubility of a metal species is of central importance, and depends on the temperature, the nature of other species present in the water, time, particle size, impurities, structural chemistry and competing chemical reactions. For example, PbS has a solubility product $K_{sp} = 8.4 \times 10^{-28}$ and the concentration of lead is 2.9×10^{-14} mol l^{-1}. But this can alter due to hydrolysis of the sulphide ion;

TABLE 11.1 Some of the Major Trace Metals in Water Effluents from Some Industries

Source	As	Ba	Be	Bi	Cd	Cr	Cu	Hg	Mn	Pb	Ni	Se	Sn	U	V	Zn
Mining and ore processing	X		X		X			X	X	X		X		X		X
Metallurgy	X		X	X	X	X	X	X		X	X				X	X
Chemical industry	X	X			X	X	X	X		X			X	X	X	X
Alloys			X							X						
Paint		X			X	X				X						X
Glass	X	X								X	X					
Pulp and paper mills						X	X	X		X	X					
Leather	X	X				X	X	X								X
Textiles	X	X			X		X	X		X	X					
Fertilizers	X				X	X	X	X	X	X	X					X
Chloro-alkali production	X				X	X		X		X			X			X
Petroleum refining	X				X	X	X			X	X				X	X
Coal burning	X		X		X	X	X	X	X	X	X	X		X		
Nuclear technology			X		X									X		

$$PbS_{(s)} \rightleftharpoons Pb^{2+}_{(aq)} + S^{2-}_{(aq)}, \tag{11.1}$$

$$S^{2-}_{(aq)} + H_2O \rightleftharpoons SH^-_{(aq)} + OH^-_{(aq)}, \tag{11.2}$$

where $$K_h = \frac{[SH^-][OH^-]}{[S^{2-}]} = 8.3 \times 10^{-2} \tag{11.3}$$

Since $[Pb^{2+}] = [S^{2-}] + [SH^-]$, and $[SH^-] = [OH^-]$, solution of these equations gives the new concentration of Pb^{2+}, i.e.;

$$K_{sp} = \{[S^{2-}] + [SH^-]\}[S^{2-}] \tag{11.4}$$

$$= \{\frac{[SH^-]^2}{K_h} + [SH^-]\}\frac{[SH^-]^2}{K_h}, \tag{11.5}$$

i.e. $$K_{sp}K_h = [SH^-]^3\{\frac{[SH^-]}{K_h} + 1\} \tag{11.6}$$

which gives: $[SH^-] = 4.1 \times 10^{-10}$ mol l^{-1}, $[S^{2-}] = 2.0 \times 10^{-18}$ mol l^{-1}, and $[Pb^{2+}] = 4.1 \times 10^{-10}$ mol l^{-1}.

Hence there is more lead in solution, and this can alter with change in pH.

Sorption processes also influence levels and availability of metal species. The sorption depends on the charge on the sorbing material (e.g. clays), and the sorbate (metal ion or compound). Ion exchange can alter the nature of the absorbed species, metal ions not strongly held may be released by another metal ion or by a species present in relatively high concentration. Sorption is also a function of pH, hydrolysis of metal ions, which is pH dependent, provides hydroxy precipitates which are good surfaces for carrying down,by co-precipitation, other ions.

Biological processes are important in the aquatic chemistry of metal species. For example, bio-amplification can be the cause of toxic metals becoming concentrated sufficiently to be dangerous. Plankton concentrate Cu 90,000 times, Pb 12,000 times and Co 16,000 times above the levels in water. This process itself is dependent on a number of factors including geographical position, time, season, temperature and salinity. Seasonal variations observed in the levels of certain metal ions may be a function of bio-activity. In summer Zn and Cd levels have been measured at 6-500 $\mu g\ l^{-1}$ and 3-80 $\mu g\ l^{-1}$ respectively, while in winter the levels drop to 4-180 $\mu g\ l^{-1}$ and 2-10 $\mu g\ l^{-1}$ respectively.

It becomes increasingly clear, that it is not easy to generalize regarding the quantitative levels and chemistry of metal ions in the hydrosphere. We will therefore outline the aquatic chemistry of some of the more serious metal pollutants, including Hg, Cd, Pb, As, Cr.

Mercury

Mercury is one of the most toxic of the metals, producing serious irreversible neurological damage, especially when in the methylmercury form. Because of this and because a number of serious episodes of mercury poisoning have occurred it could be considered the most dangerous of the metal pollutants. Up to the 1960's environmental mercury was thought to be no problem, but the situation has changed since the discovery of the methylation of mercury in the environment.

Mercury inputs, cycle and concentrations in the environment. The major input of mercury into the environment is from natural sources, particularly from the degassing of the earth's crust (Table 11.2), around 25 and 150 x 10^3 tonnes yearly. Anthropogenic inputs around 16-20 x 10^3 tonnes a year, are often localised, and high concentrations produced have had serious consequences for health. An estimate of the global mercury cycle is presented in Fig. 11.1. The major transfers of mercury are the degassing of the earth and precipitation over the oceans. Representative concentrations of Hg in various areas and materials are given in Table 11.3. The figures are a guide to the levels (1 ng l^{-1} = 1 ppb), and can vary considerably with geographical position and distance from industrial sources.

Uses of mercury. The metal has a number of uses and the world's yearly production is of the order of 1 x 10^4 tonnes yr^{-1} (1973), increasing at a rate of around 2% yr^{-1}. However, this may rapidly decline as resources (HgS and Hg) are worked out. The main uses of Hg (expressed as % of the

TABLE 11.2 Mercury Inputs into the Environment (1971 - 1974) (x 10^3 tonnes yr^{-1})

Source	Quantity	Source	Quantity
Natural		Anthropogenic**	
Continents to atmosphere	25-150*	From fossil fuel combustion	3-5
River transport to oceans	~5	From cement manufacture	1.0
In precipitation	~30	Industrial and agriculture losses	4.0
World Production			
1968	8.0		
1973	10.0		

* From natural degassing of the earth's crust

** Total could reach 20 x 10^3 tonnes yr^{-1}

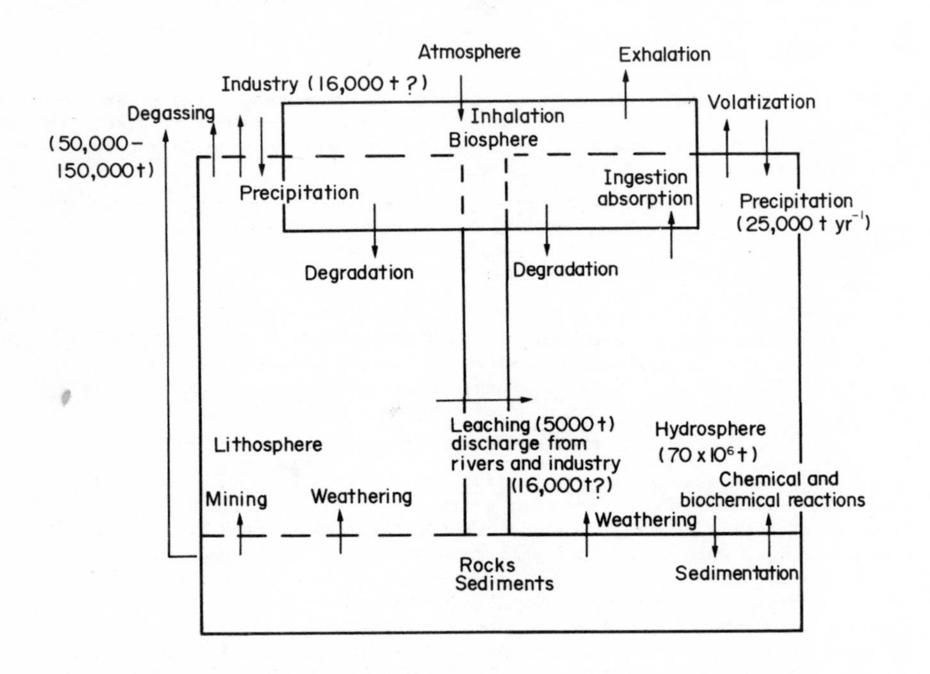

Fig. 11.1. The global mercury cycle (figures in tonnes)

total) are listed in Table 11.4. Loss of mercury, when used, as an electrode in the chloro-alkali electrolyte cell, has reached 150-200 g per tonne of Cl_2 produced. However, modifications of the plant, recycling and settling tanks, can reduce losses by 99%. Air pollution from the cells is now the main problem, but this can also be reduced by 50%.

TABLE 11.3 Concentration of Mercury in Selected Areas

Area or material	Concentration	Area or material	Concentration
Atmosphere	0.5-50 ng m^{-3} ~20 ng m^{-3} (mean)	Rainwater	0-200 ng l^{-1}
Aerosol (Hg on dust)	0.02 ng m^{-3} (background)	Continental rocks	80 ng g^{-1} (mean)
(approx. 5% of total atmospheric)	1-40 ng m^{-3} (industrial areas)	Earth's crust	500 ng g^{-1}(mean)
Freshwater	10-50 ng l^{-1} (uncontaminated) 150-700 ng l^{-1} (industrial areas)	Top soil	10-2000 ng g^{-1}
Oceans	20-270 ng l^{-1} (depends on depth) 600 ng l^{-1} (Minamata Bay)	Coal	100 ng g^{-1} or greater (mean)

TABLE 11.4 Principal Uses of Mercury

Industry	Proportion (%)*	Industry	Proportion (%)*
NaOH/Cl_2 production	25	Catalysts	4
Electrical apparatus	20	Agriculture	5
Paints	15	Dental	3
Instruments	10	Pulp and paper	1
		Other	17

* Approximately 74% of mercury can be recycled.

Mercury is used in the pulp and paper industry, as NaOH and Cl_2 are often produced on site, but also mercury slimicides are used, e.g. phenylmercuric-dimethyldithiocarbamate, methylmercury phosphate, and phenylmercury acetate. However, the use of Hg-compounds in this way is being phased out and are banned in some countries (e.g. Sweden and Finland). Organomercury compounds are also widely used as fungicide coatings on seeds. The very toxic methyl-derivatives are now widely banned but numerous phenyl-derivatives are still used.

Mercury is widely used as a catalyst, for example in the production of acetaldehyde;

$$C_2H_2 + H_2O \xrightarrow[HgSO_4]{} CH_3CHO. \quad (11.7)$$

The catalyst undergoes the following changes;

$$Hg^{2+} \rightarrow Hg \xrightarrow[Fe^{3+}]{} Hg^{2+}. \quad (11.8)$$

Mercury is used in the formation of vinyl chloride;

$$C_2H_2 + HCl \xrightarrow[HgCl_2]{} CH_2{=}CHCl \tag{11.9}$$

and in the dye industry;

$$\text{anthraquinone} \xrightarrow[HgSO_4]{\text{oleum}} \text{anthraquinone-}SO_2OH \tag{11.10}$$

Mercuric oxide is used in a drycell battery;

$$\underset{\text{cathode}}{HgO} + \underset{\text{anode}}{Zn} \rightarrow ZnO + Hg \tag{11.11}$$

the system is in a solution of KOH saturated with zinc oxide.

Mercury compounds are added to paints as antifouling agents. Finally the metal is used widely in a variety of equipment, such as thermometers, barometers and electrical switches, because as a liquid metal, it has useful physical properties.

Chemical and biochemical pathways of mercury. In the mid-sixties it had become clear that mercury in waterways was not as inactive as originally thought, but can undergo a number of reactions leading to the formation of toxic methylmercury. Bacteria feature prominently in these reactions (Fig. 11.2). Mercury enters the aquatic environment in three principal forms, Hg metal, Hg^{2+} ion and organomercury compounds. The anthropogenic sources give localized high concentrations where the reactions in Fig. 11.2 are more likely to occur. The Hg_2^{2+} ion is of less importance and will not be considered.

Reaction (a) (Fig. 11.2). The oxidation of mercury to Hg^{2+} in water occurs in the presence of O_2, assisted by organic substances which presumably complex with the Hg^{2+} ion. The reaction is important for producing Hg^{2+} in the aqueous environment.

Reaction (b). Reaction (a) may be reversed in reducing conditions by the influence of bacteria (of the *Pseudomonas* genus) and may be the principal process by which degassing of Hg from the earth occurs.

Reaction (c). Under anaerobic conditions the Hg^{2+} ion will react with H_2S to give HgS;

$$Hg^{2+} + H_2S \rightarrow HgS + 2H^+ \tag{11.12}$$

The low solubility of HgS ($K_{sp} = 1 \times 10^{-53}$) suggests that the Hg^{2+} ion is almost completely removed from the water. However, the solubility will be greater due to influence of hydrolysis of the S^{2-} ion, the competition for Hg^{2+} by organic complexing agents, as well as by reaction (d).

Reaction (d). The reverse reaction, i.e. liberation of Hg^{2+} from HgS, is of considerable importance as it indicates that the sulphide in the upper sediments of waterways, is not inactive. The Hg^{2+} is liberated by oxidation of the S^{2-} by bacteria to $SO_3{}^{2-}$ and SO_4^{2-};

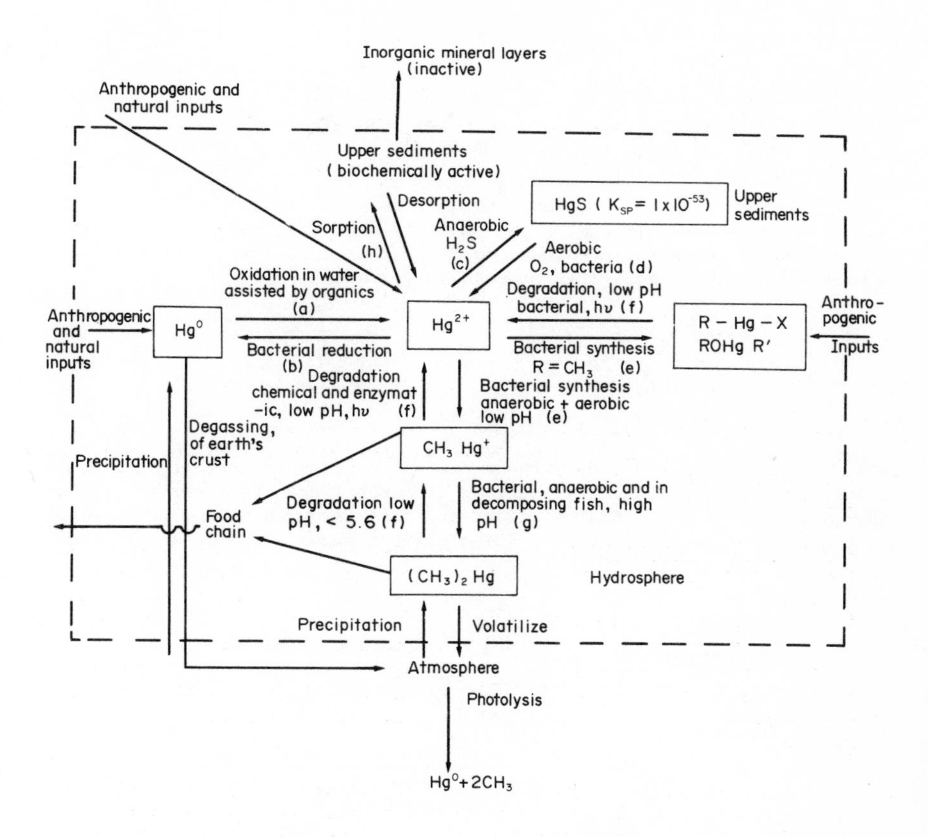

Fig. 11.2. Chemical and biochemical pathways of mercury in the hydrosphere.

$$S^{2-} + 3H_2O - 6e \rightarrow SO_3^{2-} + 6H^+, \quad -E^o = -0.59V \qquad (11.13)$$

$$SO_3^{2-} + H_2O - 2e \rightarrow SO_4^{2-} + 2H^+, \quad -E^o = -0.17V \qquad (11.14)$$

the oxidation will be achieved with dissolved O_2;

$$O_2 + 4H^+ + 4e \rightarrow 2H_2O, \quad E^o = +1.23V \qquad (11.15)$$

Reaction (e). The methylation of mercury to give CH_3Hg^+ occurs in the aquatic environment and is significant as regards Hg toxicity. Serious incidences of Hg poisoning from this source have been recorded, the methyl-mercury being consumed from the eating of fish, which accumulate the material in fatty tissue. The methylation is achieved by at least two biochemical pathways. One route is non-enzymatic under anaerobic conditions. The Hg^{2+}

ion is methylated by methylcobalamine, a methyl derivative of Vitamin B_{12}, formed by methane producing bacteria;

$$\begin{array}{c}CH_3\\ N_4Co^{3+}\\ |\\ N\end{array} + Hg^{2+} \rightarrow \begin{array}{c}CH_3\\ N_4Co^{3+}\\ |\\ N\text{-}HgOAc\end{array} \rightarrow \begin{array}{c}OH_2\\ N_4Co^{2+}\\ |\\ N\end{array} + CH_3Hg^+ \qquad (11.16)$$

As the reaction requires anaerobic conditions, and since in these conditions, the main source of Hg is HgS, it is unlikely that the reaction occurs to any great extent. The aerobic pathway occurs within cells, in which methionine is normally synthesized. This process is enzymatic;

$$\begin{array}{c}CH_3\\ N_4Co^{3+}\\ |\\ \text{enzyme}\end{array} + Hg^{2+} \longrightarrow \begin{array}{c}OH_2\\ N_4Co^{2+}\\ |\\ \text{enzyme}\end{array} + CH_3Hg^+ \qquad (11.17)$$

Both reactions are possible in the upper sediments, particularly sediments in suspension, where the outer surface is aerobic and the inner surface anaerobic (Fig. 11.3). The reactions are also favoured by low pH. Methylation may also occur for other species present in water, e.g. metal ions such as Pb, Sn, Si, Tl, Pt.

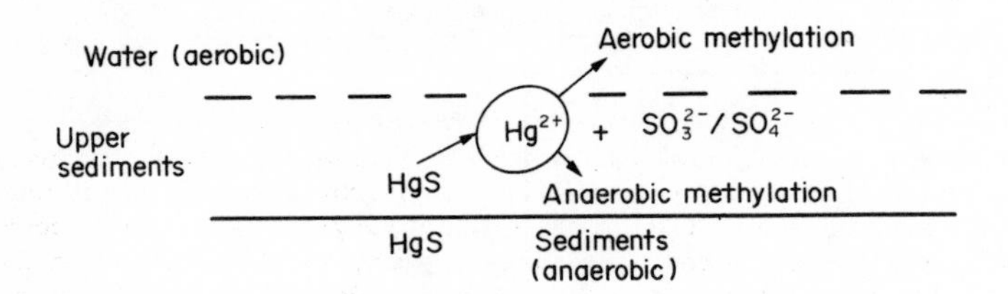

Fig. 11.3. Methylation of mercury.

Reaction (f) (Fig. 11.2). The degradation of CH_3Hg^+ back to Hg^{2+} is possible, involving bacteria, low pH and light.

Reaction (g). The further step, of producing $(CH_3)_2Hg$, is achieved by similar biochemical routes to reaction (e), and is favoured at high pH. Dimethylmercury is also produced in the bodies of decomposing fish. As $(CH_3)_2Hg$ is volatile and not soluble in water it can escape into the atmosphere, and either be returned in precipitation or undergo photolysis.

Reaction (h). It would appear that mercury in water sediments is in two forms. In the upper sediments the material is biochemically active, while in the inorganic mineral layers there is little or no activity and weathering may be the only way the Hg^{2+} is released again. The upper sediments are involved in the sorption of cations, and Hg^{2+}, one of the poorly sorbed ions, is easily replaced by other cations. A desorption order is Hg > Cu > Zn > Pb > Cr > As. Desorption may also depend on decomposing organic material giving ligands, which through coordination to the metal ions assists desorption.

The series of reactions (Fig. 11.2) tend to lead to the end products CH_3Hg^+ and $(CH_3)_2Hg$, both of which are concentrated in fish, and so become involved in the food chain. Though not much methylmercury is produced, it is bio-amplified and since a great deal of mercury is present in water sediments one can expect it to continue to form for a long time.

Mercury Toxicity. Fish appear to be the principal carriers of methylmercury, and when heavily contaminated are a threat to the general public. In fresh water, levels in fish can be 100 - 200 $\mu g\ kg^{-1}$ (wet weight), while in contaminated water levels within the range 100 - 5000 $\mu g\ kg^{-1}$ are reported (most around 500 - 700 $\mu g\ kg^{-1}$). For marine fish (especially sword fish, tuna and halibut) similar levels are attained, most around 200 - 1500 $\mu g\ kg^{-1}$.

Many people have suffered from methylmercury poisoning, either from consuming fish or seeds covered with a methylmercury fungicide. A large number of people have died or suffered irreversible brain damage, especially if exposed during the fetus stage. Mercury vapour, of which 80% is retained by the body, has similar neurological effects, getting to the brain via the blood stream. Inorganic mercury compounds are also dangerous, but to a lesser extent than organo-mercury compounds. The Hg^{2+} ion is a soft acid, and bonds strongly to sulphur, and is one way it poisons by bonding to proteins and enzymes. The signs of mercury poisoning are depression, irritability, paralysis, blindness, insanity, birth defects and chromosome breakdown, as well as death.

The WHO recommendation for the tolerable weekly intake is 0.3 mg of which no more than 0.2 mg should be mercury from methylmercury. These figures are equivalent to 5 μg and 3.3 μg per kg body weight respectively.

Mercury Poisoning. The record of mercury poisoning over the centuries is not good. In 1953 the first major disaster of methylmercury poisoning occurred at Minimata, Japan, with at least 43 deaths and 116 irreversibly affected by 1960. The mercury came from a nearby chemical plant where it was used as a catalyst in the production of vinyl chloride and acetaldehyde. A similar situation occurred at Niigata, Japan, with 5 deaths and 26 irreversibly poisoned. In Iraq (1971-72) 500 deaths occurred and around 7000 people were poisoned from eating flour made with grain treated with methylmercury fungicide. It has been estimated that around 60,000 people were affected in some way. Other episodes occurred at Lake St. Clair and the Wabigon - English - Winnipeg river systems in Canada. The mercury came from pulp and paper mills and chloroalkali plants. In New Mexico (1969-70) a family were seriously poisoned from eating pork from pigs which had died from being fed contaminated seeds.

Lead

Lead, the most widely distributed of the three metals, Pb, Cd and Hg, is less of a problem in the hydrosphere. Estimates of fluxes and reservoirs, in the hydrosphere, are listed in Fig. 11.4. More lead is deposited on the land and oceans than lost to the atmosphere from sea-spray. This comes from anthropogenic sources, i.e. industry, and lead in petrol. The run-off, from English and Welsh mining areas, has produced the pattern of concentrations of lead in the Irish Sea as shown in Fig. 11.5. A similar pattern is found for cadmium.

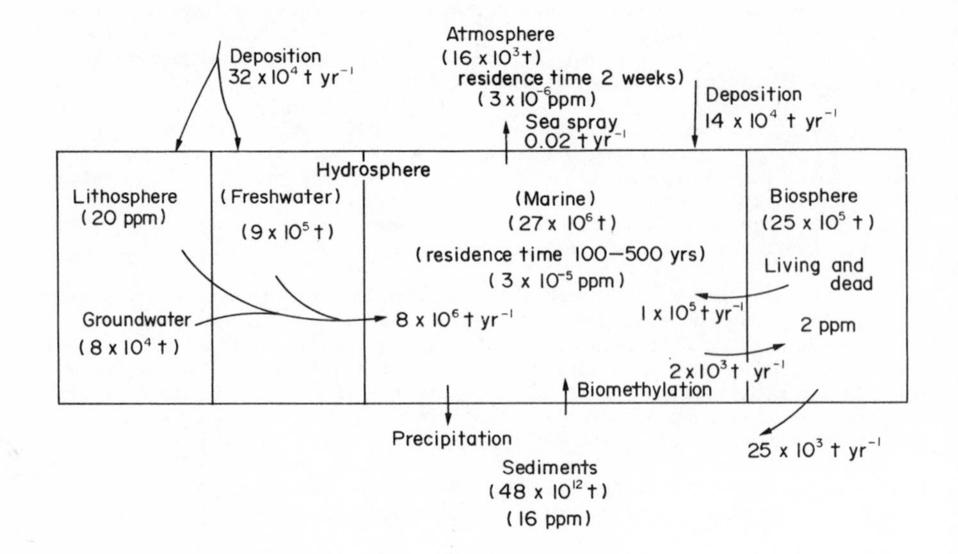

Fig. 11.4. Lead in the aqueous environment.

The forms of lead in water vary with the type of water, and can be classified in terms of the size of the particles (Fig. 11.6). In natural waters most lead appears to be associated with inorganic colloids. Since there is a range of particle sizes, chemical analysis of lead in water (and metals in general) depend on the methods used to separate small and large particles. The separation achieved is also pH dependent as the lead species, and hence particle size, can alter considerably by varying the pH. For example, lead absorbed on to clay can become a free metal ion in strongly acid conditions.

$$Pb - clay + 2H^+ \rightarrow H_2\text{-clay} + Pb^{2+} \qquad (11.18)$$

In deep water the concentration of lead is around 1 x 10^{-11} mol l^{-1}, less than expected suggesting the metal is tied up with other materials. For example, PbO_2 can form solid solutions with MnO_2. The major lead species in sea-water, and their approximate proportions are;

at pH 7		at pH 8		at pH 9	
$PbCl^+$	40%	$PbCO_3$	80%	$PbCO_3$	90%
$PbCO_3$	30%	$PbCl^+$	15%	$PbCl^+$	5%
$PbCl_2$	15%	$PbCl_2$	5%	$PbOH^+$	5%

Around 5 to 30% of lead in urban runoff - i.e. lead arising from industry, paint and its use in petrol - is soluble. The rest occurs as low soluble solids, mainly as phosphates, carbonates, sulphates and oxy-hydroxy species.

The three principal problem areas for lead in water are, the use of lead plumbing, lead in containers, e.g. tins and glazed pottery, and the biomethylation of lead.

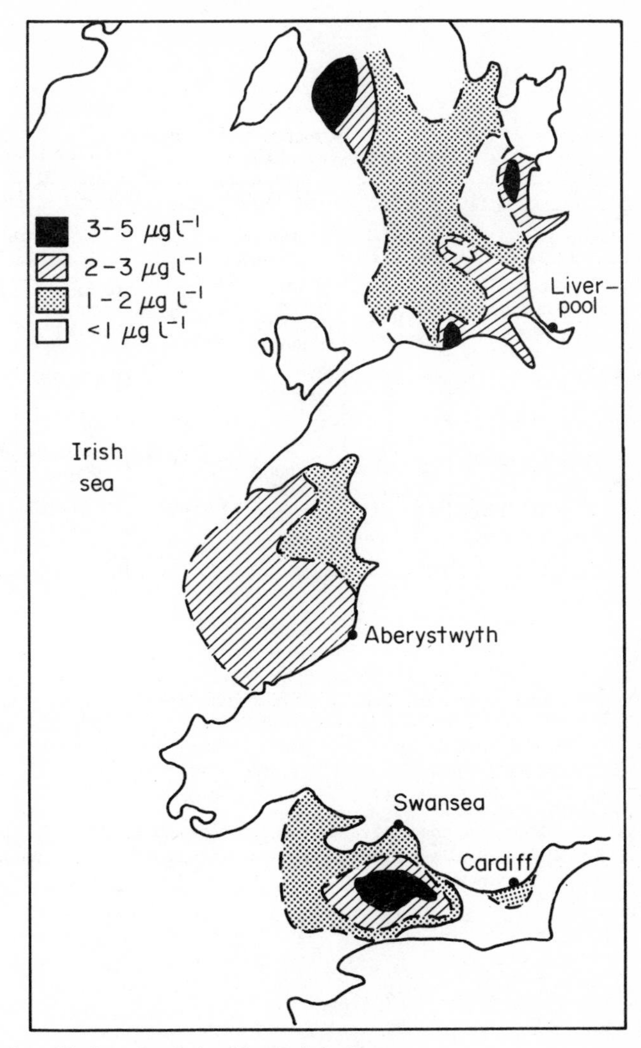

Fig. 11.5. Lead in the Irish Sea

In a recent survey (3000 homes) of lead in tap water, in the United Kingdom, it was found that 10.3% of the households had lead levels > 50 μg l^{-1} and 4.3% > 100 μg l^{-1}, while 25.3% of the homes had lead levels between 10 and 50 μg l^{-1}. Levels tended to be higher in Scotland than England and Wales. A suggested upper limit for drinking water is 50 μg l^{-1}. Water that is in contact with lead for some time can dissolve appreciable quantities of the metal. Soft, acid water appears to have the greatest "plumbosolvency". A four-fold increase in solubility is reported with a pH change from 6 to 4. Some hard waters have a significant "plumbosolvency", but to a lesser extent than soft water, also water at a pH 8 to 10 has improved solvent properties. Lead, as well as other metals, can be transferred from cooking water to food. The solubility products of common lead compounds found in the environment, are reasonably low; $PbCO_3$ (1.0×10^{-13}), $Pb(OH)_2$ (1.2×10^{-15}), $PbSO_4$ (1.3 x

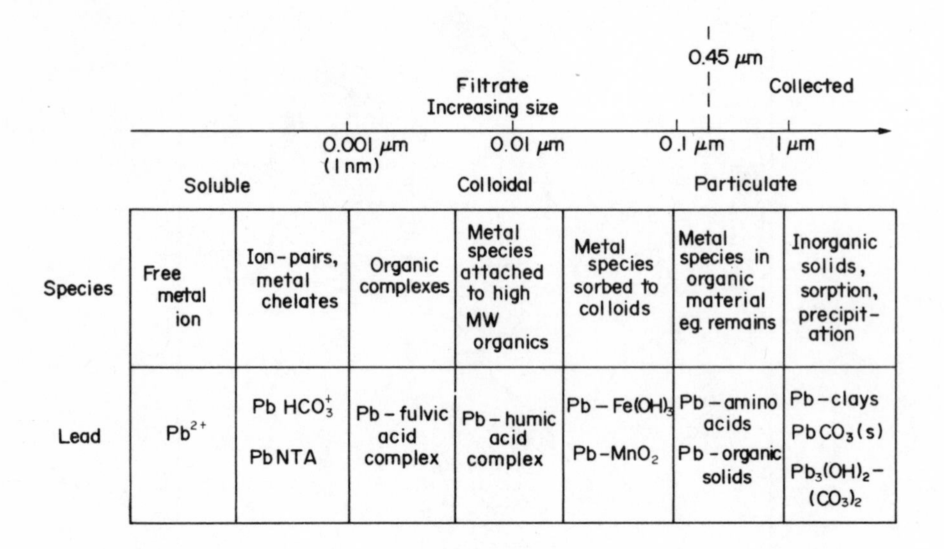

Fig. 11.6. Forms of Lead in Water

10^{-8}), $Pb_3(OH)_2(CO_3)_2$ (1.6×10^{-10}) and PbS (1.3×10^{-29}). However, all of these compounds become more soluble in acid conditions because equilibria such as;

$$2PbCO_3.Pb(OH)_{2(s)} \rightleftharpoons 3Pb^{2+}_{(aq)} + 2OH^-_{(aq)} + 2CO^{2-}_{3(aq)} \tag{11.19}$$

increase in acid. Also in the presence of complexing agents the lead becomes soluble. For $Pb(OH)_2$, in the presence of say nitrilotriacetic acid, $N(CH_3COOH)_3$,(TH_3), we can write;

$$Pb(OH)_{2(s)} + HT^{2-} \rightleftharpoons PbT^- + OH^- + H_2O, \tag{11.20}$$

where $$K = \frac{[PbT^-][OH^-]}{[HT^{2-}]} = 2.1 \times 10^{-5}. \tag{11.21}$$

At a pH of 8, i.e. $[OH^-] = 1.0 \times 10^{-6}$ mol l^{-1}, we have;

$$\frac{[PbT^-]}{[HT^{2-}]} = \frac{2.1 \times 10^{-5}}{1.0 \times 10^{-6}} = 21, \tag{11.22}$$

Most of the lead is present as the soluble complex, provided sufficient of complexing agent (or a similar agent) is available, and provided the equilibrium (11.20) has had time to establish.

Lead glazed pottery has been the cause of a number of lead poisonings. If the glaze is incorrectly fired (generally at too low a temperature) the lead ions can be leached out, particularly by acid materials such as fruit juices. Limits of extractable lead, using weak acetic acid solutions, are laid down by the health authorities. Lead solder used in tin cans is also a problem, particularly if the food is left in an opened can. Canned food can contribute up to 15% of the total dietary intake of lead. The lead content of canned food may be 10 times the lead content of fresh food, for example, analytical studies on fresh peas indicated a mean lead content of 0.02 ppm (range 0.01 - 0.04 ppm), while for canned peas the figures were 0.21 ppm (mean) and a range of 0.01 - 0.72 ppm. Tin solder is now being used more often, especially for canned baby foods. Old pewter (Sn 89%, Pb 2%, Sb 7%, Cu 2%) contains lead, though modern pewter is lead free.

Biomethylation of lead at the sediment-water interface, is not as well understood or studied as the methylation of mercury. However, evidence for methylation, producing methyl lead species, $(CH_3)_3Pb^+$ and $(CH_3)_4Pb$ has been found. This could well be a long term problem.

Cadmium

Cadmium, like Hg and Pb is considered as a serious metal toxin. But unlike Hg and Pb, it is not a neurotoxin, but has deliterious effects on bone structure and kidneys. The "classic" example of Cd poisoning occurred in Japan in the 1960's, when the disease (called Itai-Itai disease) caused large scale demineralization of bones leading to shrinkage, distortion and embrittlement (see Chapter 12, Fig. 12.10).

Cadmium is closely associated with zinc geochemically (approximately, Zn:Cd = 350:1) and zinc will contain some cadmium except when very pure. The close association between Cd and Zn stems from their similar chemical properties, being in the same periodic group. This similarity is also a reason why Cd is a toxic metal. Zinc is an important trace metal in biological systems, involved in a number of enzymes. Zinc can be replaced by cadmium, particularly when it is bonded to sulphur. The chemical similarity does not carry over to similar biochemical activity, and the Cd-enzymes are inactive.

$$\text{Enzyme}\langle S_2 \rangle \text{Zn} + Cd^{2+} \rightleftharpoons \text{Enzyme}\langle S_2 \rangle \text{Cd} + Zn^{2+} \qquad (11.23)$$

Cadmium-sulphur bonds are stronger than zinc-sulphur bonds the Cd^{2+} ion is the more polarizable of the two and bonds more strongly to the polarizable sulphur atoms.

Cadmium gets into waterways from zinc mining areas and from its industrial uses, such as metal plating and the use of CdS as an orange pigment. While most cadmium compounds that get into, or are formed in, water have low solubility; e.g. $CdCO_3$, $K_{sp} = 1.8 \times 10^{-14}$; $Cd(OH)_2$, $K_{sp} = 1.6 \times 10^{-14}$; and CdS, $K_{sp} = 6.3 \times 10^{-28}$, this is no guarantee that they remain insoluble (due to pH changes), or that they are biochemically inactive. In acid the following equilibria lie to the right.

$$Cd(OH)_2 + 2H^+ \rightleftharpoons Cd^{2+} + 2H_2O, \quad \log K = 13.61 \qquad (11.24)$$

$$CdCO_3 + 2H^+ \rightleftharpoons Cd^{2+} + CO_2 + H_2O, \quad \log K = 3.8 \qquad (11.25)$$

Also, under aerobic conditions, the concentration of Cd in sea-water can be quite high due to the formation of species such as $CdCl^+$ and Cd(OH)Cl.

$$Cd^{2+} + Cl^- \rightleftharpoons CdCl^-, \quad \log K = 2.08 \tag{11.26}$$

$$Cd^{2+} + OH^- + Cl^- \rightleftharpoons Cd(OH)Cl, \quad \log K = 5.87 \tag{11.27}$$

In anaerobic conditions the Cd concentration will drop as CdS forms;

$$CdCl^+ + H_2S \rightleftharpoons CdS + 2H^+ + Cl^- \tag{11.28}$$

Levels of cadmium in harbour sediment have been reported at 130 $\mu g\ g^{-1}$, hence high levels may well occur in the water.

The insolubility of $Cd(OH)_2$ is a function of pH (equation 11.24). This is demonstrated with a distribution diagram (Fig. 11.7). The formation constants for the cadmium-hydroxy and oxy species $CdOH^+$, $Cd(OH)_2$, $HCdO_2^-$ and CdO_2^{2-} from Cd^{2+} are of the form;

$$\beta_i = \frac{[\text{species}]}{[Cd^{2+}][OH^-]^i} \quad (i = 1 \text{ to } 4). \tag{11.29}$$

The values are $\beta_1 = 1.4 \times 10^4$, $\beta_2 = 2.5 \times 10^8$, $\beta_3 = 1.2 \times 10^9$ and $\beta_4 = 5.8 \times 10^8$. The total concentration of cadmium in solution is given by;

$$C = [Cd^{2+}] + [CdOH^+] + [Cd(OH)_2] + [HCdO_2^-] + [CdO_2^{2-}] \tag{11.30}$$

The fraction of species "i" in this solution is given by;

$$\alpha_i = \frac{[\text{species } i]}{C} \tag{11.31}$$

From equations (11.29 to 11.31) one can calculate each α_i as a function of the pH of the solution.

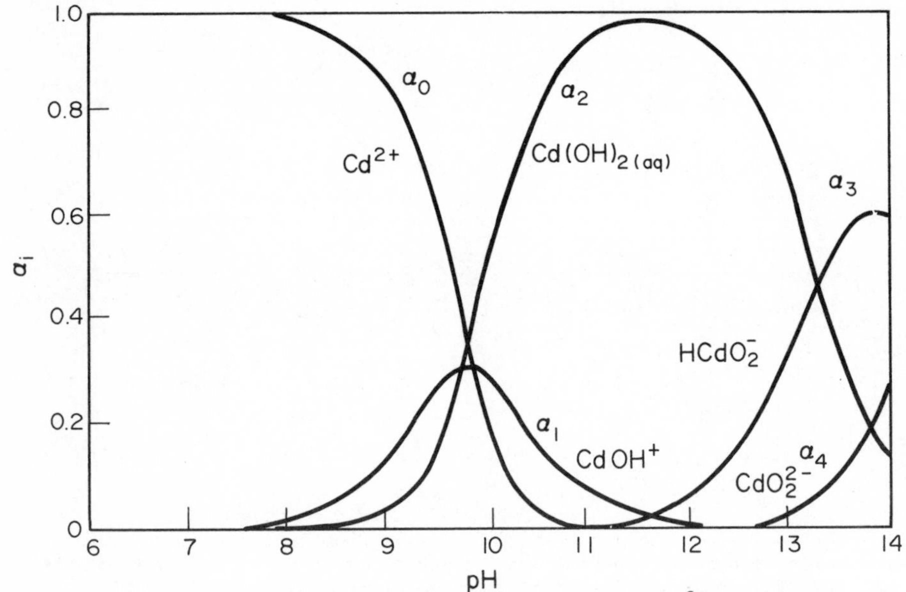

Fig. 11.7. The distribution diagram for the Cd^{2+}/OH^- system

For $$\alpha_o = \frac{Cd^{2+}}{C}, \tag{11.32}$$

the denominator (C) can be simplified so that it is expressed in terms of one cadmium species, in this case Cd^{2+}, i.e.

$$C = [Cd^{2+}] + \beta_1[Cd^{2+}][OH^-] + \beta_2[Cd^{2+}][OH]^2 + \beta_3[Cd^{2+}][OH^-]^3 + \beta_4[Cd^{2+}][OH^-]^4 \tag{11.33}$$

Hence simplifying equation (11.32) and making use of $K_w = [H^+][OH^-]$ we get;

$$\alpha_o = \frac{[H^+]^4}{[H^+]^4 + \beta_1 K_w[H^+]^3 + \beta_2 K_w{}^2[H^+]^2 + \beta_3 K_w{}^3[H^+] + \beta_4 K_w{}^4} \tag{11.34}$$

Similar expressions can be obtained for α_1, α_2, α_3 and α_4, which when plotted against pH give Fig. 11.7. It is clear from the figure that at a pH < 9 the principal species in solution is Cd^{2+}, and it is the only species at pH < 7.5, provided no other materials in solution influence the formation of the oxy and hydroxy species (e.g. carbonate). Similar distribution diagrams can be determined for other systems. For example for the Cd-chloro system at a chloride concentration of say 0.06 mol l^{-1}, the Cd species that exist in solution are Cd^{2+}, $CdCl^+$ and $CdCl_2$.

Dissolved carbonate affects the solubility of cadmium. At a pH = 8.3 with no carbonate present the solubility of Cd is 637 mg l^{-1} while at a total carbonate concentration ($[CO_2] + [H_2CO_3] + [HCO_3^-] + [CO_2^{2-}]$) of 5 x 10^{-4} mol l^{-1} the solubility drops to 0.11 mg l^{-1}. Addition of carbonate reduces the Cd^{2+} levels in water, provided the various equilibria involved have had time to establish.

Corrosion of galvanised pipes or galvanised tanks, can add cadmium to water, because of the Cd in the zinc. Over the pH range, 8.3 - 10.5, oxidation of cadmium metal occurs with O_2;

$$Cd - 2e \rightarrow Cd^{2+}, \qquad -E^o = +0.4 \text{ V} \tag{11.35}$$

$$O_2 + 4H^+ + 4e \rightarrow 2H_2O \qquad E^o = 1.23 \text{ V} \tag{11.36}$$

hence $$2Cd + O_2 + 4H^+ \rightarrow 2Cd^{2+} + 2H_2O. \qquad E = 1.63 \text{ V} \tag{11.37}$$

The corrosion decreases with an increase in both $[OH^-]$ and $[CO_3^{2-}]$ by virtue of increasing the pH. A method of reducing corrosion, and hence levels of Cd^{2+}, is to precoat the pipes with $CaCO_3$ or calcium metasilicate, and exclude O_2 from the metal surface. Because of its high toxicity, and the relative ease with which it goes into solution, it has been recommended that the upper limit of cadmium in drinking water be 10 μg l^{-1}.

Arsenic

Arsenic, whose toxicity is well known, occurs in the earth's crust at levels around 2 - 5 μg g^{-1}, and is commonly associated with phosphate rock. For this reason arsenic appears as an impurity in phosphate fertilizers and phosphate detergents. The principal arsenic minerals are FeAsS, As_2S_3, AsO_2 and $FeAs_2$. Anthropogenic sources are combustion of fossil fuels (coal contributes around 5000 tonnes yr^{-1} and oil around 10 tonnes yr^{-1}), from cement production (3200 tonnes yr^{-1}), and from mine tailings. The enrichment factor for arsenic in aerosols over the crustal concentration is approximately 310. This material eventually precipitates and accumulates in waterways. The two main forms in water are arsenite, AsO_2^- and AsO_3^{3-}, and arsenate AsO_4^{3-}, the latter form being the less toxic.

Background levels of arsenic in water (both fresh and salt) are around 2 μg l^{-1}, the recommended upper limit in drinking water is 50 μg l^{-1}. Arsenic is concentrated in the food chain and levels of 6.0 μg g^{-1} (dry weight) in zooplankton and 43 - 188 μg g^{-1} (dry weight) in fish have been reported.

As for mercury, arsenic in sediments is mobilized by bacteria and the evidence suggests that the element is methylated;

$$\underset{\text{As(III)}}{H_3AsO_3} \underset{\text{methyl cobalamin}}{\rightleftharpoons} \underset{\substack{\text{As(V)} \\ \text{methylarsenic acid}}}{CH_3AsO(OH)_2} \underset{\text{methyl cobalamin}}{\rightarrow} \underset{\substack{\text{As(V)} \\ \text{Dimethylarsenic acid}}}{(CH_3)_2AsO(OH)} \quad (11.38)$$

followed by reduction to As(III).

$$\underset{\text{As(V)}}{(CH_3)_2AsO(OH)} + 4e + 4H^+ \rightarrow \underset{\text{As(III)}}{(CH_3)_2AsH} + 2H_2O \quad (11.39)$$

producing very toxic $(CH_3)_2As^+$, and also volatile $(CH_3)_3As$. These compounds are readily oxidised to As(V), and so an As(III)-As(V) redox cycle is established.

Chromium

The chief chromium ore is chromite $FeO.Cr_2O_3$ from which ferro-chrom alloys, and chromium metal are obtained. Chromium is used in the tanning industry and in corrosion protection of steel. Treatment of steel with CrO_4^{2-} may form an oxide film on the metal rendering it resistant to corrosion.

Hexavalent chromium has two main oxy-anion forms, CrO_4^{2-} and $Cr_2O_7^{2-}$, which are involved in the equilibrium;

$$2\,CrO_4^{2-} + 2H_3O^+ \rightleftharpoons Cr_2O_7^{2-} + 3H_2O \quad (11.40)$$

Chromium (VI) is toxic, but its lifetime in the aqueous environment is limited as it is readily reduced to chromium(III) by organic material. In fact chromium(VI) is used for the estimation of total organic carbon in environmental studies of materials such as soil and water, according to the reactions;

$$Cr_2O_7^{2-} + 14H^+ + 6e \rightarrow 2Cr^{3+} + 7H_2O \quad (11.41)$$

$$\{CH_2O\} + H_2O - 4e \rightarrow CO_2 + 4H^+ \quad (11.42)$$

i.e. $$2Cr_2O_7^{2-} + 3\{CH_2O\} + 16H^+ \rightarrow 4Cr^{3+} + 3CO_2 + 11H_2O \quad (11.43)$$

The Cr(III) state, which is said to be non-toxic (though this may be because of its low concentration in water), is mobilized in acid as the Cr^{3+} hydrated ion, while in alkaline conditions low soluble $Cr(OH)_3$ is produced. The hydroxy species of chromium(III) dominates in natural water systems. The recommended limit for chromium in potable water is 50 $\mu g\ l^{-1}$.

Manganese

Manganese, an essential trace metal, occurs in nature principally as the oxide MnO_2 or Mn(II) compounds originally formed under reducing conditions (see the E_h - pH diagram, Fig. 2.17). The major sources of marine manganese are river outflows and submarine volcanic activity. The metal is used as a scavenger for oxygen and sulphur in steel production, and the organo-metallic compound $Mn(h^5 - CH_3C_5H_4)(CO)_3$, is being used as an anti-knock reagent in place of tetraalkyl lead compounds.

The end product of manganese in the oceans is precipitated MnO_2, formed with the assistance of bacteria, producing manganese nodules. Large quantities occur on the ocean floor, particularly in the Pacific Ocean (approximately 1.7×10^9 tonnes). The concretions range from a few millimeters in diameter to a few centimeters and can weigh up to 1 kg. They consist principally of hydrated oxides of manganese and iron (MnO_2 and Fe_2O_3) with elevated levels of metals such as Co, Ni, Cu, Zn and Pb, where the Mn^{4+} and Fe^{3+} ions have been replaced by the M^{2+} ions. A typical analysis of concretions is 20% Mn, 20% Fe and 1-3% Ni, Co and Cu.

The nucleus around which the nodule forms can be a clay or rock particle or bony animal material. The nodules consist of concentric spherical shells, the colour of which vary with the proportion of manganese and iron. In deep water the growth rate is slow, around 1 mm in 100,000 years, while in shallow coastal waters the rate may reach 10 cm in 100 years.

Under reducing (and low pH) conditions, the principal form of manganese in water is the hydrated Mn^{2+} ion. This mobile form of manganese oxidises under alkaline conditions, as found in sea water with a pH around 8, and rich in O_2. The whole process is assisted by bacteria.

$$5Mn^{2+} + 12OH^- - 8e \rightarrow 4MnO_2.Mn(OH)_2 2H_2O + 6H^+ \quad (11.44)$$

$$O_2 + 4H^+ + 4e \rightarrow 2H_2O \quad (11.45)$$

i.e. $$5Mn^{2+} + 2O_2 + 10OH^- \rightarrow 4MnO_2.Mn(OH)_2 2H_2O + 2H_2O \quad (11.46)$$

and $$4MnO_2.Mn(OH)_2 2H_2O + \tfrac{1}{2}O_2 \rightarrow 5MnO_2 + 3H_2O \quad (11.47)$$

The sum of equations 14.46 and 14.47 is:

$$Mn^{2+} + \tfrac{1}{2}O_2 + 2OH^- \rightarrow MnO_2 + H_2O \quad (11.48)$$

The dioxide and also Fe_2O_3 precipitate and aggregate together and initially by surface absorption concentrate other metal ions, which end up as isomorphous replacements of the Mn^{4+} and Fe^{3+} ions. The nodules, if easily mined, could become a valuable source of some of these trace metals.

SPECIES IN WATER - ANIONIC

The main "anionic" species that occur in water are species of the p-block elements, such as PO_4^{3-}, NO_3^-, SO_4^{2-}, CO_3^{2-}. Since a few of these are important macro-nutrients, their concentration affects the growth of aquatic life. The amounts of phosphate and nitrate in water also have a bearing on the levels of dissolved dioxygen. Therefore we will first discuss the dioxygen content of water and the problems arising from its depletion.

Oxygen in Water

Dioxygen is not very soluble in water and its solubility decreases with increase in temperature and salinity (Table 11.5). The amount also depends

TABLE 11.5 Solubility of O_2* in Water (mg l^{-1})

Temperature	Salinity g kg^{-1}				
°C	0	10	20	30	35
0	10.22	9.54	8.91	8.32	8.05
5	8.93	8.36	7.83	7.33	7.09
10	7.89	7.41	6.95	6.52	6.32
15	7.05	6.63	6.24	5.87	5.69
20	6.35	5.99	5.64	5.32	5.17
25	5.77	5.45	5.15	4.86	4.73
30	5.28	4.99	4.73	4.47	4.35

* 20.95% O_2 in air, 101.3 kPa pressure and 100% relative humidity.

on the levels of bacterial, plant and animal life in the water and on the turbulence of the water. Turbulence provides a method of mixing of air and water. The minimum level of dioxygen in water, for fish survival, is 3 mg l^{-1}, however, some fish, such as trout and salmon, need higher levels. If the dioxygen level falls very low, bacteria can obtain their oxygen by reduction of dissolved SO_4^{2-} and NO_3^-. In the former case H_2S is produced.

Dioxygen is consumed rapidly in water by organic material, with the assistance of micro-organisms.

$$\{CH_2O\} + O_2 \xrightarrow{\text{micro-organisms}} CO_2 + H_2O \qquad (11.49)$$

This is the principal reaction involving O_2, but other processes also consume dioxygen.

$$NH_4^+ + 2O_2 \xrightarrow{\text{micro-organisms}} 2H^+ + NO_3^- + H_2O \qquad (11.50)$$

$$4Fe^{2+} + O_2 + 4H^+ \rightarrow 4Fe^{3+} + 2H_2O \qquad (11.51)$$

$$2SO_3^{2-} + O_2 \rightarrow 2SO_4^{2-} \qquad (11.52)$$

In addition to the total dioxygen content of water there are two other estimates of dioxygen capacity. The Biological Oxygen Demand, BOD, represents the amount of O_2 required by bacteria to achieve aerobic oxidation of organic material to CO_2 and H_2O. Experimentally, a system is allowed to stand for five days and the O_2 level measured before and after. The Chemical Oxygen Demand, COD, is determined chemically by measuring the amount of oxidisable material in the water. In general COD $>$ BOD.

Dioxygen levels in water vary in two main ways, one as a consequence of added pollutants and the other a natural daily variation. When oxidisable pollutants are added to water the O_2 level will decrease represented by a "dioxygen sag-curve"(Fig. 11.8), the concentration of O_2 plotted against either time or

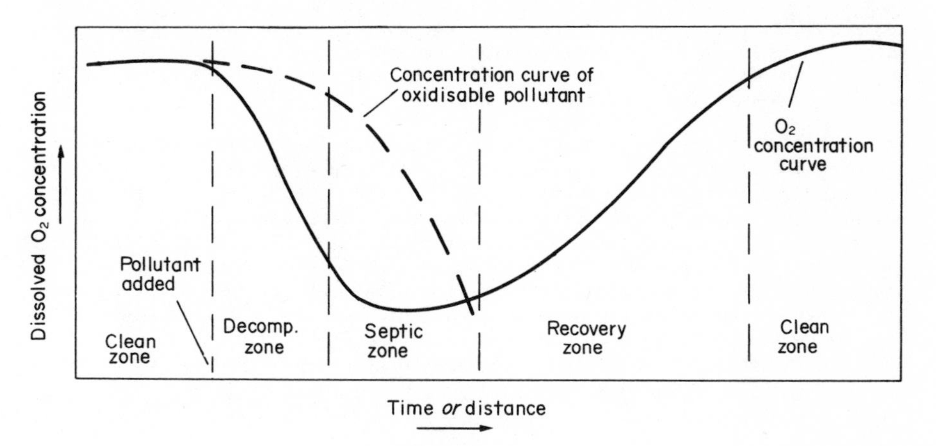

Fig. 11.8. A dioxygen sag-curve

distance downstream. The recovery time will, among other things, depend on the degree of mixing at the air-water interface. The daily variation in O_2 levels (Fig. 11.9) is a function of respiration, which removes O_2, and photosynthesis, which adds O_2.

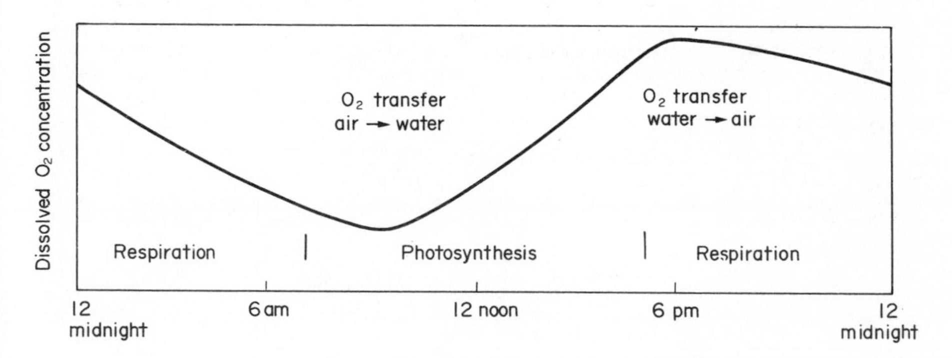

Fig. 11.9. Daily variations in the levels of dissolved dioxygen.

Eutrophication is a condition that leads to the depletion of dissolved O_2. This has occurred naturally when a body of water, for example a lake, has become enriched with nutrients (water poor in nutrients is called oligotropic) producing excessive biological activity, especially plant and algae growth. When this material dies and falls to the bottom of the lake it undergoes partial decay, using up some dissolved O_2, and recycling some of the CO_2, N, P and K. Eventually the area becomes marshy and finally becomes dry. Coal and peat deposits were produced in this way. Anthropogenic activity speeds up this process, by providing extra nutrients, especially N and P from sewage, agriculture run-off and industry. Excess algal growth as a consequence of an increase in nutrients, is at a maximum just below the surface of the water where photosynthesis is most pronounced. Phosphorus is often the limiting factor in the process of eutrophication as it occurs in forms which are the least soluble of the macro-nutrients. However, the increase in use of phosphate detergents is changing this. Phosphorus and nitrogen levels greater than 0.015 mg l^{-1} and 0.3 mg l^{-1} respectively are sufficient to produce nuisance blooms of algae.

The water temperature also plays an important part in the eutrophication process, because layers of water at different temperatures do not mix readily. Since the maximum density of water occurs at 4°C, warm water can float on colder water producing thermal stratification in the summer (Fig. 11.10a). A different situation occurs in the winter (Fig. 11.10b), while in the spring and autumn the temperature can be the same at all levels. In the summer dissolved O_2 in the epilimnion zone will not get into the hypolimnion zone. If the water contains large amounts of organic material sunlight penetration will be reduced cutting down photosynthesis and O_2 production. When the dead organic material falls into the hypolimnion it will use up the dissolved O_2 and anaerobic conditions will soon form.

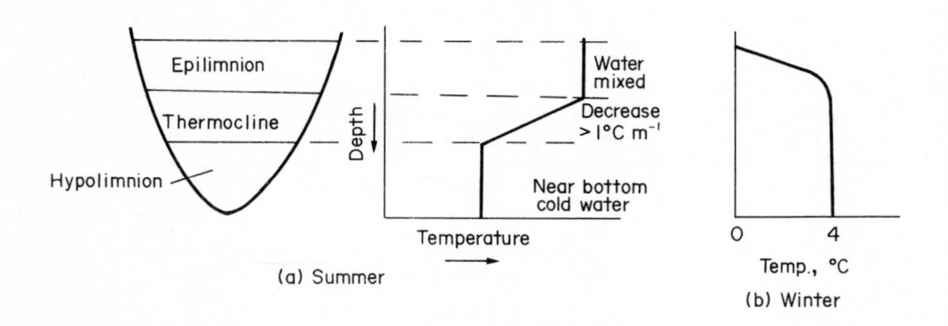

Fig. 11.10. Thermal stratification in a body of water

Whereas natural eutrophication is, broadly speaking, irreversible, this need not be the case for pollution eutrophication, if the flow of nutrients is curtailed or stopped.

Source of Anionic Pollutants

Aquatic anionic pollutants are either nutrients, or not, and they arise from point sources or diffuse sources. Point sources are mainly man made, e.g.

sewage effluent, industrial and farm effluents, while diffuse sources; precipitation, land run-off, drainage, leaf fall and bird faeces are partly influenced by man. Some representative levels of the nutrients N and P, from different sources, are listed in Table 11.6. The data shows that domestic

TABLE 11.6 N and P Levels in the Environment

Source	N	P
Precipitation	0.2 mg N l^{-1} (NO_2 av. in rain) 0.5 mg N l^{-1} (NH_3 av. in rain)	97 μg P l^{-1} (Av. across Canada)
Run off	10 mg N l^{-1} (As NO_3^- from well farmed arable land in U.K.) 5 mg N l^{-1} (from non-fertilised land)	0.05 mg P l^{-1} (Average, much less leached than NO_3^-)
Domestic Sewage (Average)	20 - 40 mg N l^{-1} (As NH_3) 10 - 30 mg N l^{-1} (organic)	10 - 20 mg P l^{-1} (Approx. half as PO_4^{3-})

sewage contributes localized high levels of N and P, and is one of the main inputs of P into waterways. Some of the main sources of anionic pollutants, are listed in Table 11.7. It is also clear from this table that domestic

TABLE 11.7 Sources of Anionic Pollutants

Major Pollutant	Metal Finishing	Iron & Steel	Chemicals	Food Processing	Agriculture	Domestic
Solids				✓	✓	✓
BOD			✓	✓	✓	✓
NH_3			✓		✓	✓
CN^-	✓	✓				
Detergents						✓
NO_3^-			✓		✓	✓
pH	✓	✓	✓			
Phosphates					✓	✓
Sulphide		✓	✓			
Pesticides			✓		✓	

waste is a significant pollution source, particularly of the type that accelerates eutrophication.

Water quality is estimated from the amount of material in it. Therefore regular monitoring is necessary, as many of the materials in water are toxic to human and animal life. In Table 11.8 some of the indicators of water quality (both non-metal and metal), are listed, together with their significance and suggested optimum safe levels. The diverse range of materials and

TABLE 11.8 Indicators of Water Quality*

Indicator	Significance	Level
Dissolved O_2	water quality	min. acceptable 4 mg l^{-1} desirable 10 mg l^{-1}
Total suspended solids	reduce light penetration, clog fish gills	depends on location
Total dissolved solids	gives mineral content, some are toxic	max. 400 mg l^{-1} for diverse fish population
BOD	water quality, dissolved oxygen	1 mg l^{-1} very clean 3 mg l^{-1} fairly clean 5 mg l^{-1} doubtful purity 10 mg l^{-1} contaminated
COD	oxidisable material	0-5 mg l^{-1} clean stream
pH	acidity	depends on location
Fe	suggests mine drainage	max. 0.7 mg l^{-1} for fish population, 0.05 mg l^{-1} drinking water
Mn	if high, due to contamination	max. 0.1 mg l^{-1} for stream quality, 0.05 mg l^{-1} for drinking water
Cu	if high, pollution or mine drainage	max. 0.02 - 0.1 mg l^{-1}
Zn and other metals	industry or mine drainage	max. 1.0 mg l^{-1}
NO_3^-	fertilizers, sewage, affects plant growth	max. 0.3 mg l^{-1}
PO_4^{3-}	fertilizers, sewage, detergents, affects plant growth	max. (inorg. P) 0.03 - 0.4 mg l^{-1}

* Microbiological indicators are also used; such as, fecal coliforms, total coliforms, fecal streptococci and algae.

the low levels being estimated means that a wide selection of analytical methods are employed. The methods used and the significance of the results, need to be continually updated in order to protect society from health risks.

Phosphates

Phosphorus occurs almost entirely in the oxidation state (V) in the environment, as ortho-, poly- and organic-phosphates. The orthophosphate forms H_3PO_4, $H_2PO_4^-$, HPO_4^{2-} and PO_4^{3-} have a distribution with respect to pH as given in Fig. 6.7. The principal species in natural water (pH = 8) are; HPO_4^{2-} (90%) and the $H_2PO_4^-$ (10%), of which the latter forms the most soluble salts. Polyphosphates (in detergents) undergo hydrolysis in water to orthophosphate;

$$P_3O_{10}^{5-} + 2H_2O \rightleftharpoons 3HPO_4^{2-} + H^+ \qquad (11.53)$$

The major species, HPO_4^{2-}, in water forms a number of low soluble salts, and so large amounts of phosphorus in the oceans end up in the sediments as phosphate salts or absorbed on to clays. The global cycle for phosphorus is given in Fig. 7.8, while some estimated pools and fluxes are given in Fig. 11.11.

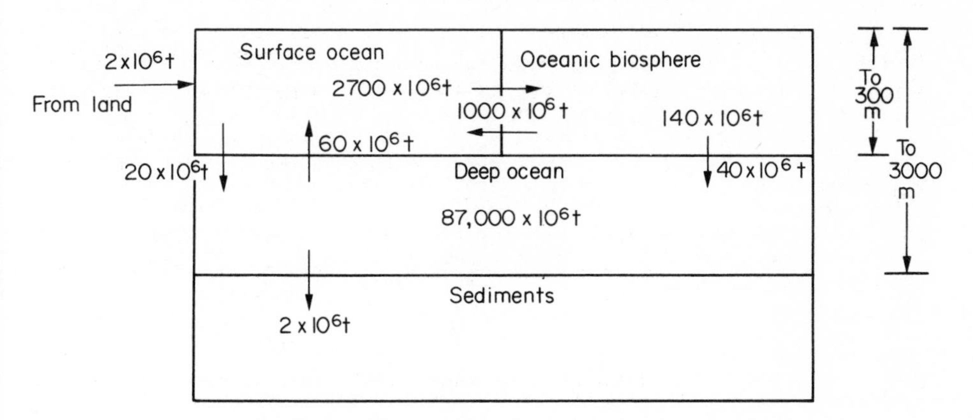

Fig. 11.11. Phosphorus pools and fluxes in the aqueous environment (quantities in metric tons, i.e. 10^3 kg)

The phosphorus concentration in water systems, such as lakes, is found to correlate with the amount of phytoplankton, as measured by the concentration of chlorophyll α. The Great Lakes in North America, are the recipients of large amounts of phosphorus from both point and diffuse sources. The estimate of the amount from point sources (sewage and industry) in 1976 was 26,000 kg day^{-1}, the proportion to each lake was approximately 62% L. Erie, 22% L. Ontario, 11% L. Michigan and 2% to each of L. Superior and Huron. Lakes Erie and Ontario also receive the greatest influx from land drainage, approximately 30% and 21% of the total phosphorus load respectively. Urban runoff from storm-waters is much less significant compared with sewage, except during storms. Phosphate that ends up in storm-water, comes from soil erosion, leachate from materials such as fallen leaves and garden soils, atmospheric fall-out, animal faeces and urban dust.

Nitrogen

Nitrogen salts being more soluble than phosphates, readily leach from soils into water-ways. The average concentration of all chemical forms of nitrogen in water-ways is around 5×10^{-5} mol l^{-1}, but varies with locality. In addition to run-off from urban and agriculture land, the species NH_3, NO_2^-, NO_3^- are produced by nitrogen fixing bacteria in the oceans. Denitrifying bacteria, that removes nitrate from water, cannot however, cope with the additions from the land. The vertical distribution of the nutrient elements (N and P) generally increase down to about 1 km and then fall off to a constant concentration at greater depth. The zone of greatest concentration, corresponds to a decrease in the dioxygen levels. The dioxygen is consumed by organic materials releasing inorganic N and P which can be recycled.

$$NO_3^-, PO_4^{3-} \rightarrow \text{plants} \rightarrow \text{animals} \xrightarrow[\text{oxidation}]{\text{decay}} \begin{matrix}\text{Organic}\\ \text{N and P}\end{matrix} \rightarrow NO_3^-, PO_4^{3-} \quad \text{(recycle)} \qquad (11.54)$$

This cycle is easily established in a moving water system (lotic), but in a lake where the water is more or less still (lentic) the recycling depends on the temperature gradients within the lake.

The recommended level of NO_3^- (as N) in drinking water is 10 mg l^{-1} (and NO_2^- nitrogen should not be above 0.1 mg l^{-1}). The nitrate level should not be too high because of the risk of reduction to NO_2^-, in the less acidic stomach of young children. If this happens a condition call methemoglobinemia may occur, where the NO_2^- ion oxidises the Fe(II) in haemoglobin to Fe(III), as well as coordinating to the metal. This can be fatal.

Other Species

Many other species end up in the world's water systems, some quite toxic, such as the cyanide ion. Cyanide is widely used for metal cleaning, electroplating and in gold extraction. In water, cyanide undergoes hydrolysis, producing the weak acid HCN ($K_a = 6 \times 10^{-10}$) which is volatile.

$$CN^- + H_2O \rightleftharpoons HCN + OH^- \qquad (11.55)$$

$$K = \frac{[HCN][OH^-]}{CN^-} \qquad (11.56)$$

Since $$K_a = \frac{[H^+][CN^-]}{[HCN]} \text{ and } K_w = [H^+][OH^-], \qquad (11.57)$$

we can write,

$$K = \frac{K_w}{K_a} = \frac{10^{-14}}{6 \times 10^{-10}} = 1.7 \times 10^{-5}. \qquad (11.58)$$

The volatility of HCN is a driving force, shifting reaction 11.55 to the right. Cyanide bonds irreversibly with Fe(II) in haemoglobin.

Both NH_3 (occurring as NH_4^+) and H_2S are the products of the decay of organic material containing N and S. The former remains in solution, but H_2S is removed as low soluble sulphides. The main form of sulphide in water is HS^-. Hydrogen sulphide is readily detected by its smell, as anyone knowns who has handled fresh sediments produced in water systems under anaerobic conditions.

The sulphite ion has a limited existence in natural water, being readily oxidised to sulphate in the presence of O_2.

From what has gone before it is clear that the pH of water is very important, and cannot be ignored. Both strongly acid and alkaline conditions are detrimental. In general acidity is the more common problem, with acid water occurring from mine-water drainage, industrial wastes and an increasing amount of acid (H_2SO_4) rain produced from coal combustion. Strongly alkaline conditions are produced by the dumping of 'red-mud' in ponds adjacent to bauxite purification plants. The red-mud, mostly insoluble iron oxides, is separated from the alkaline solution obtained when bauxite is treated with NaOH solution. The mud deposited in low-pond-areas can cause high pH contamination of water systems.

The salinity of water is a problem when it occurs in natural fresh water systems. It mainly arises from a joint increase in the use of fertilizers and irrigation. In the end, ground water may become quite salty and dangerous for use as drinking water. Plant life is also affected, especially if the saline water is used for irrigation.

The Great Lakes

The five Great Lakes in North America are situated adjacent to urban and industrial areas as well as agriculture land. Pollution has down-graded the lakes from a source of high quality water to areas of eutrophication, especially Lakes Erie and Ontario. The principal offending species are phosphorus, mercury, lead, halocarbons and micro-organisms.

The phosphorus loading of the lakes (Table 11.9) arises from atmospheric fallout, diffuse tributary sources and point sources. The point sources are

TABLE 11.9 Phosphorus Loading in the Great Lakes

Lake	Total load tonnes yr^{-1}	Atmospheric % Total	Diffuse (tributary) % Total	Point Sources % Total	Contributions to Diffuse %		
					Agric.	Urban	Forest & Other
Superior	4200	37	53	10	7	7	86
Michigan	6300	26	30	44	71	12	17
Huron	4850	23	50	27	68	12	20
Erie	17,450	4	48	48	66	21	13
Ontario	11,750	4	28	63	66	19	15

From Int. Joint Commission Int. Ref. Group on Great Lakes Pollution from Land Use Activities (PLUARG) 1978.

as, or more, important as diffuse sources for Lakes Erie, Ontario and Michigan and, except Lake Superior, agriculture runoff is the main component of diffuse tributary phosphorus. A large proportion of tributary phosphorus (36 - 80%) is sorbed by Fe and Al materials and clays in sediments, and approximately 60% of this material is not biologically available.

The main sources of mercury in the Great Lakes are industrial (chlor-alkali plants and pulp and paper mills) and urban air. Tributary sources and about 2300 kg yr^{-1} (most to Lake Erie) and over 95% of this is associated with sediments. Lake St. Clair, between Huron and Erie, is also heavily contaminated. Even after a number of years of reduced mercury inputs the metal is still being transferred from L. St. Clair to L. Erie. The Lakes Erie and Ontario have areas where sediments contain more than 2000 ppb of Hg, and large areas where the levels $>$ 1000 ppb. These high levels reflect in the levels found in fish tissue, especially for Lakes St. Clair and Erie (Table 11.10).

TABLE 11.10 Mercury and Lead in Fish

Lake	Mercury* mg kg^{-1} (wet weight)	Lead** mg kg^{-1} (wet weight)
Superior	0.07 - 0.78	0.012 - 0.066
Michigan	0.22 - 0.54	0 - 0.54
Huron	0.06 - 0.18	0.04 - 0.10
St. Clair	0.06 - 3.8	0.47 - 0.63
Erie	0.03 - 1.52	0.04 - 0.12
Ontario	0.06 - 0.49	< 1.0

* guideline concentration 0.5 mg kg^{-1}

** guideline concentration 10 mg kg^{-1}

From PLUARG 1978.

The majority of lead (85 - 99%) in the Lakes comes from atmospheric fallout of automobile emissions. The majority ends up in the sediments, and except for possible methylation is not considered a problem. Levels of lead in fish tissue (Table 11.10) are less than for mercury even though the levels in the sediments are greater (some areas in L. Erie and Ontario have $>$ 150 ppm of lead).

Other species, NO_3^-, Cl^-, Cr, Cd, Zn, Se and As, as yet in low levels, need to be continually monitored and further work needs to be carried out on these species.

HARDNESS OF WATER

The degree of hardness of water varies widely, and when high is expensive to remove. Hardness is mainly due to dissolved Ca^{2+} and Mg^{2+} ions (permanent hardness) and the HCO_3^- ion (temporary hardness). Dissolved Fe(III) and Mn(II) also contribute to a lesser extent. Dissolved CO_2, and the weathering of carbonate rocks are the main source of hardness.

$$CO_2 + H_2O \rightleftharpoons HCO_3^- + H^+ \text{ (temporary hardness)} \quad (11.59)$$

$$\underset{\text{limestone}}{CaCO_3} + H_2O + CO_2 \rightleftharpoons Ca^{2+} + 2HCO_3^- \quad (11.60)$$

$$\underset{\text{dolomite}}{CaCO_3.MgCO_3} + 2H_2O + 2CO_2 \rightleftharpoons Ca^{2+} + Mg^{2+} + 4HCO_3^- \quad (11.61)$$

(permanent hardness) — equations (11.60) and (11.61)

Hardness may be removed by three basic processes; removal of Ca^{2+} and Mg^{2+} ions as insoluble precipitates, formation of soluble metal ion complexes reducing the free metal ions in solution, and by ion exchange. Details of the application of these three methods are listed in Table 11.11. The increasing tendency is to remove hardness by the second method. However,

TABLE 11.11 Methods for the Removal of Hardness of Water

Method	Reactions	Comments
1 Soap	$Ca^{2+} + 2RCOO^-Na^+ \rightarrow (RCOO)_2Ca\downarrow + 2Na^+$ (in excess)	Wasteful if water is quite hard
Boiling	$Ca^{2+} + HCO_3^- \rightarrow CaCO_3\downarrow + H_2O + CO_2$	Get $CaCO_3$ deposits
Washing soda Na_2CO_3	$Na_2CO_3 + Ca^{2+} \rightarrow CaCO_3\downarrow + 2Na^+$	For permanent hardness only
Slaked lime $Ca(OH)_2$	$Ca^{2+} + 2HCO_3^- + 2OH^- \rightarrow 2CaCO_3\downarrow + 2H_2O$ ($2OH^-$ from $Ca(OH)_2$)	For temporary hardness only. Also removes Mg^{2+} as $Mg(OH)_2$
Lime plus soda	Both reactions above	For both temporary and permanent hardness, pH needs to be reduced using CO_2
2 Polyphosphates $Na_5P_3O_{10}$	$2P_3O_{10}^{5-} + 5Ca^{2+} \rightarrow Ca_5[P_3O_{10}]_2$ (soluble) $PO_4^{3-} + H_2O \rightarrow HPO_4^{2-} + OH^-$ OH^- + grease → soap	Keeps calcium in solution.
Detergents $ROSO_3^-Na^+$	$2ROSO_3^-Na^+ + Ca^{2+} \rightarrow (ROSO_3)_2Ca + 2Na^+$ (soluble)	R-linear biodegradable, R-branched not biodegradable
Sequestering agent EDTA, NTA	$Ca^{2+} + T^{3-}$ (NTA) $\rightarrow CaT^-$ (soluble)	
3 Ion-exchange	$2\ \text{Resin}\ SO_3^-H^+ + Ca^{2+} \rightarrow (\text{Resin}\ SO_3^-)_2Ca^{2+} + 2H^+$ $2\ \text{Resin}\ NR_4^+OH^- + 2X^- \rightarrow 2\ \text{Resin}\ NR_4^+X^- + 2OH^-$	

this process contributes to water pollution, as a consequence of the chemicals used, viz., phosphates and sequestering agents.

The principle of water softening, using a complexing agent such as $Na_5P_3O_{10}$, is clear from the following analysis. In water at a pH around 7 to 8, the tri-phosphate exists in the forms H_2P^{3-}, HP^{4-} and P^{5-} (where P stands for the P_3O_{10} group). Reaction of these species with calcium ions gives soluble CaP^{3-}. Assuming Ca^{2+} reacts primarily with H_2P^{3-} we may write;

$$Ca^{2+} + H_2P^{3-} \rightleftharpoons 2H^+ + CaP^{3-}, \tag{11.62}$$

i.e.

$$K = \frac{[H^+]^2[CaP^{3-}]}{[Ca^{2+}][H_2P^{3-}]} \tag{11.63}$$

The value of K can be obtained from consideration of the following equilibria;

$$H_2P^{3-} \rightleftharpoons H^+ + HP^{4-}, \quad K_1 = 3.2 \times 10^{-7} \tag{11.64}$$

$$HP^{4-} \rightleftharpoons H^+ + P^{5-} \quad K_2 = 6.3 \times 10^{-10} \tag{11.65}$$

$$Ca^{2+} + P^{5-} \rightleftharpoons CaP^{3-}, \quad K_3 = 1.3 \times 10^{8} \tag{11.66}$$

where

$$K = K_1K_2K_3 = 2.6 \times 10^{-8} \tag{11.67}$$

If excess triphosphate (one mole) is added to the water, containing Ca^{2+} at 40 mg l^{-1} (i.e. 1 x 10^{-3} mol l^{-1}), and assuming that most of the Ca^{2+} ions are complexed, then at a pH = 7, from equation (11.63) we have;

$$[Ca^{2+}] = \frac{(10^{-7})^2(1 \times 10^{-3})}{(2.6 \times 10^{-8})(1 \times 10^{-3})} = 3.8 \times 10^{-7} \text{ mol } l^{-1} \tag{11.68}$$

At this level the concentration of CO_3^{2-} required to exceed the solubility product of $CaCO_3$ (7.2 x 10^{-9}) is;

$$[CO_3^{2-}] = \frac{7.2 \times 10^{-9}}{3.8 \times 10^{-7}} = 1.9 \times 10^{-2} \tag{11.69}$$

Without the triphosphate the amount of carbonate required would be 7.2 x 10^{-6} mol l^{-1}. Hence the triphosphate has effectively removed Ca^{2+} ions from solution without precipitation. The more phosphate used the lower will become the Ca^{2+} concentration.

Sequestering agents have the same effect as the polyphosphates. For example nitrilotriacetic acid (NTA),

```
HOOC - CH2
          \
           N - CH2 - COOH,
          /
HOOC - CH2
```

has been used as a water softener. Its use is now discontinued as it complexes other metal ions, and extracts them from precipitates or sediments, taking them into solution. Such reagents accelerate the transport of metals within the hydrosphere. The polyphosphates, while producing significant pollution, do not act as transport agents, because in the hydrosphere they eventually hydrolyse to orthophosphate

$$H_5P_3O_{10} + 2H_2O \rightarrow 3H_3PO_4 \rightleftharpoons 2HPO_4^{2-} + H_2PO_4^- + 5H^+ \tag{11.70}$$

Many heavy metal ions then form low soluble precipitates with the orthophosphate.

COORDINATION COMPLEXES IN WATER

Both inorganic and organic species in natural and polluted waters can coordinate to metal ions. In the presence of chloride ions the Cd^{2+} ion forms chloro-complex species. The metal also bonds to polyphosphates and to organic reagents, such as carboxylate. The donor atoms of the ligands are p-block elements;

C	<u>N</u>	<u>O</u>	F
Si	P	<u>S</u>	<u>Cl</u>
	As	Se	Br
Sn	Sb	Te	I

in particular N, O, S and Cl in natural waters.

The stability of a complex is expressed by the formation constant, which is often a function of the pH of the solution, and competition between metal ions for the donor sites. For the reaction;

$$HL^{n-} + M^{m+} \rightleftharpoons ML^{(m-(n+1))+} + H^{+}, \qquad (11.71)$$

the formation constant is;

$$K_f = \frac{[ML^{(m-(n+1))+}][H^+]}{[HL^{n-}][M^{m+}]}. \qquad (11.72)$$

The species in natural water that act as ligands are Cl^-, S^{2-}, carboxylates, amino acids, amines, phenols and humic substances. In polluted waters additional ligands occur, polyphosphates, EDTA, NTA, citrate and anions such as CN^-. Complex formation reduces the concentration of the free metal ions in solution. Coordination can alter the redox potentials of metal ions. In the case of sewage treatment, complexes can alter the chemical processes involved.

Citric acid is produced biologically (in the Krebs cycle), it comes from citrus fruit, soft drinks and is used as a food additive. Therefore it is present in both natural and polluted waters, probably at a concentration range of 1×10^{-4} to 1×10^{-8} mol l^{-1}. The acid is triprotic (H_3L).

```
       CH2 - COOH
        |
HO  -   C  - COOH
        |
       CH2 - COOH
```

The three acid dissociation equilibria are;

$$H_3L \rightleftharpoons H^+ + H_2L^-, \quad K_1 = 1.3 \times 10^{-3} \qquad (11.73)$$

$$H_2L^- \rightleftharpoons H^+ + HL^{2-}, \quad K_2 = 5.0 \times 10^{-5} \qquad (11.74)$$

$$HL^{2-} \rightleftharpoons H^+ + L^{3-}, \quad K_3 = 2.0 \times 10^{-6} \qquad (11.75)$$

The total amount of citrate is given by;

$$C = [H_3L] + [H_2L^-] + [HL^{2-}] + [L^{3-}], \tag{11.76}$$

and the fraction of any component (say L^{3-}) is;

$$\alpha_3 = \frac{[L^{3-}]}{C} \tag{11.77}$$

Equation (11.77) can be expressed in terms of $[H^+]$ and the three dissociation constants K_1, K_2 and K_3 by suitable manipulation of equations (11.73) to (11.76). For example α_3 becomes;

$$\alpha_3 = \frac{K_1K_2K_3}{[H^+]^3 + K_1[H^+]^2 + K_1K_2[H^+] + K_1K_2K_3} \tag{11.78}$$

and for various values of $[H^+]$ the values of α_3 are found, similarly for α_0, α_1 and α_2. The results are represented in Fig. 11.12. Above a pH = 7 citrate is present entirely as L^{3-}, therefore complex formation involves the reaction;

$$M^{2+} + L^{3-} \rightleftharpoons ML^- \tag{11.79}$$

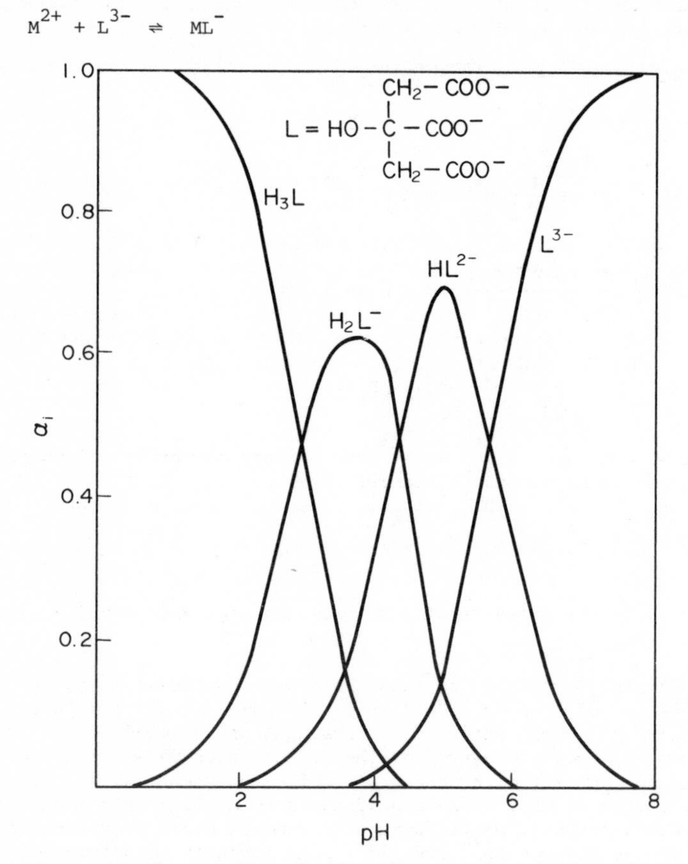

Figure 11.12. pH distribution diagram for citric acid, H_3L.

For a range of metal ions the relative order of K_f is;

$$Fe^{3+} >> Pb^{2+} > Cu^{2+} > Ni^{2+} > Co^{2+} \approx Zn^{2+} > Mn^{2+} > Mg^{2+} > Fe^{2+} > Ba^{2+} \approx Sr^{2+} > Cd^{2+} \quad (11.80)$$

The order for the divalent transition metals ions closely follows the well known Irving-Williams order of stability of metal complexes. The much higher value for Fe^{3+} is because of its greater charge. In water, at a pH of 6 to 9 and citrate concentration approximately 1×10^{-8} mol l^{-1}, calculations give the major citrate complexes present as $Fe(OH)L^{2-}$ and $Cu(OH)L^{2-}$, while at higher concentrations ($\sim 1 \times 10^{-4}$ mol l^{-1}) the complexes CuL^-, PbL^-, ZnL^-, MgL^- and free L^{3-} are also present.

Solid $PbCO_3$ dissolves in citrate, in the pH range 7 to 10. In this pH range, citrate exists as L^{3-} and the principal carbonate species is HCO_3^-. The various reactions involved are;

$$PbCO_{3(s)} + L^{3-} + H^+ \rightleftharpoons PbL^- + HCO_3^-, \quad (11.81)$$

$$PbCO_{3(s)} \rightleftharpoons Pb^{2+} + CO_3^{2-}, \quad (11.82)$$

$$CO_3^{2-} + H^+ \rightleftharpoons HCO_3^-, \quad (11.83)$$

and

$$Pb^{2+} + L^{3-} \rightleftharpoons PbL^-; \quad (11.84)$$

where

$$K_s = [Pb^{2+}][CO_3^{2-}] = 1.5 \times 10^{-13}, \quad (11.85)$$

$$K_b = \frac{[HCO_3^-]}{[CO_3^{2-}][H^+]} = 2.1 \times 10^{10}, \quad (11.86)$$

$$K_f = \frac{[PbL^-]}{[Pb^{2+}][L^{3-}]} = 3.2 \times 10^6, \quad (11.87)$$

and

$$K = \frac{[PbL^-][HCO_3^-]}{[H^+][L^{3-}]} = K_s K_b K_f = 1.0 \times 10^4 \quad (11.88)$$

At a pH of 7, and $[HCO_3^-] = 1 \times 10^{-3}$ mol l^{-1} the proportion of Pb citrate complex formed to free citrate ion is;

$$\frac{[PbL^-]}{[L^{3-}]} = \frac{K[H^+]}{[HCO_3^-]} = 1.0, \quad (11.89)$$

i.e. citrate, in the presence of $PbCO_3$ will be approximately 50% utilized in a complex.

Humic substances, formed by the decay of organic material in soils, are an important source of ligands. Both insoluble humic acid and soluble fulvic acid, are high molecular materials, containing around 40-60%C and 30-50% O. Many of the compounds isolated from the acids are aromatic with functional groups such as carboxylic, ketonic, alcoholic and phenolic, all of which are good coordinating agents. Fulvic acid takes metal ions into solution, and aids in their transport, while humic acid removes metal ions from solution. The trivalent metal ions Fe^{3+} and Al^{3+}, bond strongly to fulvic acid and the stability of the divalent metal ion complexes follows the Irving-Williams order.

The amount of organic material in water is normally in the region 0.1 to 10 mg l^{-1}. Of this it is reasonable to expect 1 mg l^{-1} is fulvic acid, which, with a molecular mass of 950, gives a concentration approximately 1 x 10^{-6} mol l^{-1}. For the reaction of fulvic acid, FH^-;

$$FH^- + M^{2+} \rightleftharpoons FM + H^+, \tag{11.90}$$

$$K_f = \frac{[FM][H^+]}{[FH^-][M^{2+}]} \tag{11.91}$$

From equation 11.91 the proportion of a metal ion complexed by fulvic acid is given by;

$$\frac{[FM]}{[M^{2+}]} = K_f\frac{[FH^-]}{[H^+]}, \tag{11.92}$$

if K_f is known, and the pH is specified.

WATER TREATMENT

Urban and rural governing bodies need to supply potable water and treat polluted water, and domestic and industrial sewage. The type of treatment depends on the end use of the water and its source. For example, potable water must be free of pathogens, but can be hard, while water used in industrial boilers needs to be soft, but does not necessarily need to be free of all bio-organic materials. River water may be highly polluted and need extensive treatment before use, while artesian water may be of such a quality it hardly needs any treatment. We will consider the three major water treatment processes, production of domestic water, treatment of sewage, and treatment of industrial waste water.

A Potable Water Supply

Potable water must be free of pathogens, have no undesirable tastes, odours, colours or tubidity, and contain no harmful organic or inorganic chemicals. The level of bacterial contamination is estimated by the "coliform count", that is the amount of *Escherichia coli* present per millilitre. Algae contribute to unpleasant taste, odour and colour. Some pathogens are removed by storage, around 50% die within 2 days and 90% within 7 days. However, some pathogens last for longer periods (>2 yr) and chemical treatment is necessary. Storage also allows large solid particles to separate out by sedimentation. Aeration of the water removes volatile materials such as H_2S, CO_2 and CH_4, as well as odorous bacteria. Oxidation of Fe(II) to Fe(III) and Mn(II) to MnO_2 is also achieved with aeration, especially at high pH, but if the metal ions are complexed, stronger oxidising agents, such as Cl_2, are necessary.

The precipitation of Ca^{2+} and Mg^{2+} as carbonates and the removal of hardness is achieved on a large scale by the lime-soda process (Table 11.11). The high pH ensures that Fe(III) hydroxy species and MnO_2 will precipitate. Precipitation is achieved for colloids and fine material by adding coagulants, such as $Al_2(SO_4)_3$ or $Fe_2(SO_4)_3$. The Al^{3+} and Fe^{3+} hydrolyse to give gelatinous hydroxide precipitates which, because of their large surface area, absorb other ions and solids and carry them down. Also the high positive charge associated with the metal ions and hydroxy species overcome the repulsion between small negative particles in the water and so the size of the particles grow, i.e. coagulate, and precipitate.

Removal of harmful bacteria is achieved by oxidation with dichlorine which, in water, gives the weak acid HOCl;

$$Cl_2 + H_2O \rightleftharpoons H^+ + Cl^- + HOCl, \quad K = 4.5 \times 10^{-4}, \tag{11.93}$$

$$HOCl + H_2O \rightleftharpoons H_3O^+ + OCl^-, \quad K = 2.7 \times 10^{-8} \tag{11.94}$$

The hypochlorite ion can be added independently as calcium hypochlorite, $Ca(OCl)_2$. Both HOCl and OCl^- are effective in killing bacteria, but it is important that the potential for oxidation lasts while the water is being distributed to its various outlets. This can be achieved by the addition of ammonia, which reacts with HOCl to produce chloramines.

$$NH_4^+ + HOCl \rightarrow NH_2Cl + H_2O + H^+, \tag{11.95}$$

$$NH_2Cl + HOCl \rightarrow NHCl_2 + H_2O, \tag{11.96}$$

$$NHCl_2 + HOCl \rightarrow NCl_3 + H_2O \tag{11.97}$$

A problem in using Cl_2 is that organo-chlorine compounds can be produced. This is overcome by either using activated charcoal to remove the organics before or after chlorination or using ozone as the oxidising agent.

Filtration of the water is also carried out to remove solids. The water is generally filtered through a sand bed or diatomaceous earth or a microscreen (stainless steel mesh 23 - 65 μm).

If the water supply is partially saline or even straight sea-water it is necessary to remove the salts before use as drinking water. The methods employed are distillation, electrodialysis or reverse osmosis. In Kuwait, for example, around 20 million litres of sea water are treated daily. Distillation is energy intensive and is not a preferred process. In electrodialysis water is added to a tank divided into compatments by alternation of cation and anion permeable membranes. A current is applied across the water resulting in some compartments being freed of dissolved ions, while others become more concentrated. The purified water is then tapped off. In reverse osmosis water is forced through a semi-permeable membrane under pressure.

Treatment of Sewage

A typical sewage consists of many types of materials. Some of the principal ones being BOD and COD materials, solids, sediments, grease, pathogens, nitrates, phosphates and heavy metals. Treatment is carried out in three stages, and sewage can be discharged at the end of any one of them. This will depend on the type of sewage being handled, the cost and the level of pollution allowed in waterways into which the treated sewage is emptied (Fig. 11.13).

Primary treatment. The primary treatment is mainly mechanical removal of solid materials by sedimentation, coagulation, flocculation, grinding and skimming processes. As some of the removed material is oxidisable primary treatment lowers the BOD and COD levels. A good primary treatment plant

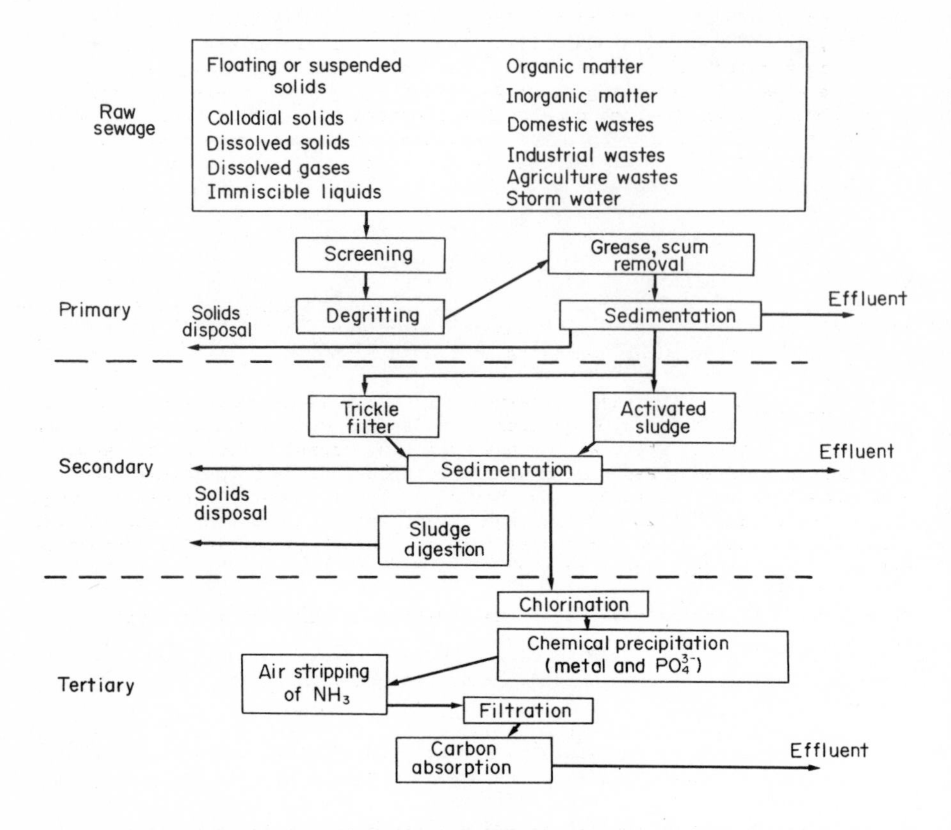

Fig. 11.13. Sewage and waste water treatment

will remove approximately 50 - 60% suspended solids, 35% BOD, 30% COD, 20% of total N and 10% of total P (in the solids). For the majority of sewage this is the only treatment given. Sometimes the effluent will be chlorinated before discharge.

Secondary Treatment. In the secondary treatment dangerous bacterial material and organic material,which if discharged into waterways would seriously deplete O_2 levels, are removed. The process is biochemical, and two methods are used. The trickle method allows the sewage to spray over beds of crushed stone coated with micro-organisms which carry out the oxidation of the organic material. The two nutrients elements N and P are necessary for the process. The second method makes use of activated sludge which contains bacteria which in the presence of dioxygen initiates aerobic decomposition of the organic materials. Also a lot of material is removed from the water by absorption onto the sludge. In both processes organic N and P are released as NO_3^- and PO_4^{3-}. The primary plus secondary treatments remove around 90% BOD, 80 - 90% COD, 90% suspended solids, 40 - 50% N, 30% P and around 5% of dissolved materials. Finally the water may be chlorinated and discharged.

The disposal of the sludge is a major problem of sewage treatment. For example, in Chicago, with 5.5 million people, approximately 6.6 x 10^6 m^3 of sewage is handled daily giving 815 tonnes of sludge. The sludge - a thin mud-like material containing about 1% solids - can be dewatered to some extent by spreading over sand, and then disposed of at this stage. However it may be treated further by first concentrating by sedimentation, coagulation and flocculation and then digesting in tanks at 35°C, where organic material is converted anaerobically to CH_4 (65%), CO_2 (30%) and NH_3, H_2S, N_2 and H_2. The methane is often burnt as an energy source on the plant. After about 60 days the sludge can be used as a humic substance, incinerated, used as land-fill, dumped in the ocean or used as a fertilizer. Sludge acts as an efficient sorption material for heavy metals, and levels of metals such as Cr, Cu, Pb, Hg and Cd can reach many thousands ppm. Hence use as a fertilizer is questionable, perhaps used with soil for growing horticulture plants which are not consumed by human beings is a safer use.

Tertiary treatment. Tertiary treatment of sewage is necessary when special problems occur, and the type of treatment depends on the contents. Wide use is made of activated charcoal for absorption of organic material and metal ions that are already associated with gelatinous hydroxide precipitates. Some phosphate is also removed by the charcoal. The spent charcoal is regenerated by heating to 925°C in air and steam, when the absorbed materials are oxidised and lost to the atmosphere (air pollution?). Persistent pathogenic materials can be oxidised with H_2O_2 or O_3

Phosphates are precipitated after the addition of lime, alum or ferric chloride;

$$5Cu^{2+} + 3HPO_4^{2-} + H_2O \rightarrow Ca_5(PO_4)_3OH + 4H^+ \quad (11.98)$$

$$Al^{3+} + HPO_4^{2-} \rightarrow AlPO_4 + H^+ \quad (11.99)$$

$$FeCl_3 + HPO_4^{2-} \rightarrow FePO_4 + H^+ + 3Cl^-. \quad (11.100)$$

Nitrogen may be removed either by ammonia stripping or denitrification of nitrates. In ammonia stripping lime is added to raise the pH and the reaction;

$$NH_4^+ + OH^- \rightleftharpoons NH_3 + H_2O \quad (11.101)$$

moves to the right as the NH_3 is flushed out in a stream of air.

Advanced biological purification may be used whereby the effluent is fed to plants which use up the PO_4^{3-} and NO_3^-. The natural biological process lowers the concentration of these nutrients before being discharged into water-ways. Shallow oxidation ponds are also used to purify water making use of aerobic bacteria.

Metal ions can be removed by precipitation as hydroxides using lime, or by sorption on to Fe and Al hydroxide precipitates. Metal-organic complexes need to be destroyed, usually with Cl_2, to liberate the metal ions prior to precipitation. Sulphide may be added to remove Pb, Cd, Hg etc. as low soluble sulphides. Scrap iron will remove Cu^{2+} ions;

$$Fe + Cu^{2+} \rightarrow Cu + Fe^{2+}. \quad (11.102)$$

The methods described above for desalination of water can also be used to remove metal ions from sewage, but only after most of the organic material (in suspension) has been removed as it clogs up the membranes. Ion exchange can also be used to remove metals.

Industrial Waste Water

Industrial waste water is either put into sewage or directly into waterways. Preferably it should be cleaned-up before either of these disposal methods, making use of some or all of the tertiary water treatment processes.

The electroplating industry produces heavy metals, acid and cyanide wastes. High levels of metals such as Cr, Cu, Ni, Zn, Fe, Cd occur in the solutions arising from the metal preparation stage, where the metal is treated with acid (called pickling) to remove rust, scale and oxide films. Electroplating is carried out in either acid or in alkaline cyanide solutions, and again high concentrations of metal ions occur in solution. When the levels get high they may cause sewage work failure, for example effluent with 700 mg l^{-1} of chromium causes a breakdown in slude digesters. Therefore the concentration of metal ions need to be low before discharge into sewers. This would involve some treatment on site. Techniques used are evaporation, incineration, reverse osmosis and ion exchange. In addition the cyanide should be oxidised and acid or base neutralized.

Iron and Manganese

The metals iron and manganese are relatively abundant and are commonly found in water. Iron in solution is present as Fe(II), normally associated with bicarbonate, or as soluble organic complexes. In acid water both Fe^{2+} and Fe^{3+} will exist, but ferrous iron is rapidly oxidised in air to ferric iron in neutral or alkaline solutions;

$$4Fe(HCO_3)_2 + O_2 + 2H_2O \rightarrow 4Fe(OH)_3 + 8CO_2, \tag{11.103}$$

turning fresh clear water turbid, and brown-red oxy-hydroxy species settle out on standing. The presence of Fe and Mn in domestic water is undesirable due to the turbidity, colour, staining of utensils and a bitter taste. With the tannin in tea iron forms a dark-inky colour. Iron must be absent in water used for dyeing, ice, paper and film making.

Iron bacteria can grow in water pipes and add to the problem of build up of iron(III) hydroxy-oxy species leading to pipe blockages. Removal of iron and manganese means producing precipitates. Aeration helps because of the oxidation of Fe(II) to Fe(III). Lime is the main chemical used to precipitate the metals and remove CO_2. Other chemicals that may be used are Na_2CO_3 and NaOH.

THERMAL POLLUTION

Thermal pollution may become a problem when water is used as a coolant in a number of industrial processes, and in thermal power-stations. The warm water is either cooled in cooling towers, transferring the heat to the air, or discharged into rivers and lakes. A small increase in the water temperature is beneficial for temperate climates, but large increases are to be avoided.

One obvious effect is that the solubility of dioxygen reduces as the temperature rises (Table 11.5). Warm water going into the epilimnion can postpone or even prevent the turn-over of lake water so that the hypolimnion will remain anaerobic all year. If the cooling water is taken from the hypolimnion it could be that the warmed cooling water will still be at a lower temperature than the epilimnion, and this would be an advantage. If the temperature of the water sink rises too high certain aquatic life may be affected. In unpolluted waters diatoms flourish best at 18 - 20°C, green algae at 30 - 35°C and blue-green algae at 35 - 40°C. It is possible that thermal pollution would favour the blue-green algae (a less useful food source) over the green algae.

WATER STANDARDS

The standards laid down for water quality vary from county to county, and economics is one factor that is considered. A list of typical standards of some of the more important inorganic components is given in Table 11.12.

TABLE 11.12 Water Quality - Domestic Water

Parameter	Permissible level mg l^{-1}	Levels considered dangerous
pH	6.5 - 8.5	<6.5, >9.2
Total hardness	30 - 500	
Ammonia	0.5	
Arsenic	0.5	>0.2
Barium	1.0	
Cadmium	0.01	
Chloride	250	>600
Chromium(VI)	0.05	
Copper	1.0	>1.5
Cyanide	0.01	>0.2
Dioxygen	>4.0	
Fluoride	0.8 - 1.7	
Iron	<0.3	>1.0
Lead	<0.05	>0.1
Magnesium	50	>150
Manganese	<0.01	>0.5
Mercury	<0.002	
Nickel	<0.05	
Nitrate(NO_3^-)	<45	
Nitrate (N)	<10	
Phosphorus	0.01 - 0.05	
Selenium	0.01	
Silver	0.05	
Sulphates	250	>400
Total dissolved solids	500	>1500
Uranyl	5	
Zinc	5	>15

The standards given are, in general, the same for the United States of America, Europe and recommended WHO figures. It is interesting to note that the permissable concentrations versus the average concentration found in many streams in the world show a reasonable correlation (Fig. 11.14). This

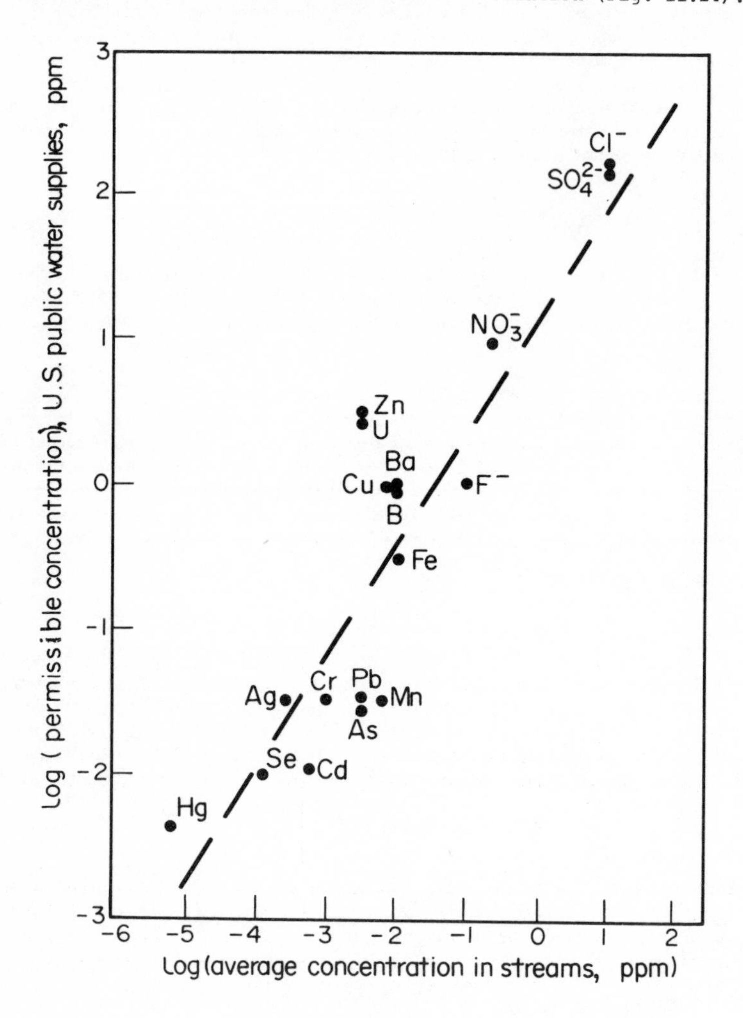

Fig. 11.14. Correlation between permissible concentration of trace species in public water supplies (USA) and the average concentrations in streams. Source: Garrels, R.M., Mackenzie, F.T., and Hunt, C., Chemical Cycles and the Global Environment (1975), Kaufmann, Los Altos

suggests that standards are, in part, determined by natural levels, the standards being around 10 to 100 times greater than the average of the levels in streams.

CHAPTER 12

The Lithosphere and Pollution

Pollution of the lithosphere is not as readily described as in the atmosphere and hydrosphere because of the absence of significant mixing. The lithosphere cannot be considered in isolation, as both air and water in contact with land have the potential to spread materials to other areas and provide some mixing. The three broad divisions of the lithosphere are rocks, soil and areas of mineralization. Of the three, soil is most directly in contact with human beings, and the part on which we depend for most of our food.

SILICATES

The basic materials of the lithosphere are the silicates and aluminosilicates, making up the rocks and clays. Silicon has a valency of four, but unlike carbon, is unable to form $p_{\pi} - p_{\pi}$ bonds, of any significance, with oxygen, because of the diffuse nature of the Si p_{π} orbitals and the large silicon radius. Hence, while carbon forms discrete species, with multiple bonding;

$$O = C = O, \quad C \equiv O \quad \text{and} \quad \left[O \text{---} C \begin{matrix} O \\ \\ O \end{matrix} \right]^{2-}$$

the bond between Si and O atoms in silicates is single (with perhaps some $Si_{d\pi} \leftarrow O_{p\pi}$ bonding). However, since oxygen is divalent it must either form a further single bond, by bridging two silicon atoms, or gain an electron for each unsatisfied valency. The silicate structures therefore arise from the different ways oxygen achieves its valency requirements. The results are summarised in Fig. 6.10.

The Si - O bond energy (468 kJ mol^{-1}) is high, and resistant to chemical attack. Hence silicate structures persist in the environment, and only gradually weather.

The replacement of some Si atoms by Al adds to the diversity of structures, and it is the alumino-silicates in particular, which are the basis of many rock structures (feldspars) and of clays. The change from Si to Al increases the negative charge associated with silicate anions, which is offset by additional counter cations.

The zeolites, a group of three dimensional framework aluminosilicates, are important chemicals, obtained both naturally and synthetically. The framework structure encloses large cavities (400 - 1100 pm diameter, with 200 pm openings) in which cations and/or water molecules may reside. Selective movement, in and out of the zeolite cavities, is made use of in ion-exchange and molecular sieve properties. Zeolites find use as catalysts in the production of high octane petrol. They are involved in the cracking, hydroisomerization and reforming processes. Conversion of methanol to petrol is also catalysed by zeolites, a process currently under investigation for commercial development.

ROCKS

There are three rock types in the earth's crust, igneous, sedimentary and metamorphic. Their interrelation is demonstrated in Fig. 12.1, which includes the important weathering step.

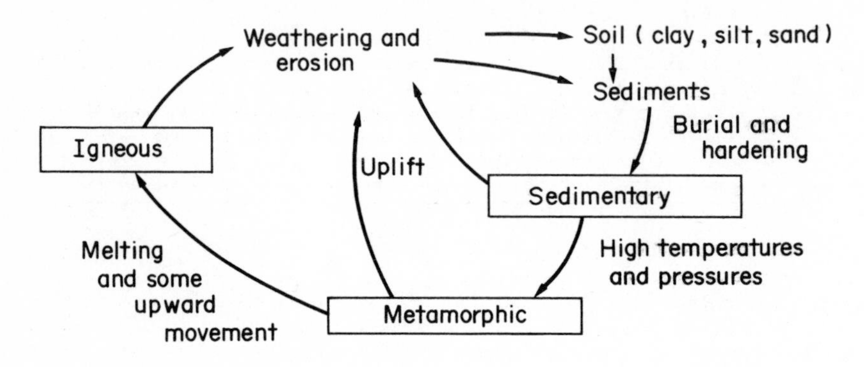

Fig. 12.1. The rock cycle

Igneous Rocks

Igneous rocks are solidified magma. Intrusive igneous rocks cooled below the earth's crust, and since cooling was normally slow the grain or crystal size can be quite large. Extrusive igneous rocks cooled on the earth's surface normally have a smaller crystal size because of more rapid cooling.

Igneous rocks are also classified in terms of the "SiO_2", or Si content. When the SiO_2 > 66% the rocks are called acidic, 52 - 66% intermediate, 45 - 52% basic and <45% ultrabasic. The most common rocks are the basalts (extrusive and basic) and granites (intrusive and acidic). This may be seen from the frequency curve of igneous rocks versus %Si (or % SiO_2) given in Fig. 12.2. Increase in the silicon content correlates well with changes in concentration of other elements (Fig. 12.3). The reason for this is because of the various silicate components of the rocks (Fig. 12.4).

The major change from one igneous rock to another is in the proportion of feldspars, pyroxenes, olivine and quartz. Therefore an order of rock formation in relation to solidification temperature may be determined from the different crystallising temperatures of the component minerals (see

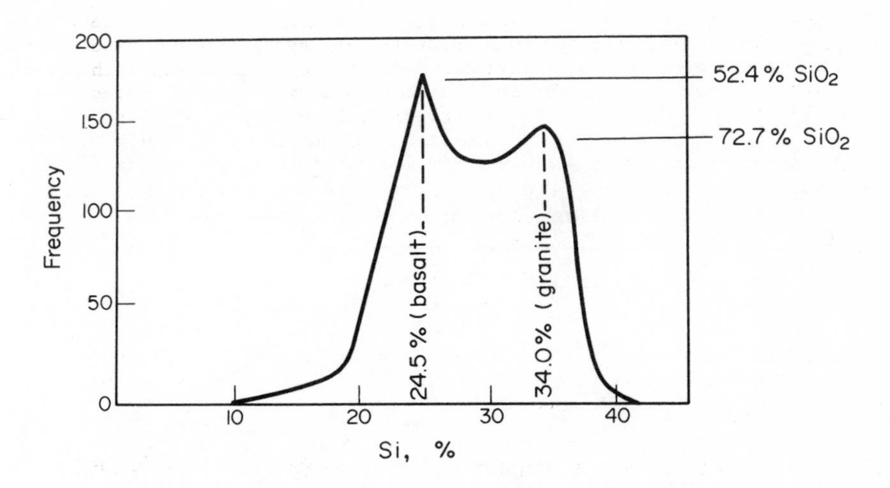

Fig. 12.2. The silicon content of igneous rocks and the frequency of occurrence.

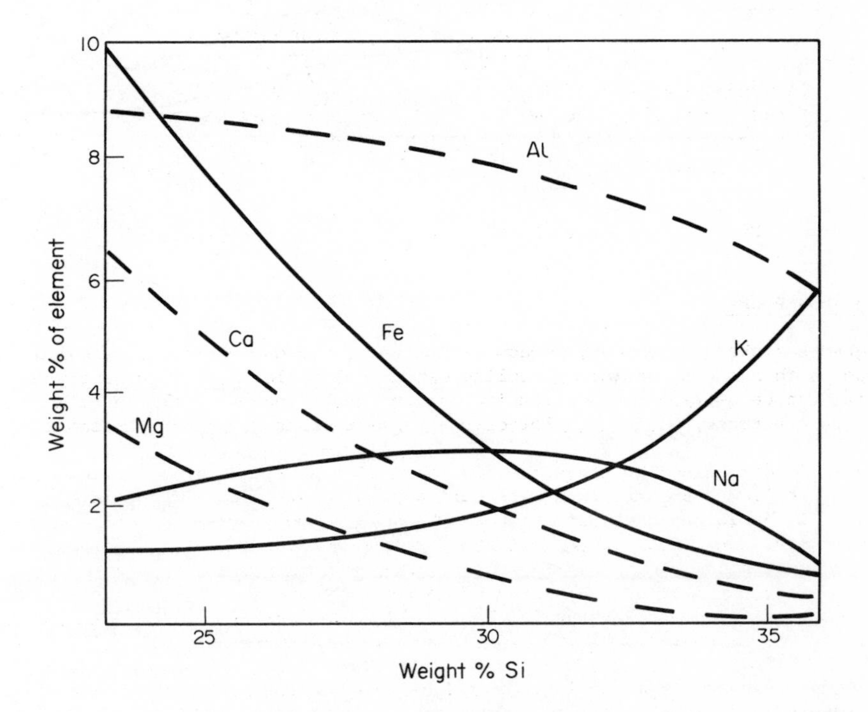

Fig. 12.3. Variation in the concentration of some elements in igneous rocks as a function of silicon content.

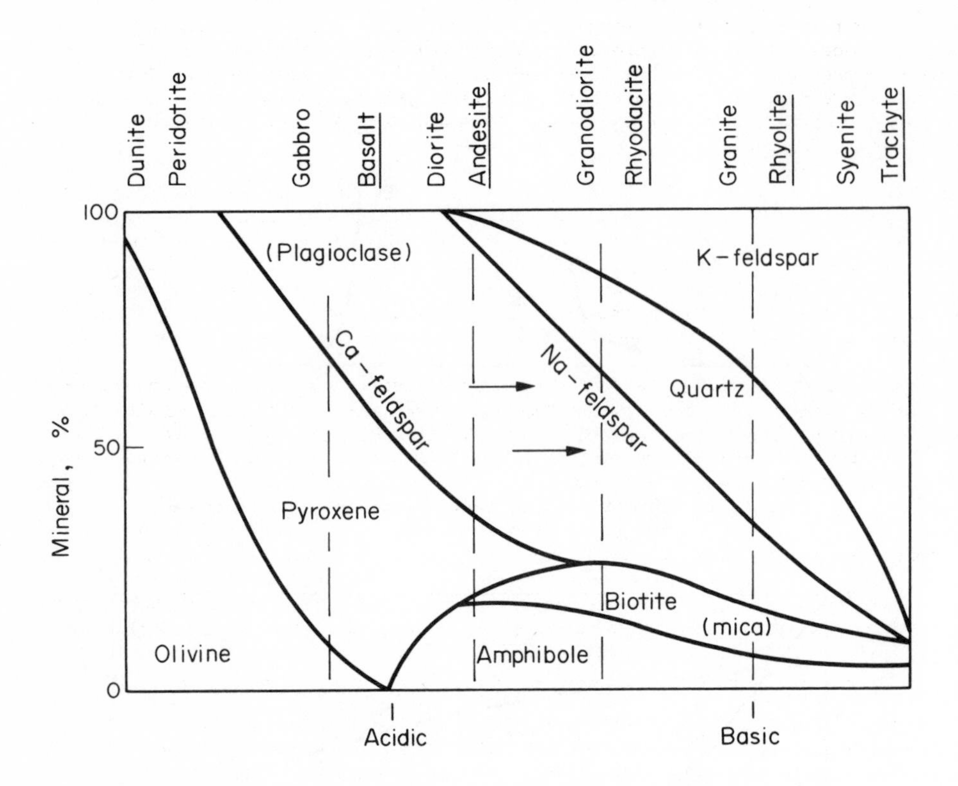

Fig. 12.4. Mineral components of igneous rocks.

Bowen reaction series, Fig. 12.5). The order of crystallation, from high to low temperatures, correspond to an increase in the SiO_2 content, and also the amount of Si to O condensation (i.e. the Si:O ratio).

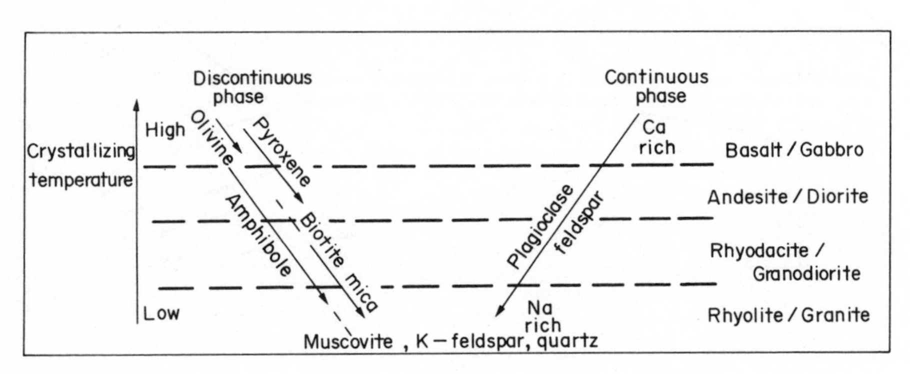

Fig. 12.5. The order of crystallisation of igneous rocks and component minerals.

Sedimentary Rocks

Sedimentary rocks slowly form from sediments by a process called diagenesis. The process which is little understood, includes compaction, loss of water under pressure, and further chemical breakdown of the sediments. The particles are cemented together with materials such as gypsum, anhydrite ($CaSO_4H_2O$, $CaSO_4$), calcite ($CaCO_3$), dolomite ($(Ca,Mg)CO_3$), silica, iron oxides and sulphides.

The principal materials of sedimentary rocks are quartz, clay, calcite and dolomite, and of lesser importance goethite (FeO.OH), hematite (Fe_2O_3), halite and gypsum. The rocks are of three main types, clastic - formed from fragments of rocks, chemical - formed from precipitates and evaporites, and organic - formed from organic material. Sedimentary limestone may occur in each type, i.e. with fragements of rocks embedded in it, chemically formed stalactites and stalagmites and limestone formed from coral and seashells.

Clastic sedimentary rocks include sandstone, siltstone, mudstone and greywacke. Examples of chemical sedimentary rocks are dolomite, rocksalt, flint, rock gypsum and organic sedimentary rocks include coal and oil shale. The most abundant sedimentary rocks are shale (mudstone) ($>50\%$), sandstone ($\sim 20\%$), limestone and dolomite ($\sim 15\%$).

Metamorphic Rocks

Sedimentary and igneous rocks can alter under the stress of high temperatures and/or pressures, to produce metamorphic rocks. Numerous new minerals are formed, also the materials may settle into bands giving banded rocks. Typical changes are, shale to slate and limestone to marble.

Metamorphic rocks are classified as slate, which is fine-grained and splits into sheets; schistose, coarser grained and has good cleavage; granular, which have no cleavage, e.g. marble and quartzite; and gneisses are coarse grained rocks with irregular banding. Some of the more important metamorphic minerals are: muscovite $KAl_2AlSi_3O_{10}(OH)_2$, biotite $K(Mg,Fe)_5(Al,Fe^{3+})_2Si_3O_{10}(OH)_8$, garnets $A_3B_2(SiO_4)_3$ where A = Ca, Mg, Fe, Mn and B = Fe^{3+}, Cr^{3+}, examples are pyrope $Mg_3Al_2(SiO_4)_3$, almandine $Fe_3Al_2(SiO_4)_3$ and pyroxenes, example jadeite, $NaAlSi_2O_6$.

Some materials in rocks weather more rapidly than others (Chapter 2), and as a result the minerals in soils either remain from weathering, or have been transported. Weathering is the most advanced, in the humid tropics, the basic materials (alkali and alkaline earth minerals) weather most rapidly, then quartz followed by iron and aluminium minerals. This has been summarized in Fig. 2.11 and some specific weathering products from minerals are given in Table 12.1. The extent of weathering, and the minerals present determine the soil formed and subsequent properties.

Contamination of a rocky environment by pollution could alter the rate of weathering and the products. Acid conditions will speed the process up. Weathering of rocks today could be considered of little concern to present day generations, but in acid conditions this may not be a valid assumption. Acid rain, for example, will speed up weathering processes and the leaching of soils, which could alter soil fertility in the future.

TABLE 12.1 Minerals and Weathering Products

Stage of Weathering	Mineral	Released Ions	Residual Material	Soil Type
Early	Biotite	K^+, Mg^{2+}	Clays, limonite, hematite	Dominated by these minerals and residual materials in young soils of world, espeically in arid regions.
	Calcite		Calcite	
	Gypsum	Ca^{2+}, SO_4^{2-}	Gypsum	
	Olivine	Mg^{2+}, Fe^{2+}	Clays, limonite, hematite	
	Feldspars (Na,Ca)	Na^+, Ca^{2+}	Clays	
	Pyroxene	$Mg^{2+}, Ca^{2+}, Fe^{2+}$	Clays, limonite, hematite	
Intermediate	Clay Minerals	SiO_2	Tending to bauxite (humid tropics)	Soils of temperate regions, wheat and corn belts.
	Muscovite	K^+, SiO_2	Muscovite, clays (e.g. montmorillonite)	
Advanced	Clay Minerals	SiO_2	Tending to bauxite and iron oxide	In humid tropics, low fertility soils
	Gibbsite		Gibbsite	
	Hematite		Hematite	

SOIL

Soil could be described as a mixture of inorganic and organic materials ranging from colloids to small particles, containing both dead and living materials, water and gases in variable proportions, but normally in dynamic balance. For a chemist soil could be considered as a nightmare, a far cry from the pure compounds in the chemical laboratory. However, a study and understanding of soil-chemistry, is important, as it is often in the area of chemistry that the clue to resolving soil pollution problems lies.

Soil Components

Sand, Silt and Clay. The principal components of soil are inorganic materials called sand, silt and clay. The division between the three is somewhat arbitrary, and is in terms of particle size. Sand particles have diameters within the range 0.05 - 2.0 mm, which is subdivided further from very fine (0.05 - 0.01 mm), to very coarse (1.0 - 2.0 mm). Sand is mainly quartz. A sandy soil is light, easily worked, but very permeable to water, so soluble salts are readily leached from it. Soil particles with diameters 0.002 - 0.05 mm are called silt, which is mainly quartz, with other silicate minerals of both primary and secondary origin. Finally clay particles have diameters <0.002mm. Clay consists of a wide range of secondary silicate and aluminosilicate minerals, which we will consider in more detail below. The proportions of sand, silt and clay in soils vary widely from soil to soil. Soil texture is determined by the relative proportions of the three materials as portrayed in the triangular chart in Fig. 12.6.

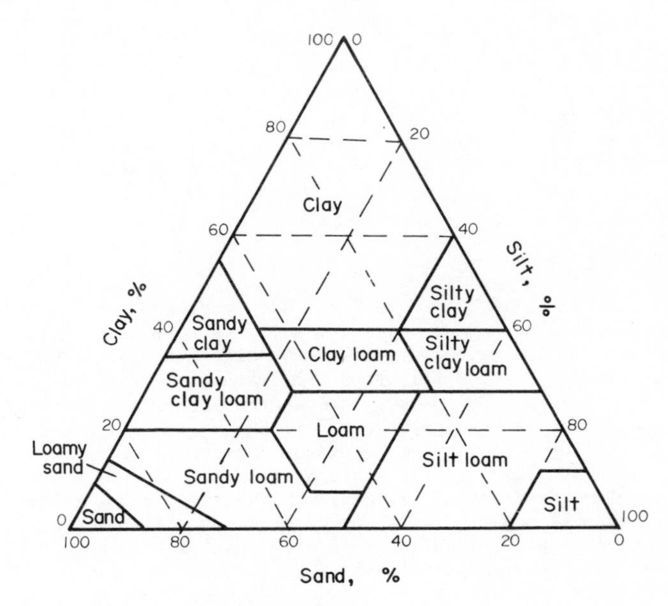

Fig. 12.6. Soil texture chart.

Organic. Soil organic material is around 2 - 5% (or less) of the total soil mass, however, it has an important and active role. The materials, which are mainly in top layers of the soil, consist of dead and living organic substances such as plant litter, decaying plants and organisms, and numerous soil organisms. The types of organisms include bacteria, fungi, algae, protozoa, worms, arthropods, mollusks and vertebrates. The amount of organic materials in a soil depends on climatic conditions, type of inorganic soil components and topography. An important intermediate product of the decay of organic substances in soil is humus. In the formation of humus most of the cellulose material has disappeared, lignins modified and protein retained. Humus influences the water-holding capacity of a soil, its ion-exchange capacity and binding of metal ions. The condition of a soil, as regards the breakdown of organic material, is indicated by a C:N ratio. A value of approximately 10:1 indicates a satisfactory condition.

Water and Atmosphere. Soil water and atmosphere fill up the interstitial spaces between soil particles. The space is determined from a knowledge of the bulk density of the soil and particle density. The movement of H_2O and gases is controlled by the amount of space and the size of the pores. This is in part determined by the size of the soil particles, i.e. the proportion of sand, silt and clay. The smaller the particles (i.e. the more clay) the greater the number in a certain volume, and this generally means more pore space. For a sandy soil, the water holding capacity is low but most of the water is available to plants, while for a clayey soil the holding capacity is greater, but a larger proportion is held tightly by minerals and in microscopic pores and unavailable to plants. The water is a solution of dissolved materials, from which plants obtain the necessary nutrients (as well as water). The amount dissolved depends on the pH of the soil solution. Acid and basic pollutants can therefore, influence the soil solution composition.

The soil atmosphere is vital, for oxidation processes involved in the decay of organic material. Absence of oxygen, as in a waterlogged soil, allows reducing conditions to occur. The soil atmosphere and water, together fill the available space. The atmosphere is similar to, but not identical to, the normal atmosphere because of a high CO_2 level from the decaying processes.

Soil-Forming Factors

Over the centuries, soil-forming factors have determined soil type. The *parent material,* the rock from which the soil has developed, determines the mineral content and the proportions of sand, silt and clay. The *typography,* i.e. the slope and exposure influences the drainage of the soil and the impact of the climate. The *climate* (temperature, rainfall and wind), affects the water content of, and evaporation from, the soil. *Organisms* influence the soils continuing fertility, as distinct from the fertility arising from the parent material. *Time* is important as regards continuing weathering and change. In quite recent times (relative to the time it has taken to form some soils), *man* has modified soils by his treament of them and the over-lying vegetation. It should be kept in mind that all these factors are interrelated, and cannot be discussed in isolation. For example, the topography (gentle undulation), the climate, alternating wet and dry seasons, bacteria and a soil rich in iron and manganese, all contribute to the transport of ions and build up of iron-manganese nodules in a soil. And this all takes time.

$$4Fe^{2+} + O_2 + 4H_2O \overset{\text{bacteria}}{\rightleftharpoons} 2Fe_2O_{3(s)} + 8H^+ \qquad (12.1)$$

$$2Mn^{2+} + O_2 + 2H_2O \overset{\text{bacteria}}{\rightleftharpoons} 2MnO_{2(s)} + 4H^+ \qquad (12.2)$$

Waterlogged, acid soil (winter). Ions move from high areas into depressions.	Dry, aerated soil (summer). Nodules accumulate in the depressions.

The intervention of man, providing drainage in the winter and irrigation in the summer, may modify the above processes.

The end result of soil formation is seen from the soil profile. The profile divides into horizons which differ in the amounts of soil components. The top horizon, the O horizon, contains the majority of the organic material and humus. Next the A horizon is the main mineral zone. Downward leaching and movement of iron, aluminium and clay (eluvial zone) leaves behind the resistant minerals. The clay, iron and aluminium, accumulate in the third, the B horizon, (illuvial zone). Finally, the C horizon contains the parent material, but not just bed-rock. Horizons are further subdivided depending on the contents.

Three of the principal processes in well drained areas are *podzolization*, in the cool humid areas, *latosolization* in the humid tropics and *calcification* in dry areas. In podzolization the soil is leached by acid water (as low as pH 4), the acidity coming from organic materials. Soluble bases, aluminium and iron oxides are transported downwards, clays peptise (i.e. particle size reduces) and also move downwards. Hence the soils have pronounced A and B horizons. In latosolization, weaker acid (pH = 5-7) leaching occurs, or even alkaline leaching. This is probably because of the rapid mineralization of humus in the warmer climate. At higher pH SiO_2 is more soluble (Fig. 2.9) and removed from the soil, leaving behind low silicate clays (e.g. kaolinite), iron and aluminium oxides. In calcification, $CaCO_3$, $MgCO_3$ and $CaSO_4$ build up in top horizons because of the lack of water. A classification of the world's soils is given in Table 12.2.

TABLE 12.2 World Soil Classification

Soil Order	Derivation of Root Word	Approximate % of World Soils	Comments	Weathering
Entisols	recent	13	no profile developed	low
Vertisols	verto (L) = to turn	2	crack badly when dry	low
Inceptisols	inceptum (L) = beginning	16	young, profile just beginning	low
Aridisols	aridus (L) = dry	19	desert soils	low to moderate
Mollisols	mollis (L) = soft	9	brown to black soils	moderate
Spodosols	spodos (G) = woodash	5	developed spodic horizon	moderate
Alfisols	Alf - from aluminium and iron	15	pronounced mineral horizon	moderate
Ultisols	Ultimos (L) ultimate	9	red-yellow mineral horizon	high
Oxisols	oxi from oxide	9	free sesquioxides deep horizons	high
Histosols	histos (G) = tissue	1	bog soils	

Clays

The clay minerals form from weathering of rocks (Fig. 2.11). Clays have a layered structure, there being three basic types, 1:1, 2:1 and 2:2. The ratios refer to the arrangement of the two different sheet structures in clays. One sheet consists of SiO_4 tetrahedra, three of the oxygen atoms are only associated with the silicon atoms, while the fourth (all pointing in the same direction) is also involved in the second sheet. This sheet consists of close-packed oxygen and OH groups, with Al in the octahedral cavities.

For the 1:1 clays, e.g. kaolinite, the repeating unit is one layer composed of one "SiO_4 sheet" and one "AlO_6 sheet" (Fig. 12.7). Hydrogen bonding holds the layers together and is sufficiently strong to prevent ions getting between them.

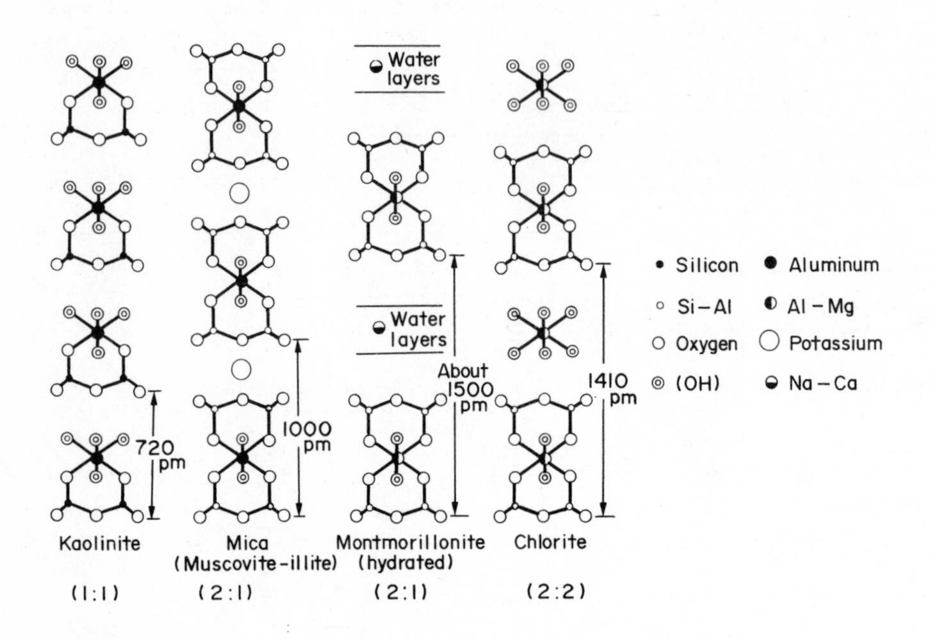

Fig. 12.7. Schematic diagrams of some typical clays.

A 2:1 clay, e.g. illites, has layers made up of three sheets, one "AlO_6 sheet" sandwiched between two "SiO_4 sheets". In muscovite one quarter of the Si atoms are replaced by Al, and in vermiculite some of the octahedrally coordinated Al atoms are replaced by Mg and Fe. These changes alters the charge on the clays and cations such as K^+, Ca^{2+} and Mg^{2+} fit between the layers, as well as water, all helping to hold the layers together. A 2:2 clay, e.g. chlorite, has the same layer as for a 2:1 clay but between these layers a discrete "AlO_6 sheet" occurs, in which some of the Al atoms are replaced by Mg and Fe. A summary of the clay types is given in Table 12.3.

Soil Properties

We will consider just two soil properties, their ion exchange capacity and soil pH. The properties are interrelated, and influence the availability of

TABLE 12.3 Some Clays

Clay Type	"SiO_4 sheet"	"AlO_6 sheet"	Between layers	Name
1:1	4Si	4Al		kaolinite
	4Si	4Al	water	halloysite
2:1	4Si	4Al		pyrophyllite
	4Si	5Al+Mg	various cations & water	montmorillonite
	3Si+Al	4Al	K^+	muscovite*
	3Si+Al	6(Al+Mg+Fe)	Ca^{2+}, Mg^{2+}	vermiculite
2:2	3Si+Al	6(Al+Mg+Fe)		chlorite

* A mica rather than a clay

plant nutrients, and both are susceptible to pollution interference.

Ion-exchange capacity. The ion-exchange capacity of a soil is its ability to hold and exchange ions. Organic matter, but particularly clays, are effective ion-exchangers. The large surface area of clays is a reason why they are good ion-exchangers.

Clays in which Si atoms are replaced by Al, are anionic, and at high pH the OH groups lose protons, or the protons are replaced by metal ions. The hydroxyl groups of the soil organic acids react in a similar way. Equations (12.3) to (12.6) summarise the ways clays and organic acids are cation exchangers.

$$\text{Clay}^- + M^+ \rightleftharpoons \text{Clay-M}, \tag{12.3}$$

$$\text{Clay-M}' + M^+ \rightleftharpoons \text{Clay-M} + M'^+, \tag{12.4}$$

$$\text{Clay-OH} + M^+ \rightleftharpoons \text{Clay-OM} + H^+, \tag{12.5}$$

$$\text{R-}\underset{\underset{\text{O}}{\|}}{\text{C}}\text{-OH} + M^+ \rightleftharpoons \text{R-}\underset{\underset{\text{O}}{\|}}{\text{C}}\text{-OM} + H^+, \tag{12.6}$$

For a reaction such as;

$$\text{Clay}^- + \text{Clay-M} + 2K^+_{(soln)} \rightleftharpoons 2\text{Clay-K} + M^+_{(soln)}, \tag{12.7}$$

if K^+ is removed from solution through plant uptake, K^+ ions are released from the clay to re-establish equilibrium. Hence clay-K is a nutrient reservoir and addition of K^+ ions builds the reservoir up. Material which competes with the nutrient for sorption on the clay may disrupt the equilibrium. Herbicides, such as paraquat and simazine, which are said to become inactive in the soil, may in fact affect the nutrient reservoir because their "inactivity" involves sorption on to the clays.

Cations are sorbed on to clays in the order, $Al^{3+} > Ca^{2+} > Mg^{2+} > K^+ > Na^+$. The order reflects the charge on the ions, i.e. $M^{3+} > M^{2+} > M^+$. The orders

$Ca^{2+} > Mg^{2+}$ and $K^+ > Na^+$ are the inverse of what one would expect on the basis of ionic size (i.e. $r_{Ca^{2+}} > r_{Mg^{2+}}$ and $r_{K^+} > r_{Na^+}$). However, the cation size which is important is not that of the free cation, but of the hydrated cation, which is greater for the smaller cations (Table 12.4).

TABLE 12.4 Hydration of Na^+ and K^+

Property	Na^+	K^+
Ionic radius (pm)	95	113
Hydrated ion radius (pm)	276	232
Number of water molecules associated with cation	16.6	10.5

The anion exchange capacity of clays is not as significant as for cation exchange, but does occur according to the reaction;

$$\text{Clay-OH} + A^- \rightleftharpoons \text{Clay-A} + OH^- \qquad (12.8)$$

Nitrate is held weakly, and easily washed out, so clay does not act as a good nutrient source of NO_3^-. Phosphate, on the other hand, is more strongly held, especially by the iron and aluminium in the clay, as well as by any Fe and Al oxides present. Sorption of phosphate is largely irreversible and much of the added phosphate rapidly becomes unavailable to plants.

The ion-exchange capacity of a clay is measured in milliequivalents 100 g^{-1} soil (where 1 milliequiv. = 1 millimole of M^+ and ½ millimole of M^{2+}, etc.) A K^+ exchange capacity of 0.8 milliequivalents per 100 g of soil means the available K^+ in one hectare of soil, to a depth of 10 cm, is around 400 kg (0.4 tonne). The 2:1 clays have the best ion-exchange capacity as they are expanding. The decreasing order of ion-exchange capacity is:

vermiculite (1.0-1.5) > montmorillonite (0.7-1.2) > chlorite(0.1-0.4) > halloysite (0.05-0.5) > kaolinite (0.03-0.15) equiv. kg^{-1} (i.e. 2:1 > 2:2 > 1:1).

Soil pH. Soil pH influences the fertility of a soil, and is affected by the cations sorbed on to the clays. For example a predominance of cations such as Ca, Mg, Na and K tends to raise the pH when they are released from a clay;

$$\text{Clay-Na} + H_2O \rightleftharpoons \text{Clay-H} + Na^+ + OH^-. \qquad (12.9)$$

Cations such as Al and Fe (and of course H^+), when desorbed from a clay, produce acid soil solutions due to hydrolysis of the cations;

$$\text{Clay-Al} + 5H_2O \rightleftharpoons \text{Clay-}H_3 + Al(OH)_4^- + H_3O^+. \qquad (12.10)$$

Acid soils are associated with humid climates, where frequent leaching tends to produce protonated clays which lower the pH;

$$\text{Clay-H} + H_2O \rightleftharpoons \text{Clay}^- + H_3O^+, \qquad (12.11)$$

whereas arid soils where Ca, Mg, Na and K salts accumulate, have a high pH, equation (12.9).

The influence of pH on the availability of nutrients is summarised in Fig. 12.8, where the band width represents the availability of the elements.

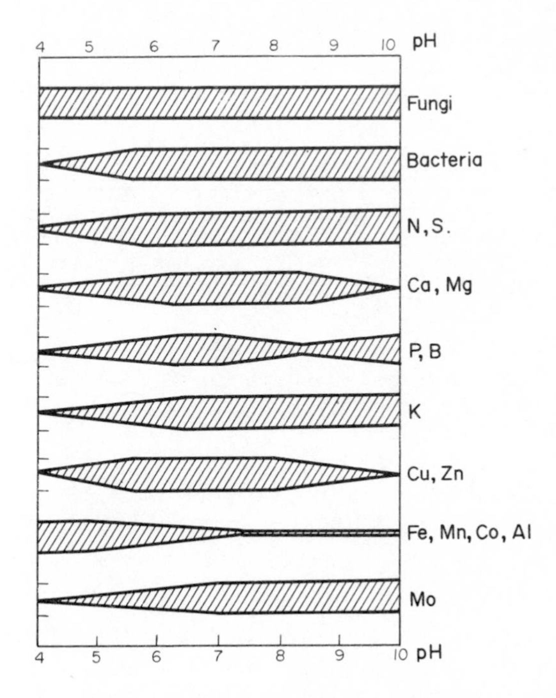

Fig. 12.8. Nutrient availability in relation to soil pH, the wider the band the more available is the nutrient.

Calcium and magnesium ions are removed from the soil solution at high pH as insoluble carbonates, this does not occur for K^+. Metals, such as Fe, Mn and Al, are not available much above a pH of 5 to 6 because of formation of insoluble hydroxides.

Phosphorus availability to plants is greatest in the narrow pH range 6 to 7. The form of phosphate most soluble and therefore assimulated by plants, is $H_2PO_4^-$, which exists over the pH range 2 - 7 (Fig. 6.7). However, below a pH of 6 insoluble Fe and Al phosphate form, and above a pH of 7 insoluble $Ca_3(PO_4)_2$ starts to form. Therefore to ensure the best conditions for the utilization of P, the soil pH needs to be regulated. Soil pH is raised by liming;

$$\text{Clay-H}_2 + CaCO_3 \rightarrow \text{Clay-Ca} + H_2O + CO_2, \qquad (12.12)$$

and the clay-Ca species increases the pH, equation (12.9). Soil pH is lowered by adding acid salts (e.g. $(NH_4)_2SO_4$) or sulphur, which is oxidised by soil bacteria to sulphuric acid;

$$2S + 3O_2 + 2H_2O \xrightarrow{\text{bacteria}} 4H^+ + 2SO_4^{2-}, \quad (12.13)$$

$$Clay^- + H^+ \rightleftharpoons Clay\text{-}H. \quad (12.14)$$

Trace Metals

Natural Levels. The natural levels of some of the more important trace elements in soils and rocks are listed in Table 12.5. For most elements the amount correlates with the levels in the parent material. However, soil

TABLE 12.5 Some Trace Elements in Soils

Element	Crustal abundance ppm	Basic Rocks ppm	Acid Rocks ppm	Sedimentary Rocks ppm	Range in Soils ppm	Approximate Mean in soils ppm
As	2	1.5	1.5	12	1-100	1
B	10	1-2	3	100	tr*-300	20-50
Co	25	50	8	20	0.05-300	10-15
Cr	100	2000	2	100-500	tr-4000	100-300
Cu	55	150	10	10-100	tr-300	15-40
Fe	50,000	100,000	25,000	35,000	·2-50%	30,000
Mn	1000	2000	1000	1000	tr-10,000	500-1000
Mo	1.5	2	2.5	2	tr-24	1-2
Ni	80	200-1000	10	52	tr-5000	20-30
Pb	16	3	24	19	tr-1200	15-25
Se	0.05	0.1	0.1	0.4	tr-12	0.01
Ti	5600	9000	2300	3800	tr-25,000	4000-5000
V	135	200	50	100	tr-400	100
Zn	70	100	60	95	tr-900	50-100

*tr = trace

forming factors can alter the pattern somewhat. Ultrabasic and basic extrusive rocks tend to be rich in the transition metals, Co, Cr, Cu, Ni, V and Zn, and poor in the alkali and alkaline earth metals, while the reverse holds for acid extrusive rocks.

Some elements, such as B, Pb and Mo are at a higher concentration in the soil than the parent rocks, suggesting some retention process has occurred. Variation in the concentration of trace metal ions down a soil profile tend to be the same for all metals ions; namely highest levels in the organic rich horizons followed by the clay rich horizons. Accumulation in the organic layers is due to the formation of metal-organic complexes, and in the clay zones due to sorption. In some tropical soils the major accumulation is in the clay-rich horizons, because of limited organic material. Titanium generally increases down a soil profile. Sandy soils are normally deficient in trace elements because of poor retention.

In acid soils the common form of most trace metals is the hydrated cation, which are soluble, and relatively easily leached, while at higher pH insoluble carbonates, oxides and hydroxides form, so the metals are retained. In semi-arid to arid soils (vertisols) with high pH trace metal levels may be high. In podsols (leached soils) levels will tend to be lower.

The available trace metal in a soil is not normally the same as the total concentration, but is the amount extracted by different reagents, such as water, ammonium acetate, oxalic acid, ammonium oxalate, mineral acids and EDTA. Depending on the reagent the amount extracted is said to be an estimate of what is available to plants. Extraction studies suggest that leached soils (e.g. podsols), high organic soils (e.g. peats), sandy soils, leached acid soils, and alkaline soils can be deficient in available trace metals. For example copper levels may be quite high in peaty soils, but unavailable to plants because of stable complexes formed with organic acids.

The availability of selenium is important as it is an essential element for animals (and maybe man). The levels in rocks range from 0.05 ppm in granites to 0.6 ppm in shales with a mean around 0.1 ppm. Selenium is often found associated with sulphur around volcanic areas and in rocks and minerals. But selenium species are less readily oxidised and are less mobile than sulphur, so it is not as readily removed by weathering. The differences in the redox chemistry of S and Se helps to explain their different chemistry in soil.

$$H_2S \rightarrow S + 2H^+ + 2e, \; -E_0 = -0.14V, \tag{12.15}$$

$$H_2Se \rightarrow Se + 2H^+ + 2e, \; -E_0 = +0.40V, \tag{12.16}$$

$$S + 3H_2O \rightarrow H_2SO_3 + 4H^+ + 4e, \; -E_o = -0.45V, \tag{12.17}$$

$$Se + 3H_2O \rightarrow H_2SeO_3 + 4H^+ + 4e, \; -E_o = -0.74V, \tag{12.18}$$

$$H_2SO_3 + H_2O \rightarrow SO_4^{2-} + 4H^+ + 2e, \; -E_o = -0.17V, \tag{12.19}$$

$$H_2SeO_3 + H_2O \rightarrow SeO_4^{2-} + 4H^+ + 2e, \; -E_o = -1.15V. \tag{12.20}$$

Relative to sulphur, elemental selenium is the more redox stable element, and for the oxyanions selenite is the favoured form for Se, while sulphate is the favoured form for S. From the selenium E_h - pH diagram, Fig. 12.9, it can be seen that the dominant selenium species is selenite in acid soils. At higher pH, since the oxidation potential of equation (12.20) is less negative the product, SeO_4^{2-}, become more important. The selenite ion in iron rich soils forms insoluble basic ferric selenite, or sorbs strongly on to iron oxides. This shows up in the acid soils of Hawaii where the selenium content can be as high as 20 ppm and yet very little is available to plants. In alkaline soils, even with low Se levels, plants, especially accumulator plants (e.g. species of *Astragalus*) can attain high levels of Se, because more soluble SeO_4^{2-} is the common form in the soil. An Australian study has found an inverse relationship between the selenium content of plants and rainfall. In high rainfall areas, where acid soils occur, the Se levels in plants are low (<0.1 ppm), whereas the reverse holds in low rainfall areas (plant Se ~0.7ppm, yearly rainfall <250mm).

Iron and manganese, which are relatively abundant in soils can influence the distribution of trace metals. Iron occurs, either as a replacement for Al in clays, or as oxides and hydroxide minerals, e.g. goethite, $\alpha FeO(OH)$;

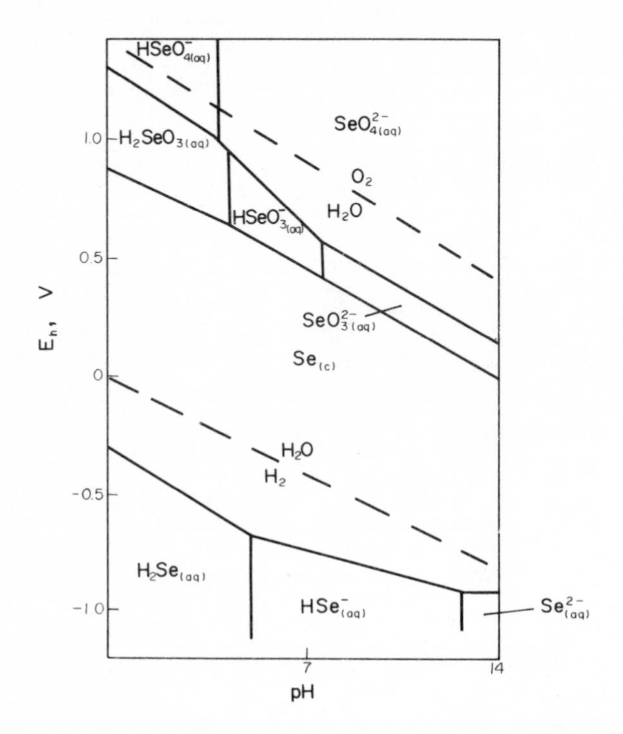

Fig. 12.9. E_h-pH diagram for selenium in aqueous conditions.

hematite, αFe_2O_3 (the main two); maghemite, γFe_2O_3; and lepidocrocite, $\gamma FeO(OH)$. Manganese also occurs as oxides, but more complex, e.g. lithiophorite, $Li_2Mn_2Al_8Mn_{10}O_{35}14H_2O$ and birnessite $(Na,Ca)Mn_7O_{14}2.3H_2O$. Manganese oxides tend to concentrate Fe, Co, Ni, Cu, Zn, Ba and Pb while iron oxides concentrate V, Mn, Ni, Cu, Zn, As, Se, Mo, P, S, Si. The last six elements associated with iron, occur as oxyanions, therefore iron oxides appear to have greater attraction for oxyanions than Mn oxides. To some extent the availability of these trace elements is controlled by the availability of iron and manganese in the soils. The tendency of Mn and Fe oxides to concentrate trace metals can be explained in terms of isomorphous replacement of ions. However, lattice energies, and maybe crystal field stabilization energies, will contribute to the overall stabilization of the replacement process.

Pollution. The addition of materials to soils by man, arises from the use of agriculture chemicals such as herbicides, fungicides and insecticides, from dustfall and precipitation and use of fertilizers and contaminated water. Whether the material becomes a problem or not, depends on the soil type and the levels. In many situations the problems occur later, when the elements are leached into waterways. Chemical fertilizers are a source of trace metals, either added deliberately, or as an impurity. For example, Cd, As and Pb, common trace metals in rock phosphate, are also in superphosphate. This source of trace metals may not be a problem yet, but this may change as more are added, with increased use of fertilizers. Around 13 Tg of fertilizers were used in 1950 in the world, this had risen to 81 Tg by 1975, and estimates for 1980 are 100-120 Tg.

Serious cadmium pollution of soil occurred in Japan. Over a 20 year period around 100 people (mostly older women) living near the Jintsu River died of a syndrome called Itai-Itai (Ouch-Ouch) disease. The effects were severe and painful bone damage. Bones were easily fractured and twisted, and the effects took around 5-10 years (and in some cases up to 30 years) to manifest themselves. In 1961 cadmium was found to be the cause, the source was traced to a zinc mine (Cd is closely associated with Zn ores) 50 km upriver. The water from the river was used by the people to irrigate the rice paddy fields. The Cd now in the soil, in a number of places >3.0 ppm, was taken up by the rice plants and consumed by the people. Studies in the area demonstrated a close correlation between levels of Cd in the soil and the percentage of women over 50 years with the disease (Fig. 12.10). Mean cadmium levels in

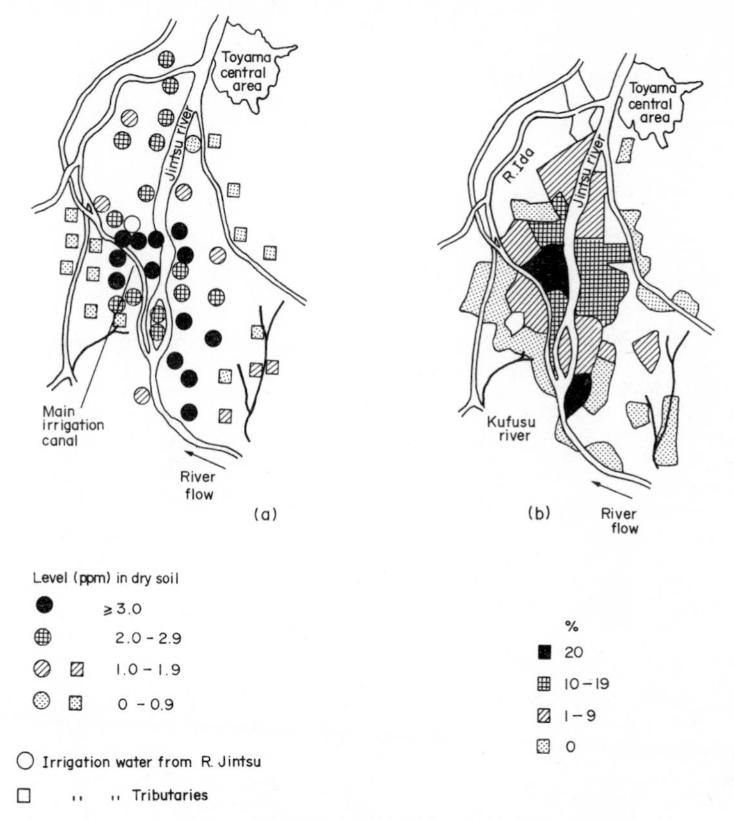

Fig. 12.10. Relationship between levels of cadmium in soil and incidences of Itai-Itai disease. (a) Cd levels in surface soils, (b) % women over 50 yrs with the disease. Source: Friberg, L., Piscator, M. and Nordberg, G. Cadmium in the Environment, (1971) Chem. Rub. Co., Cleveland.

rice of 0.68 μg g^{-1} (wet weight) was ten times greater than from other areas of Japan. Cadmium water levels were found to be 1 μg l^{-1} in well water, 5-61 μg l^{-1} (mean 17 μg l^{-1}) in mine waste water and 1-9 μg l^{-1} in the river.

A possible reason why older women were mainly affected was that they also suffered from a deficiency of vitamin D (from a lack of sunshine). This, together with the high cadmium intake, gave rise to the very painful and often fatal disease. This situation highlights the complexity of the inter-relation of more than one contributing factor.

Transport of Trace Metals. If a metal species is soluble, or remains in suspension, it is clear how it is transported. Normally the movement is downwards. Transport is achieved biologically when trace elements taken into a plant are redistributed when a plant dies and decays, the enrichment being in the top layers. Micro-organisms, especially those that move (e.g. earthworms) also translocate trace elements. Finally volatile forms of trace elements, e.g. $(CH_3)_2Hg$, will be transported via the soil atmosphere. However the principal mode of transport is with water-dissolved or suspended materials.

Suspended materials are usually associated with colloidal material (<0.45 μm), i.e. clays, humic acid, iron and manganese hydroxide/oxides and bacteria. The colloids have a surface charge, positive at low pH and negative at high pH, i.e.;

$$\underset{\text{(colloidal)}}{\text{solid-OH}} + H^+ \rightleftharpoons \underset{\text{(positive, acid medium)}}{\text{solid-OH}_2^+}, \quad (12.21)$$

$$\underset{\text{(colloidal)}}{\text{solid-OH}} \rightleftharpoons \underset{\text{(negative, alkaline medium)}}{\text{solid-O}^- + H^+}, \quad (12.22)$$

Clays can also be charged as a result of their ion exchange capacity.

Transport may be stopped by coagulation of colloidal particles. If the electrostatic repulsion between the particles is removed, coagulation occurs, the process being pH dependent. Species attached to the colloids may then be carried down as co-precipitates or as occluded material.

Precipitation of soluble trace element species is probably not significant, except in a few special circumstances, because the concentrations are low and solubility products are unlikely to be exceeded. Exceptions may occur when fertilizers, or lime, or concentrated pollutants, are added to the soil raising the levels of certain species. Precipitation occurs when a soil dries out and the material may not redissolve readily when the soil becomes wet again.

Sorption on to large particles is another method by which material can be removed from solution. This process, as for sorption on to colloids, is pH dependent.

Copper and Zinc in Soils. The levels of copper and zinc in soils reflect the concentrations in the parent rocks. Igneous basaltic rocks (90 ppm), sedimentary shale and clay (50 ppm), and black shale (70 ppm) contain the highest levels of copper. Soils from granites, gneiss and quartzite may be deficient in zinc, while soils from calcerous rocks normally have the highest levels of zinc. Soils around copper ores are anomalously high, e.g. soils around Sudbury, Canada reach copper levels of 2000 μg g^{-1}. Even though zinc is an essential metal, large amounts can be toxic, and soils with added sewage sludge may have dangerous levels of zinc.

Copper and zinc pollution in soils comes from mining and smelting activities, industrial emissions, urban development, western living styles, dumped waste solids, dustfall and precipitation, sewage sludge, fertilizers and for copper, pesticides. Around solid waste dumps Cu levels of 90 $\mu g\ g^{-1}$ have been reported, whereas nearby the levels are as low as 3 $\mu g\ g^{-1}$. The copper concentration in sewage sludge can reach 6000 $\mu g\ g^{-1}$.

Copper and zinc are concentrated in the soil organic horizon, through coordination to phenolic, carboxylic and hydroxylic groups. Some fungi concentrate Cu, and may influence its mobilization. Podsolization raises the Cu and Zn levels in the B or clay rich horizon. The accumulation of Cu and Zn by inorganic soil fractions is in the order sand < silt < clay < Mn, Fe hydrous oxides. For zinc 20-45% is associated with clay and 30-60% with the Mn-Fe hydrous oxides. Zinc replaces magnesium in silicate structures, their ionic radii being similar (Zn^{2+}, 83 pm and Mg^{2+}, 78 pm).

In rocks, copper occurs mainly in ferromagnesium silicates (e.g. olivine and hornblende) and as sulphides, while in soils the major forms are: $Cu(OH)_2$, $CuCO_3$, CuS, Cu_2S and $Cu_2P_2O_7$. The principal zinc minerals in soils are sphalerite (ZnS), smithsonite ($ZnCO_3$) and hemimorphite ($Zn(OH)_2Si_2O_7H_2O$), while the main weathered forms of zinc are Zn^{2+} and $ZnOH^+$, the latter occurring at higher pH.

SEDIMENTS

The third major group of solids on the earth's crust is sediments. Many chemical and physical processes that occur in sediments are similar to those described for soils, but there are significant differences. One reason is that sediments are more intimately involved with water, and for the majority of sediments the water is salty.

Formation and Components

Sediments from soil are river borne clay, silt, sand, organic material, and other materials. It has been estimated that, prior to man's influence on the earth's crust, the oceans received annually around 9-10 x 10^9 tonnes of sediments, while estimates for today are around 25 x 10^9 tonnes yr^{-1}. Even so, the concentration of suspended sediments in the oceans is only 10-20 ppb (10-20 $\mu g\ l^{-1}$).

Sediments eventually settle out on the ocean floor or at the bottom of rivers, lakes and estuaries. The smaller particles such as clay particles, can remain suspended for months, assisted by currents and turbulence. Clay particles settle out more rapidly in salt water than fresh water. The surface of a clay particle can be negatively charged, which keeps the small particles apart. In salt water, due to the presence of dissolved ions, cations are absorbed on to the surface neutralising the charge. This destabilises the colloid and allows coagulation to occur. The neutral particle with a negative surface layer ana an absorbed positive layer is called an "electric double layer".

The flow of river water entering the ocean is reduced and particle size fractionation occurs; the larger particles settle out first and the finer particles being deposited further from the shore. Hence ocean sediments are mainly finely-grained aluminosilicates. The next major component is $CaCO_3$, produced biologically, then silicon and finally minerals, especially those of iron and manganese.

Unlike a soil solution, where most trace element concentrations are low, in the ocean higher concentrations mean that for some species solubility products may be exceeded. Some of the precipitation reactions which contributing to sediment build-up are;

$$5Ca^{2+} + 3PO_4^{3-} + OH^- \rightleftharpoons Ca_5(OH)(PO_4)_3\downarrow, \qquad (12.23)$$

$$Ca(HCO_3)_2 \rightleftharpoons CaCO_3\downarrow + CO_2 + H_2O, \qquad (12.24)$$

$$4Fe^{2+} + O_2 + 10H_2O \rightleftharpoons 4Fe(OH)_3\downarrow + 8H^+, \qquad (12.25)$$

$$Fe^{2+} + H_2S \rightleftharpoons FeS\downarrow + 2H^+. \qquad (12.26)$$

Organic material, and the availability of dioxygen, lead to two zones in the sediments, an aerobic zone, on top and in contact with the water, and an anaerobic zone where oxidation of the organic material is achieved with sulphate or by fermentation. In the aerobic zone the oxidation reaction is;

$$\{CH_2O\} + O_2 \xrightarrow{} _{\text{micro-organisms}} CO_2 + H_2O. \qquad (12.27)$$

Sulphate reduction under anaerobic conditions is represented by the equations;

$$SO_4^{2-} + 10H^+ + 8e \rightarrow H_2S + 4H_2O, \qquad (12.28)$$

and $$\{CH_2O\} + 2H_2O - 4e \rightarrow HCO_3^- + 5H^+ \qquad (12.29)$$

which add up to;

$$SO_4^{2-} + 2\{CH_2O\} \xrightarrow{} _{\text{micro-organisms}} H_2S + 2HCO_3^-. \qquad (12.30)$$

The H_2S produced will react with iron to give FeS and FeS_2;

$$2Fe(OH)_3 + 6H^+ + 2e \rightarrow 2Fe^{2+} + 6H_2O, \qquad (12.31)$$

$$H_2S - 2e \rightarrow S + 2H^+, \qquad (12.32)$$

$$2Fe^{2+} + 2H_2S \rightarrow 2FeS + 4H^+, \qquad (12.33)$$

which gives overall;

$$2Fe(OH)_3 + 3H_2S \rightarrow 2FeS + S + 3H_2O, \qquad (12.34)$$

and, $$FeS + S \rightarrow FeS_2. \qquad (12.35)$$

If no sulphate is present oxygen is obtained from the organic material itself;

$$2\{CH_2O\} \xrightarrow{} _{\text{micro-organisms}} CO_2 + CH_4. \qquad (12.36)$$

If the CO_2 cannot escape, and under the prevailing reducing conditions, $FeCO_3$ may be produced;

i.e. $$Fe^{3+} + e \rightarrow Fe^{2+}, \tag{12.37}$$

and, $$Fe^{2+} + H_2O + CO_2 \rightarrow FeCO_3 + 2H^+. \tag{12.38}$$

The water in sediments can be considered as interstitial, which can contain higher concentrations of metal ions than occurs in soil solutions. Dissolved gases may also occur. Determination of the amounts of CH_4 in interstitial water shows that the levels increase down the sediment profile.

Properties

The properties of sediments similar to those of soils, are ion-exchange and sorption capacity. However, sediments are mostly in an anaerobic environment and trace metal mobility is potentially greater in sediments than in soils. Sediments are continually leached, while soils have wet and dry periods. There is more organic material in sediments, which has a significant effect on chemical processes.

Sediments are good ion-exchangers with a capacity range of 20-100 milli-equiv/100g. To study this property it is necessary to avoid contact between the sediments and dioxygen, otherwise the levels of exchangeable cations, especially Mn^{2+} and Fe^{2+}, will alter.

Trace Elements

The trace metals found in sediments, in significant amounts include Cr, Cd, Cu, Mo, Ni, Co, Mn and Pb. They occur in discrete compounds, or sorbed on to clays and Mn/Fe oxides or as low soluble organic complexes. Of all the copper transported to the oceans approximately 1% (6×10^4 tonnes) is in solution, the rest is particulate, around 85% as crystalline material, 6% as metal oxide coatings, 5% in insoluble organic material and 4% absorbed to particles. Levels of copper in stream sediments correlate reasonably well with the levels in the parent material from which the sediments have derived. Because of this, stream sediments may be used to search for a trace metal anomaly. For example, around Sudbury, Canada, copper levels in sediments reach 3000 $\mu g\ g^{-1}$, as a result levels in the water are elevated. Copper levels of 1000 $\mu g\ g^{-1}$ in sediments will produce around 50 ppb of the metal in water. Trace metal levels in esturine and near shore sediments are useful in pollution studies. For example in Sorfjord, Norway, high levels of sediment zinc, 800-118,000 $\mu g\ g^{-1}$ come from smelting wastes (natural background 130 $\mu g\ g^{-1}$). Levels generally decrease with depth in sediments, and since the depth is age related this indicates increased inputs, especially over the last 80 years, for example a 10 fold increase in Hg and Cd, 5 fold in Cr, 4 fold in Cu. The types of trace metal compounds that occur will depend on the redox conditions of the environment. Some typical compounds are listed in Table 12.6, oxy-species in oxidising conditions and mainly sulphides in reducing conditions. The solubility of these materials is dependent on pH, and on the nature of available ligands, e.g. amino acids, citrate and Cl^-. Solubility also depends on the age of the sediments; aged material is generally less soluble than new.

TABLE 12.6 Some Metal Compounds in Sediments

Metal	Oxidising Conditions	Reducing Conditions
Cd	$CdCO_3$	CdS
Cu	$Cu_2(OH)_2CO_3$	CuS
Fe	$Fe_2O_3nH_2O$	FeS, FeS_2, $FeCO_3$
Hg	HgO	HgS
Mn	MnO_2nH_2O	MnS, $MnCO_3$
Ni	$Ni(OH)_2$, $NiCO_3$	NiS
Pb	$2PbCO_3.Pb(OH)_2$, $PbCO_3$	PbS
Zn	$ZnCO_3$, $ZnSiO_3$	ZnS

The transport of elements in rivers via the water and sediment has been studied along the Rhine and other rivers. The ratio of the metal concentration in water to metal in sediments (<16 μm), is <1 for Pb, Cr, Cu, As and Hg, and >1 for Cd, Zn and Ni, indicating the medium which is the principal carrier of the metals along the river. In order to estimate the degree of mobilization of metal ions from sediments, the material is treated with a variety of leaching agents. One scheme uses acetic acid in hydroxylamine-HCl which is said to remove trace metals from carbonates and some sulphides (reducible fraction). Hydrogen peroxide dissolves metals from sulphides and metal organic compounds, (oxidisable fraction), and $HF-HClO_4-HNO_3$ mixtures remove residual metals, for example from detrital silicate minerals. The proportion obtained from each reagent varies considerably with the type of sediment, and the schemes are only useful if their limitations are recognised.

The hydrous oxides of Mn(IV) and Fe(III)(and also of aluminium) play a big role in holding trace metals in sediments. The oxides occur as particles or coatings on clay surfaces. In the latter case they are efficient, especially if freshly formed, for fixing trace elements. As mentioned earlier the surface charge on the oxides is positive in acid solutions, and negative in basic solutions. The pH of the zero point charge (zpc), i.e. when the oxide surface is neutral or there are equal numbers of positive and negative charges, varies with the particular material, as shown in Table 12.7. Sorption on an oxide is represented by the two equations;

$$-\overset{|}{\underset{|}{Fe}} - OH + Zn^{2+} \underset{\text{high pH}}{\rightleftharpoons} -\overset{|}{\underset{|}{Fe}} - OZn^{+} + H^{+}, \qquad (12.39)$$

$$-\overset{|}{\underset{|}{Fe}} - OH + HPO_4^{2-} \underset{\text{low pH}}{\rightleftharpoons} -\overset{|}{\underset{|}{Fe}} - OPO_3H^{-} + OH^{-}. \qquad (12.40)$$

For phosphates the sorption is quite efficient, even though stability constant data would suggest otherwise;

$$Fe^{3+}_{(aq)} + OH^{-} \rightleftharpoons Fe(OH)^{2+}, \log K = 11.8, \qquad (12.41)$$

$$Fe^{3+}_{(aq)} + HPO_4^{2-} \rightleftharpoons FeOPO_3H^{+}, \log K = 8.4. \qquad (12.42)$$

TABLE 12.7 pH of Zero Point Charge

Particle	pH	Particle	pH
SiO_2	2.0	γFe_2O_3	6.7
kaolinite	4.6	"$Fe(OH)_3$" (amorphous)	8.5
montmorillonite	2.5	$Al(OH)_3$	5.0
MnO_2	2-4.5		

However, the sorption depends on the ratio $[OH^-]/[HPO_4^{2-}]$ in solution, and this favours phosphate sorption at low pH.

The chemical processes that can occur at the sediment/water interface have been mentioned for mercury (Fig. 11.4). Similar processes may occur for other elements, for example, arsenic, present as AsO_4^{3-} in water, can undergo the following reactions in top sediments.

$$\underset{\text{arsenate}}{AsO_4^{3-}} \xrightarrow[\text{reduction}]{\text{bacteria}} \underset{\text{arsenite}}{AsO_3^{3-}} \xrightarrow{\text{bacteria}} \underset{\text{methylarsenic acid}}{CH_3-As\text{-}O(OH)_2} \xrightarrow{\text{bacteria}} \underset{\text{dimethylarsenic acid}}{(CH_3)_2AsO(OH)} \qquad (12.43)$$

Each one of these materials may then be released into the water.

URBAN DUST

The climate of urban areas produce an urban dome (Fig. 8.9) which entraps material within the urban area. Therefore the particulate material produced within cities is retained giving around 10 times more than in rural areas. The range of copper levels in urban atmospheres has been reported as 30-200 ng m^{-3} compared with 5-50 ng m^{-3} in rural areas.

Industrial and commercial land use produce elevated levels of inorganic species in street dust. The principal sources being incinerators, motor vehicles, oil burners, soil erosion and industry. The material that settles out as urban dust, especially on paved areas, is removed in part by stormwater runoff, i.e. the soluble material and the fine fraction that gets into suspension. Lead compounds in urban street dust are mainly in the dense, but fine fraction, and so a portion is washed away during rain storms. Provided an area has a regular rainfall, it is likely that in the long term, a steady state of the levels of materials in city street dust will be reached, where inputs are balanced by removals. However this would not apply to soils. The amount of material estimated to be removed by a storm is a function of the intensity, and duration, of the storm, the length of the dry period before, land use, method of sampling for analysis, and effectiveness of the storm water drainage system, i.e. fall, and size etc. Results suggest that during a storm in urban areas the contribution of heavy metals in stormwater runoff, can in some cases, be greater than from secondary sewage treatment effluent. For example for Lodi, N. Jersey, U.S.A. (in kg yr^{-1}):

	Zn	Pb	Cu	Ni	Cr
Stormwater	1777	3289	665	215	102
Sewage treatment	278	696	174	348	174

The removal by stormwater reflects in a reduction of levels in street dust. For example, lead levels in a dust were 4820 $\mu g\ g^{-1}$ and 2470 $\mu g\ g^{-1}$ before and after a storm, respectively. Within one week of the storm, levels had risen to 4210 $\mu g\ g^{-1}$.

Sources of the materials in street dust appear to be natural soil, automobile emissions, tyre wear, metals from corrosion (of car parts), cement particles from buildings and in coastal and areas of severe winter, salt particles. Of these materials soil is the major component (~75%), but the amount depends on the land use, i.e. city centres where little soil is around compared with residential areas.

The component of street dust that has received most attention is lead, produced mainly from the burning of petrol containing tetraalkyllead. Evidence that the motor vehicle is the major source is given in Fig. 12.11. The lead levels in soil fall off away from the roadside, and also down the soil profile. Lead levels in street dust mainly lie in the 500 to 10,000 $\mu g\ g^{-1}$ (0.05 - 1.0%) range. The levels tend to be higher at street intersections (e.g. 3020 $\mu g\ g^{-1}$) compared with straight stretches of road (e.g. 2400 $\mu g\ g^{-1}$). This is because lead compounds held up in the exhaust system are ejected during acceleration.

Some typical lead species found in street dust are: $PbSO_4$, Pb, PbS, $PbSO_4.(NH_4)_2SO_4$, $PbO.PhSO_4$(trace), PbO_2 (trace) and $2PbCO_3.Pb(OH)_2$ (trace). The presence of metallic lead would suggest that PbO is reduced by CO in the exhaust system. The ultimate product, after prolonged weathering, is probably $2PbCO_3.Pb(OH)_2$. The primary lead compounds in the exhaust gases $PbCl_2$, $PbBr_2$ and PbBrCl, rapidly hydrolyse in the atmosphere to oxy-compounds. Frequently the levels of lead in street dust correlate with the levels of bromide.

High levels of lead in the top-soil (Fig. 12.11) suggest organic acids complex the metal preventing its rapid removal by leaching. It is likely that similar complexes also occur in street dust. Lead will also be sorbed on to clay particles.

An area of dust accumulation, which is only beginning to be investigated, is dust in homes. Again, lead has been the metal most frequently studied. The major sources of lead in the home are weathered lead paint, and street dust carried inside as well as aerosols. In older areas of cities house dust levels are reported to be greater than for newer housing areas (e.g. 960 $\mu g\ g^{-1}$ and 470 $\mu g\ g^{-1}$ respectively). Current methods of cleaning houses may have the effect of just redistributing the finer particles.

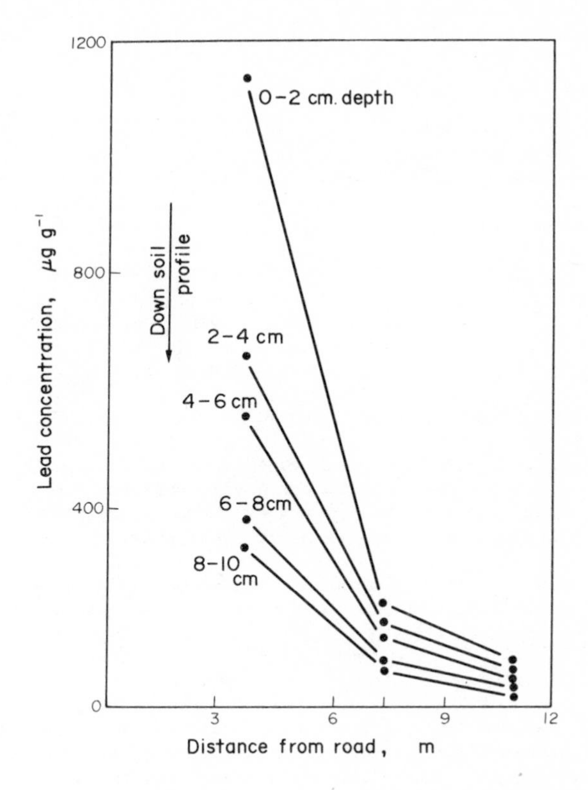

Fig. 12.11. Variation in lead concentration in soil as a function of distance from a highway and depth of soil.

MINE TAILINGS

A mine tailing is the solid waste material left over from a mining operation, in which the levels of the required elements were considered too low for economic recovery. Such material is a potential source of trace elements in the environment as ore bodies are often rich in a number of elements. For example the maximum enrichment factors of trace metals in effluents from gold mine tailings in South Africa have been reported as Mn(30,000), Fe(5500), Cr(4000), Co(19,500), Ni(15,900), Cu(1800), Zn(2600), Cd(105), Pb(580) and SO_4^{2-}(1000).

The material in the tailings has been disarranged, it is often dumped in structurally unsafe mounds, and the material is more open to water and air, than was the case before mining. As a result leaching, disturbance by wind and erosion are common. These processes continue to liberate bound metal ions long after the mining has stopped. In addition to the physical transport of tailings, biogeochemical processes also occur, often accelerated because of the free access of H_2O and O_2.

One common component of mine tailings is FeS_2, which has an important role in the formation of acid mine waters (Chapter 2). With suitable thio-bacteria elements may be released from sulphide enriched mine tailings;

$$2FeS_2 + 2H_2O + 7O_2 \rightarrow 2FeSO_4 + 2H_2SO_4, \quad (12.44)$$

for example, arsenic from arsenopyrites will be released by oxidation;

$$4FeAsS + 13O_2 + 6H_2O \rightarrow 4SO_4^{2-} + 4AsO_4^{3-} + 4Fe^{2+} + 12H^+. \quad (12.45)$$

Bacterial oxidation of Fe(II) can lead to further reactions;

$$12FeSO_4 + 3O_2 + 6H_2O \xrightarrow{bacteria} 4Fe_2(SO_4)_3 + 4Fe(OH)_3, \quad (12.46)$$

$$CuFeS_2 + 2Fe_2(SO_4)_3 + 2H_2O + 3O_2 \rightarrow CuSO_4 + 5FeSO_4 + 2H_2SO_4, \quad (12.47)$$

$$2ZnS + 2Fe_2(SO_4)_3 + 2H_2O + 3O_2 \rightarrow 2ZnSO_4 + 4FeSO_4 + 2H_2SO_4. \quad (12.48)$$

The ions Cu^{2+} and Zn^{2+} are liberated and mobilized in the acid environment.

In time mine tailings are covered with vegetation, often of a type that can tolerate high metal concentrations in the soil. However, seepage of water is on-going and contaminated water continues moving into the outlying environment and may pollute such things as nearby well water.

Uranium mining adds a further dimension to these problems because of the radio-activity associated with the uranium and its decay products.

SOLID WASTES

The problems in the handling and control of solid wastes generated by the activity of human beings lie in their variety and wide range of densities. In the U.S.A., around 4450 x 10^{-6} tonnes are produced per year from the sources listed in Table 12.8. Though Municipal wastes are a relatively small amount (5%), it is the area which has received most attention because they occur in urban areas, affecting many people. The composition of

TABLE 12.8 Solid Wastes (U.S.A.)

Source	Percent	Source	Percent
Municipal	5	Animal	39
Industrial	3	Crop	14
Mineral	38		

municipal wastes has local characteristics, but for an average Western urban society it approximates to the figures given in Table 12.9. Approximately 1.4 to 2.3 kg of urban solid waste are produced per person per day.

The chemical composition of wastes is an indicator of methods of disposal, for example purely organic wastes may be composted. Bearing in mind that the figures are only approximate, it is estimated that the composition of solid waste is: 28% H_2O, 25% C, 21% O and 3 to 4% H. The principal disposal

TABLE 12.9 Composition of Municipal Wastes

Component	Approximate amount (% dry weight)	Component	Approximate amount (% dry weight)
Metals[a]	9	Garden Refuse	5
Glass	9	Wood	4
Paper[b]	55	Plastic	1
Food	14	Other	3

[a] Ferrous metals (7), [b] Newsprint (12), Cardboard (11).

methods, at present, are dumping and land-fill, followed by incineration, specialized disposals such as dumping at sea, and recycling. Adequate dumping or land-fill requires frequent compacting and covering with soil. Initially the decomposition is aerobic (and the temperature may rise), but as the O_2 is used up the decomposition becomes anaerobic. Hence gases, such as CH_4, H_2S, CO_2 and H_2, are produced and need to be vented. In some places (e.g. Southern California) the CH_4 produced is collected and used as a fuel. Organic acids are produced, which when leached can transport heavy metals. A poorly maintained land-fill can lead to high concentrations of a variety of species in the leachate, which continues long after the fill has been completed. Some typical concentrations of species in leachates are given in Table 12.10.

TABLE 12.10 Concentration of Species in Land-fill Leachate

Material	Concentration ($mg\ l^{-1}$, ppm)	Material	Concentration ($mg\ l^{-1}$, ppm)
Cu	0 - 9	Zn	0 - 1000
Cl	34 - 2800	N(NO_3)	0 - 1300
Fe	0.2 - 5500	PO_4^{3-}	0 - 154
Pb	0 - 5	alkalinity ($CaCO_3$)	0 - 20850
Mn	0.06 - 1400	pH	3.7 - 8.5

Incineration is becoming more popular for disposal. The problem is now altered to one of air pollution, and the disposal of a greatly reduced bulk of solid (viz. ash), more homogenous in nature. However incineration is of little use if air pollution controls are not also in operation. For complete combustion temperatures around 800 - 1000°C are required. Some typical products of incineration of municipal refuse are given in Table 12.11, the figures include the O_2 and N_2 from the air, introduced into the incinerator (approximately three times greater than necessary). The amount of gas produced per tonne of refuse (excluding the remaining O_2 and N_2 introduced) is around 800 kg, the major component being CO_2.

Increased use of plastics over the last 15-20 years add a further dimension to the problems of decomposition in land-fills. Bio-degradable plastics have been advocated, but they need to last as long as they are needed. Incineration of plastics produce mainly CO_2, H_2O, CO and some NO_x, and,

TABLE 12.11 Incineration Products of Municipal Refuse

Product	Amount by Gas volume (dry)
CO_2	6%
SO_2	22 ppm
HCl	5-500 ppm
CO	0.06%
O_2	14.3%
NO_x	93 ppm
N_2	79.6%
water vapour	640 kg $tonne^{-1}$
solids ash and residue	220 kg $tonne^{-1}$

depending on the plastic (e.g. PVC or polyacrylonitriles) some HCl and HCN. Pyrolysis (combustion in the absence of O_2) rather than incineration may be a better approach. Temperatures of 850 - 1050°C are required and the gaseous products contained by the system are H_2 (54%), CO_2 (25%), CO (10%), CH_4 (10%), traces of other hydrocarbons, N_2, HCl and NH_3. Some of gases may be used for heating.

Solid wastes are potentially available for recycling, provided problems of separation of the components can be overcome. Glass, paper, plastics and ferrous metals are relatively easy to separate. The thermal energy can also be recycled. Recycled materials nearly always have to be used in lower grade products because of the problems and expense of purification. For example many recycled metals are used to produce alloys, or chemicals which can then be purified (e.g. the use of silver in photography). Aluminium accounts for 0.5 to 0.7% by weight of municipal wastes, which is worth recovering because of the high energy costs in its extraction. However, the market price is an influencing factor. The approximate amounts of materials recycled in the U.S.A. are given in TAble 12.2. The amounts are not large, especially for glass and rubber. The problem of separating the components is a major

TABLE 12.12 Amounts of Materials Recycled in the U.S.A.

Substance	Amount Recycled % of total used	Substance	Amount Recycled % of total used
Al	20	Ag	47
Cu	41	Zn	27
Au	16	Rubber	4
Pb	40	Paper	18
Hg	21	Glass	3

barrier, for example the de-inking and removal of other additives from paper is a problem restricting the amount of paper recycled. Also, the fibre in recycled paper is of inferior quality and it is said the public do not like it. In 1970 recycled paper in the U.S.A. was used as follows: 3% in newsprint, 5% in printing and writing paper, 69% paperboard and 12% in construction paper and moulding pulp.

The automobile poses particular recycling problems, though some recycling goes on, for example in second hand spare parts, battery lead for new batteries, copper from electric motors and radiators and cast iron from the engine block. But even so the wide range of materials used in motor car construction aggravates the separation problem.

Hazardous and toxic wastes pose an addition problem of disposal, which will depend on the actual material to be disposed. No satisfactory method has yet been found for the disposal of long-lived radioisotopes, partly because the life times are greater than man's experience of handling storage facilities. In the case of short-lived isotopes suitable storage until the activity is down to background is satisfactory. Then the problem becomes one of handling inactive solid wastes.

Pyrolysis of organic hazardous and toxic solid wastes would appear to be one of the safer methods, as everything is contained within the pyrolysis vessel. However, such a method does not allow for recovery of useful products, except thermal energy.

More effort is required to find methods of disposal which reduce the volume of the waste, separate components, recover useful materials, and detoxify the wastes.

CHAPTER 13

Inorganic Materials in the Biosphere

We will briefly consider the interactions of inorganic species with the biosphere. This is a vast topic, and this account can only be an introduction and restricted in its coverage.

ELEMENTS IN THE BIOSPHERE

From the information in Chapter 7, it is clear that inorganic species are intimately linked with the biosphere. With some exceptions (Ca, N, P, K, Na) the inorganic species are present at trace levels. The elements considered essential to life (plant and animal) are listed in Table 13.1, as well as

TABLE 13.1 Essential and Toxic Elements

Essential Elements	
Macro	Micro (average daily intake for man, μg)
C, Na, Ca H, K, N O, Mg, P Cl, S	Co (150-600), Gr (50-120), Cu (2000-5000), I (60-500), Mn (2000-3000), Mo (100-400), Se (unknown), V (1000-4000)*, Zn (10,000-15,000), B (plants), Fe (large amounts), F (710-3400)*, Ni (300-600)*, Si*, Sn (1500-3500)*
Non-Essential and Toxic Elements	
Cd, Hg, Pb, Tl, Be, Ba, Al, As, Sb, Bi	

* probably essential, but uncertain (from experiments with chickens)

some non-essential elements considered to be toxic to life, particularly animal life. Such a table, can not be complete because of uncertainty, for some elements, whether they are essential or not. For example there is evidence that the elements F, Ni, Si and Sn are essential to life, and it is reasonably certain the list will grow. As the accuracy and detection limits of trace element analysis improve it will be possible to be more positive regarding the levels of elements considered essential for health.

Inter-element effects are a common feature of biochemistry, where the level of one element influences the activity of another. For example excess intake of Co can reduce the availability of iodine for the production of the thyroid hormone thyroxin for some people. The hormone controls cellular metabolism. Also excess Mo can lead to copper deficiency in mammals. The consumption of large doses of one element, prescribed for a particular problem, may alter levels of other elements in other areas producing a different health problem. The effects of some elements are additive (the synergistic effect) and therefore they need to be considered together. Since the chemistry of Cd, Hg and Pb with regard to forming bonds to sulphur is similar, their toxic effect, inhibiting sulphur containing enzymes, should perhaps be considered as additive.

TOXICITY AND DEFICIENCY

Most of the essential elements (Co, Cr, Cu, I, F, Mn, Ni, Se, Si, Sn, V) listed in Table 13.1 are also toxic when taken in excess. Too much copper (100-150 times the essential amount) causes necrotic hepatitis and hemolytic anaemia. Selenium is the most toxic of the essential elements, and intakes greater than 4 $\mu g\ g^{-1}$ (4 $mg\ kg^{-1}$) is serious. Too much manganese causes psychic and neurological disorders. The dual action of trace elements, in relation to toxicity and deficiency, is illustrated by a dose-response curve (Fig. 13.1). The cut-off points, between deficiency, normal health and toxicity, vary from nutrient to nutrient, and from person to person. The range for normal health is narrow for Se, but quite large for Zn. Both insufficient and an excess of iodine produces thyroid problems. Deficiency

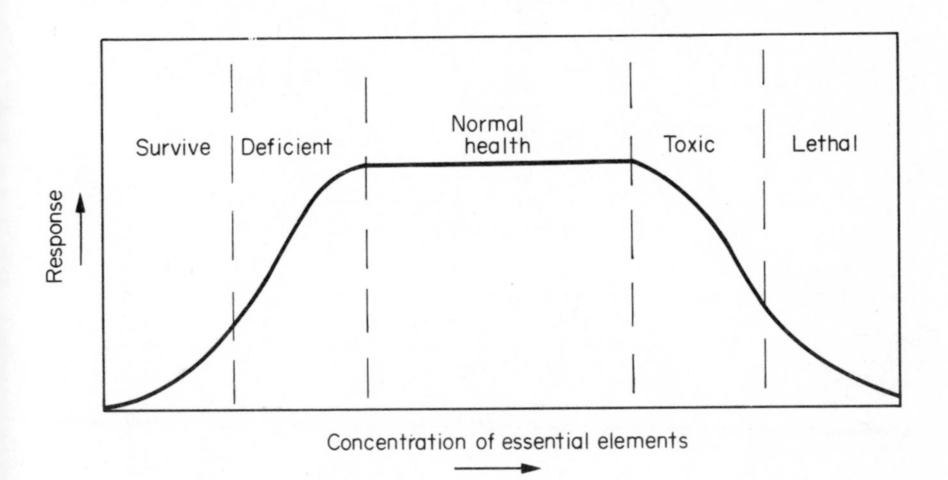

Fig. 13.1. Dose-response curve for essential elements.

of Cu and Se may contribute to kwashiorker. Deficiency in Co, a component of vitamin B_{12}, can lead to anaemia.

The nature of the chemical form of an element influences the dose-response relationship. For example the cations Sn^{2+} and Ni^{2+} are not particularly toxic, the chemical species $Sn(CH_3)_3^+$ and $Ni(CO)_4$ are dangerously toxic. In these cases the toxicity is due to the whole species, not just the trace element. In the case of $Ni(CO)_4$ the toxicity arises from the dissociation of the CO groups. Chromium VI is more toxic than chromium III, which may be because of its strong oxidising properties.

The dose-response curve is person dependent, i.e. not everyone is affected equally. This situation is portrayed by an S type cumulative percentage frequency curve (Fig. 3.2) relating the proportion of people affected with

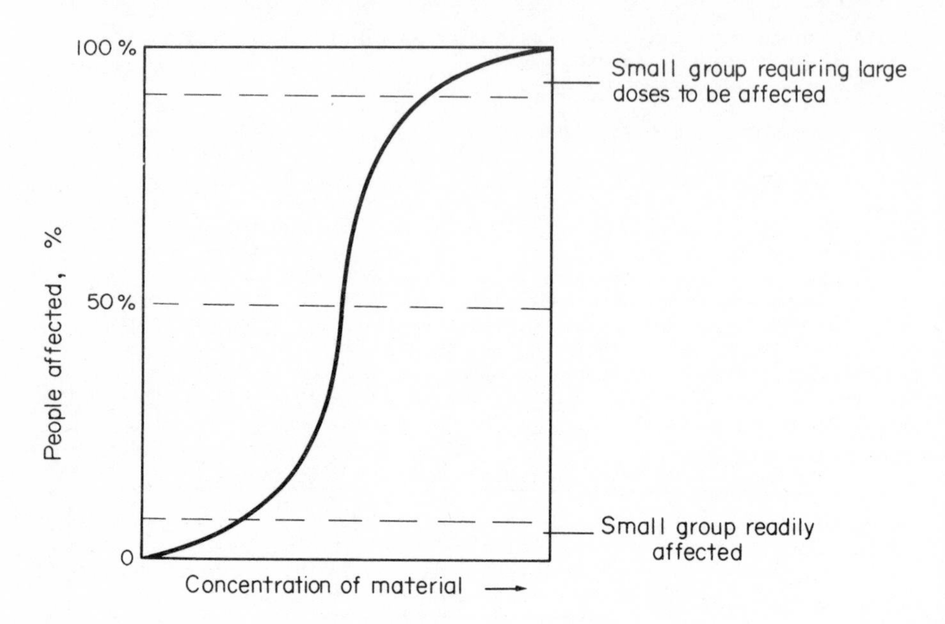

Fig. 13.2. The relationship between the concentration of a material and its toxic effect on living things.

concentration of the toxic material. The inverse curve may be drawn for a deficiency of essential elements. The concentration of toxic material to give 50% mortality, is used as a measure of toxicity and is given the symbol LD50, i.e. lethal dose to kill 50% of the population.

SOME ASPECTS OF BIOCHEMISTRY

Most trace inorganic species, such as metal ions, are involved in biological molecules by way of coordination. For example, Co is coordinated to N atoms of a corrin ring in vitamin B_{12}, Fe to nitrogen atoms of a porphyrin ring in haemoglobin. To understand these interactions some brief details will be given of some biochemicals.

The basic building unit of living material is the *cell*, which is around a few μm in size. A mammal cell is surrounded by a *membrane*, permeable to some materials and not to others, and therefore protects the cell from foreign substances. The cell *nucleus* contains deoxyribonucleic acid, *DNA*, which determines cell reproduction.

Proteins, one of the main structural chemicals of living material, consist of sequences of amino acids. There are twenty naturally occurring amino acids, and a particular sequence in a protein gives the protein its particular properties, e.g. an enzyme. The amino acids are linked together through the *peptide bond:*

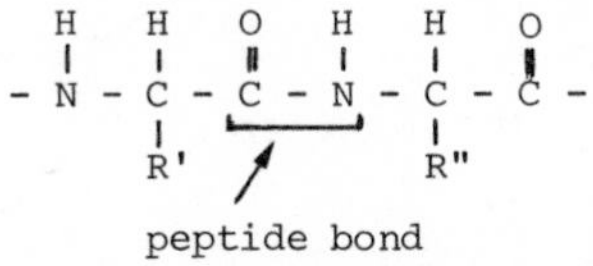

```
     H   H   O   H   H   O
     |   |   ‖   |   |   ‖
   - N - C - C - N - C - C -          (13.1)
         |   └───┘   |
         R'    ↗     R"
         peptide bond
```

Carbohydrates, $(CH_2O)_n$, are the principal source of energy in living materials, and are produced in plants by photosynthesis. The energy is provided by combustion of the carbohydrate;

$$C_6H_{12}O_6 + 6O_2 \rightarrow 6CO_2 + 6H_2O + \text{energy} \quad (13.2)$$

Lipids are fats or oils, which are water insoluble, and non-polar. Hence non-polar materials such as alkyl metallates may dissolve in lipids. They are part of the structural material of cell membranes, and protect the interior of cells from polar substances such as metal ions or hydrated metal ions. Two important cell membranes in the body are the blood-brain barrier and the placental barrier. The brain and fetus are therefore protected from polar materials, but not necessarily from non-polar substances.

An *enzyme* is a biochemical catalyst, a protein with active sites, which are sometimes metal atoms. Enzymes achieve sophisticated biochemical reactions at ambient temperatures and pressures. The catalyst acts upon a substrate and sometimes a co-enzyme is necessary for an enzyme to operate.

DNA, the genetic material of living substances, consists of two helical polymers linked by hydrogen-bonds. The method of mutual twisting of the two strands, either a double helix or side-by-side, is yet unresolved. Each strand is made from heterocyclic amines, phosphate and deoxyribose (a pentose). The ordering of amines establishes the genetic code, which determines a particular amino acid sequence in, say, an enzyme.

The nervous system consists of nerve cells called *neurons*, the main component of which, is called an *axon*. A nerve impulse moves along an axon under the influence of changes in the cell membrane permeability to Na^+ and K^+ ions. When the pulse reaches the end of the axon a gap (50 nm) called the *synapse* is bridged by a neuro-transmitter such as acetylcholine $CH_3 - \overset{O}{\overset{\|}{C}}OCH_2 - CH_2 - N(CH_3)_3^{\ +}$, and norepinephrine $(HO)_2C_6H_3 - \overset{OH}{\overset{|}{C}H} - CH_2 - NH_2$. To restore nerve sensitivity, so it can respond to the next impulse, the neuro-transmitter is removed. In the case of acetylcholine, this is done by hydrolysis with an enzyme acetylcholinesterase.

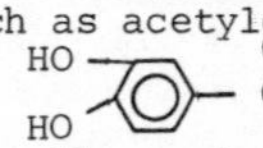

TOXICITY AND TRACE ELEMENTS

Trace elements are taken in through food, water and air. Of these three, air intakes are the most readily assimilated into the body. The various routes are shown schematically in Fig. 13.3.

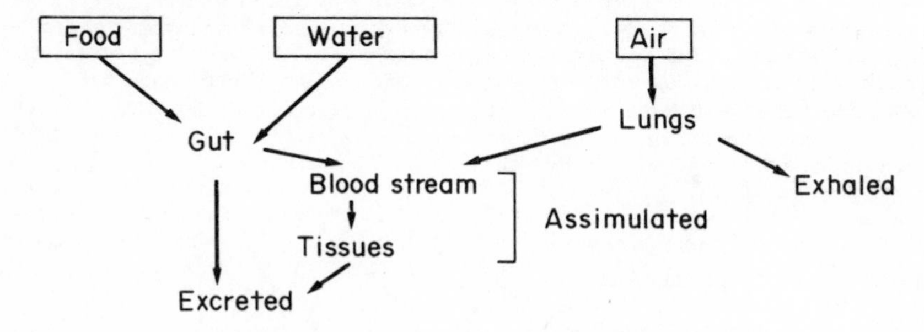

Fig. 13.3. The routes by which trace elements enter mammals

The concentration, c, of a species in a particular organ, at a particular time, t, are related, using a highly simplified model by the expression;

$$\frac{dc}{dt} = A - L \tag{13.3}$$

where A = assimilated rate and L = loss rate.

Also, $$L = kc, \tag{13.4}$$

where k is the rate constant for the loss of material. The loss follows an exponential rate law, hence the half-life for an element in an organ is;

$$t_{\frac{1}{2}} = \frac{0.693}{k} . \tag{13.5}$$

Under normal conditions an assimilated element reaches a steady state in the body; when that state is upset problems may arise.

However, foreign material may be metabolised to less harmful forms, such as converting a non-polar compound into a polar one, e.g. $R_4Pb \rightarrow R_3Pb^+$. But the metabolism process may itself be detrimental, and the products still toxic.

Enzyme Inhibition

A major mode of toxicity is the inhibition of enzymes. Elements, such as Pb^{2+}, Cd^{2+}, Hg^{2+}, As(III) and As(V), can block the active sites of enzymes, or replace an essential element such as Zn^{2+}. This occurs mainly by bonding to the sulphur of the amino acids cysteine and methionine;

$$\text{Enzyme}\begin{smallmatrix}\diagup SH \\ \diagdown SH\end{smallmatrix} + M^{2+} \rightleftharpoons \text{Enzyme}\begin{smallmatrix}\diagup S \diagdown \\ \diagdown S \diagup\end{smallmatrix}M + 2H^+, \tag{13.6}$$

$$\text{Enzyme}\langle{}^{S}_{S}\rangle\text{Zn} + M^{2+} \rightleftharpoons \text{Enzyme}\langle{}^{S}_{S}\rangle\text{M} + Zn^{2+} \qquad (13.7)$$

The reactions may be reversed under suitable conditions, for example excess Zn^{2+} could drive reaction (13.7) to the left.

The metals Cd, Hg and Pb are soft acids on the Hard and Soft Acids and Bases (HASB) scheme, i.e. they are relatively large and polarizable (compared with zinc). Sulphur is a soft base and also polarizable. Since each atom in the M-S bond polarizes the other,bonds between sulphur and Cd, Hg and Pb are stronger than with zinc. The inhibition of enzymes by one metal ion replacing another is because of similar chemistry, but different biochemistry. Cadmium and zinc, in the same periodic group have similar chemical properties but zinc is biochemically active while cadmium is not. Some of the enzymes inhibited by toxic metals are listed in Table 13.2. Lead inhibits a number of enzymes involved in the biosynthesis of haem (Fig. 13.4), the porphyrin

TABLE 13.2 Enzymes Inhibited by Toxic Metals

Metal Ion	Enzyme
Cd^{2+}	adenosine triphosphatase, amylase, alcohol dehydrogenase*, carboxypeptidase (peptidase activity), carbonic anhydrase*, glutamic-oxaloacetic transaminase.
Hg^{2+}	alkaline phosphatase*, glucose-6-phosphatase, lactic dehydrogenase.
Pb^{2+}	acetylcholinesterase, alkaline phosphatase*, adenosine triphosphatase, carbonic anhydrase*, cytochrome oxidase, haem synthetase, ALA dehydrase*.
As(III), As(V)	pyruvate dehydrogenase.

* Contain zinc as an essential element

in haemoglobin and cytochromes. The two best documented steps are the inhibition of ALA-dehydrase and haem synthetase. In both cases the substrates accumulate, ALA (δ-amino-levulinic acid) in serum and urine and protoporphyrin IX in red blood cells. Elevated levels of these products are diagnostic of lead poisoning.

Other Toxic Effects

Some toxic materials are mutagenic, i.e. they produce changes in the base sequence in DNA, so incorrect genetic information is transmitted. The outcome is that wrong proteins and enzymes are produced, leading to a change in the organism. Terato-genesis, or birth defects, can also be triggered by toxic materials, again by interfering with the DNA molecule. The list of carcinogenic chemicals is increasing, i.e. materials that stimulate uncontrolled cell growth, called cancer. These three areas, involve complex biochemical processes, many of which are not understood.

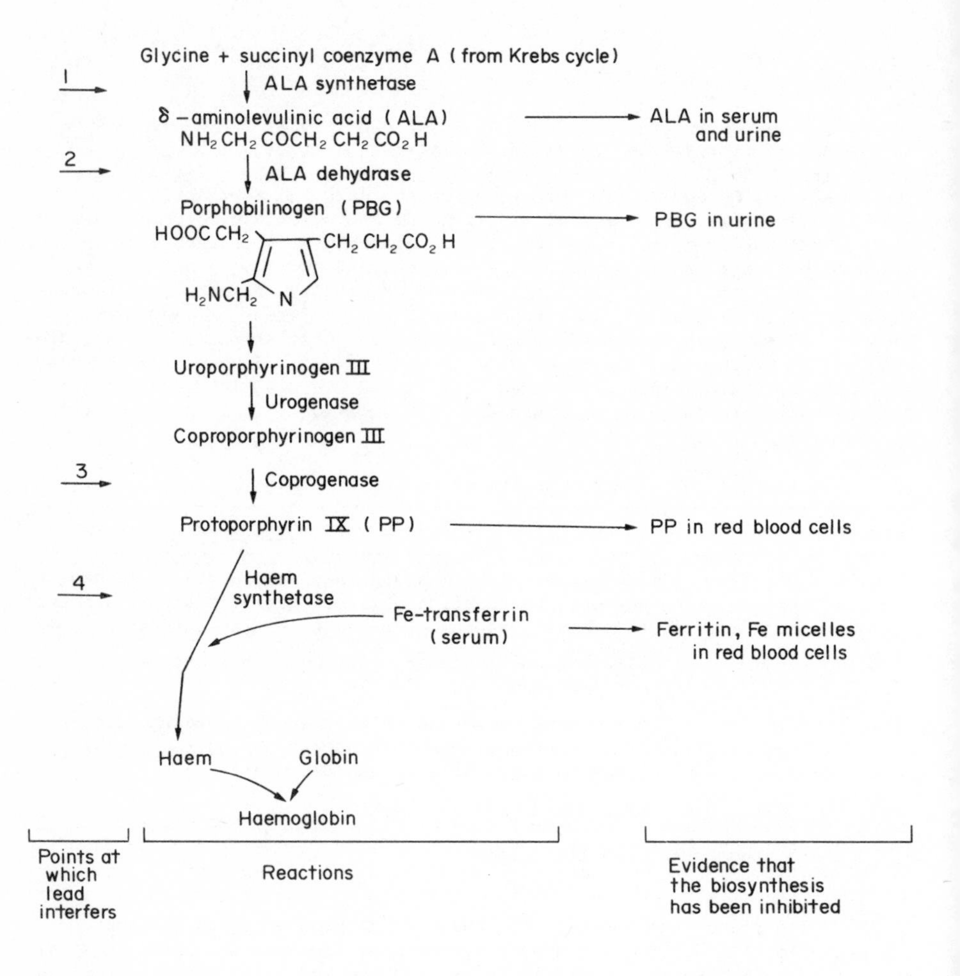

Fig. 13.4. Points in the biosynthesis of haem where lead inhibits the synthesis.

Trace Metal Characteristics in Relation to Toxicity

As mentioned above, replacement of zinc by heavy metals such as cadmium, is a function of the stronger Cd-S bond. Generally soft acids are expected to be more toxic than hard acids, because of their greater bonding capacity to soft ligands, such as -SH. Therefore the toxicity of metals tend to increase down a periodic group, i.e. Zn < Cd < Hg, Cu < Ag < Au, Al < In < Tl and Mg, Ca < Sr, Ba.

Solubility of a compound can influence the toxicity of an element. For example $BaSO_4$ is used internally to enable X-rays to be taken of the stomach and intestines, even though the Ba^{2+} ion is toxic. The low solubility of $BaSO_4$, even in stomach acid, means it is relatively safe. The availability of a toxic element depends on the solubility of its salts in water. The most soluble salts are nitrates and halides (except fluorides) while less soluble salts are, normally, sulphates, phosphates, carbonates, perchlorates, oxides and hydroxides. Some metal ions hydrolyse at the pH of biological fluids producing insoluble hydroxides.

Many biochemical molecules offer potential chelating sites for metal ions, e.g. amino acids, heterocyclic amines and phosphates. Phosphate groups of the ribose and deoxyribose-phosphate backbone of RNA and DNA respectively are coordination sites for metal ions. Coordination of certain metal ions stabilise the structures by reducing intramolecular repulsions. Metal ions that readily bond to phosphate are; Mg < Co < Ni < Mn < Zn < Cd < Cr.

INTERPRETATION OF TRACE ELEMENT DATA

It is not always possible to measure directly the amount of a trace element in an organ of a living person, and indirect approaches have to be used. Some readily obtainable materials for analysis are urine, faeces, hair, finger and toe nails, teeth, blood and sweat. The value of data obtained from these materials depends on the knowledge of how an element gets into the substance, and how the levels relate to the body burden. Considerable care must be exercised in obtaining the sample for analysis, for example it is easy to contaminate blood samples, even by the sampling method, if due care is not taken. Some materials are contaminated, before sampling, from contact with the external environment. Analytical results from these materials need careful interpretation. Hair is such a material, where trace elements originate from internal deposition in the hair fibre, and also absorption from the external environment and sweat. It has still not been resolved how to separate out the contribution from the various sources, especially when the hair is subjected to different washing techniques. It could be that hair roots are the best material to use for the estimation of body burden from trace element levels in hair.

Some materials when analysed give the current status of trace elements in the body, these are blood, urine, faeces and sweat. Others, such as hair, finger and toe nails, give an historical record over a few months, while materials like teeth and bones, which store trace elements give an historical record over a longer time. However the record does not provide accurate dating.

Frequently it is not necessary to measure the amount of a trace element directly, but substances produced, or lost as a result of the trace element levels, may be measured. The ALA in urine, or protoporphyrin IX in red blood cells mentioned above are used as indicators of lead toxicity. Correlation analysis may be used between different sets of data, to investigate inter-related effects, and to provide the justification for using data from one measurement to infer facts about something else. As an example the correlation (-0.841) between ALA-dehydrase activity and blood lead levels is given in Fig. 13.5. Statistical methods must be used to help interpret the data, because of the S-type frequency versus concentration curve given in Fig. 13.2. Some data lie well off the regression line, because some subjects respond to a little toxic material while others require a lot before responding.

Correlations can be made between analytical data from biological material and possible environmental sources, such as the levels of an element in blood and levels in the air that people are in contact with. Also

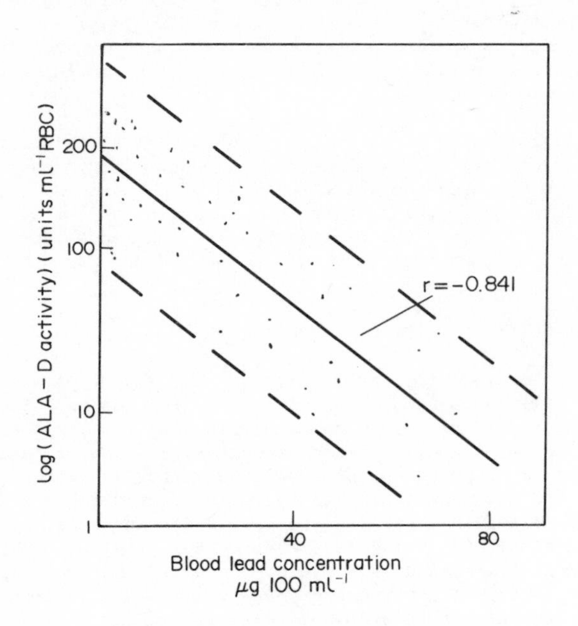

Fig. 13.5. Red blood cells (RBC) ALA-dehydrase activity in relation to blood lead concentrations.

correlations can be made with some characteristic of the subjects, such as age or sex. However, care must be taken in the interpretation of the numerous correlations that can be made as they can be misleading. In fact some correlations can be invalid, if, for example, the data is not normally distributed. In general the more parameters that are measured the better; clearly multielement analysis is important here.

BIOCHEMICAL EFFECTS OF TRACE ELEMENTS

In this section we will consider some of the more important trace elements, and those most commonly encountered in the environment.

Lead

Though it is not considered as toxic as mercury and cadmium, lead is so widespread in our consumer society, it is probably the most serious toxic metal. The many unique properties of lead and its compounds make it a very useful metal. However, it must be treated carefully, and in some areas, particularly where young children are involved, it should be excluded, for example in paints.

Lead production in 1975 was 3500 x 10^3 tonnes of ore, and 4200 x 10^3 tonnes of metal were used. The main uses of lead are in the lead accumulator (44%), tetraalkyl lead compounds, cable sheathing, lead pigments, alloys and in manufacturing - all around 9-11% each. The uses of lead most commonly associated with poisoning are lead glazed pottery, solder, plumbing, in paint and an additive to petrol.

Numerous attempts have been made to estimate the lead consumption per person per day, in an urban environment. Any estimation must be considered as approximate because of difficulties in deciding on an "average" consumption for a person, and because of the wide variations in the levels of lead in food, water and air. The figures given in Table 13.3 (which must be treated cautiously), represent the usual levels of intake considered. One important

TABLE 13.3 Daily Lead Intakes

(a) Adults

Material	Daily Intake μg	Absorption Factor	Assimilated Lead μg	% of Total
Food	200 - 300	0.1	20 - 30	69 - 71
Water[1]	10 - 20	0.1	1 - 2	3 - 5
Air[2]	20 - 25	0.4	8 - 10	29 - 24
Total	230 - 345		29 - 42	100

(b) Children

Material	Daily Intake μg	Absorption Factor	Assimilated Lead μg	% of Total
Food	60 - 160	0.1	6 - 16	70 - 76
Water[3]	3 - 7	0.1	0.3 - 0.7	<0.1 - 3
Air[4]	4 - 15	0.4	1.6 - 6.0	24 - 27
Total	67 - 182		7.9 - 22.7	100

[1] Assume a consumption of 1.5 l per day. [2] Assume air intake 20 - 25 m^3 per day and lead concentration in the air of 1 μg m^{-3}. [3] Assume a consumption of 0.5 l per day. [4] Assume air intake of 4 - 15 m^3 per day and lead concentration in the air of 1 μg m^{-3}.

feature that comes from the data is the significant amount of lead assimilated into the body from air intakes. This is because of the high absorption factor of 0.4, compared with 0.1 for food and water. For people living close to busy roads, where the average lead in the air could be 2 or 3 times higher, the contribution from air intake can become quite significant (up to 50%). An air lead level of 1 μg m^{-3} is said to raise the blood lead level by 2 μg 100 ml^{-1}. The data for children is even more tentative because of their wide range of intakes of food, water and air, and variations with age. Also, children have a greater metabolism rate and absorption factor. The limit suggested by WHO for a tolerable intake of lead per day for an adult is 430 μg. No figure has been given for a child, but a value of 200 μg day^{-1} seems reasonable.

Around 70 to 90% of the lead assimilated goes into the bones, the next major accumulator is the liver, then the kidneys. Divalent lead readily replaces Ca^{2+} in bone, and either becomes firmly fixed or reversibly fixed. Lead in the latter form may be released into the blood stream and a person who has been purged of lead may have a reoccurrence of lead poisoning at a later date due to release from bone tissue.

The principal indicator of lead levels in human beings is lead in blood, expressed as μg 100 ml^{-1} or μmol l^{-1}, (80 μg 100 ml^{-1} = 3.86 μmol l^{-1}). The lead concentration can be given for whole blood, or in the red cells (packed cell volume), where it mostly resides. The latter is more useful, because the red cell level may be low if the subject has anaemia. Blood lead levels and some related health effects are given in Table 13.4. While the occupational blood lead level is around 70 - 80 μg 100 ml^{-1}, considerable dispute exists as to what is considered a high blood lead level for the general population. People in urban areas have levels around 10 - 20 μg 100 ml^{-1}, but if levels reach 30 - 40 μg 100 ml^{-1}, especially in children, there is reason to be concerned about the lead intake.

TABLE 13.4 Blood Lead Levels and Related Health Effects

Whole Blood Lead Level		Health Effect
μg 100 ml^{-1}	μmol l^{-1}	
10	0.48	ALA-dehydrase inhibition
15 - 20	0.72 - 0.97	Protoporphyrin elevation in erythrocytes
40	1.93	Increased ALA in urine Anaemia Coproporphyrin elevation in urine
50 - 60	2.42 - 2.90	Effect on central nervous system Peripheral neuropathics
>80	>3.86	Encephalopathic (Brain) symptoms.

Symptons of lead poisoning are many and not always the same for every person. A brief list and the areas affected are given in Table 13.5. The principal

TABLE 13.5 Lead Poisoning: Symptons

System	Symptons
General	Pallor, anaemia, drawn look.
Digestive	Colic, constipation, loss of appetite, metallic taste, pain in abdomen.
Muscular	Loss of coordination, loss of strength, tiring.
Nervous	Peripheral motor paralysis, atrophy of most used muscles, headache, insomnia, tremor, dizziness, encepthalopathic condition
Vascular	Diminished haemoglobin, arteriosclerosis, hypertension.
Other organs	Lead line in gums, miscarriages (repeated), loss of vision, joints painful.

danger areas are, the haemopoietic system, leading to anaemia in serious cases, the nervous system, leading to irreversible brain damage, i.e. acute encephalopathy, and the renal system. The onset of brain damage appears to be associated with high levels of blood lead, >80 μg 100 ml^{-1}. Young children are particularly susceptible, the damage showing up in behavioural and educational abnormalities, and maybe mental retardation. In mild form, lead effects the central nervous system, manifested in hypertension, irritability and impaired motor skills. It is in this latter area where the debate has been fiercest. There are many who say that there is little or no effect with blood levels around 30 μg 100 ml^{-1}, and there are many who say there is detrimental effects. From the information in Table 13.4 it is clear that even "normal" levels (<20 μg 100 ml^{-1}) affect the human biochemistry, and so lead cannot be taken lightly. The effect of lead on the renal system may be reversed if caught in time, otherwise irreversible damage may be done to the kidneys.

Alkyl lead compounds are not as common in the environment as inorganic lead. Being lipid soluble their main affect is on the central nervous system. It is thought that the principal toxic metabolite is R_3Pb^+, produced in the liver.

Methods for removal of lead from the body all depend, to some extent, on coordination of the metal to give a soluble species that may be excreted from the body. The most common method used today is chelation using $CaNa_2$ EDTA. Other approaches are to administer a low calcium high phosphate diet, or to prescribe sodium citrate and 2,3-dimercaptopropanol (British Anti-Lewisite, BAL). BAL was developed as an antidote for arsenic, but it also works for lead.

$$2HSCH_2 - \underset{\displaystyle SH}{CH} - CH_2OH + Pb^{2+} \rightleftharpoons \left[\left(\begin{array}{l} H_2C - S \\ \;\;| \\ HC - S \\ \;\;| \\ CH_2OH \end{array} \right)_2 Pb \right]^{2-} + 4H^+ \qquad (13.8)$$

It may be necessary for a person to have a number of removal treatments, because of continued release of lead from the bones. It is also necessary to watch for removal of other materials, such as essential trace elements, by the reagents.

Cadmium

Cadmium is less used than lead (i.e. production 17 x 10^3 tonnes yr^{-1}), but it is more toxic, i.e. lower levels have an adverse effect on human beings. The metal is used for plating (30-50%), in pigments (20-30%), in batteries (Ni-Cd and Ag-Cd cells), stabilizers for plastics and in alloys. Cadmium is association with zinc, and the Cd/Zn ratio for a number of ores and ore concentrates lies in the range 0.0021 - 0.0073 (mean 0.0042). The cadmium continues in association with zinc as an impurity in the metal and its compounds. For example, rubber tyres can contain 20-90 μg g^{-1} of cadmium, due to the use of zinc compounds such as zinc oxide and zinc dialkylcarbamates in the vulcanisation process. The abrasion of tyres on the road, adds Cd to street dust and mean levels of 2 μg g^{-1} and greater have been reported. Cadmium plated utensils are dangerous, especially if used in an acid medium.

The total biomass contains around 7.5×10^{11} g of Cd, of which anthropogenic additions are 2×10^8 g. The total intake of Cd by human beings, is approximately 2.2×10^8 g yr^{-1} and the residence time spans 1 to 40 years. Food is the principal source of Cd, and the average intake is around 50 μg day^{-1} $person^{-1}$ (range 40 to 190 μg day^{-1} $person^{-1}$, the lower amount for rural non-smokers, and the higher amount for urban smokers). Tobacco contains around 1 μg g^{-1} of Cd, so up to 2 to 4 μg of metal are inhaled from smoking one packet. This is the major source of inhaled Cd, as air levels are low (~0.025 μg m^{-3}). Approximately 25 to 50% of inhaled cadmium is absorbed into the body, while from food the absorption is about 6%. The WHO tolerable intake per day per person is given as 57 - 71 μg. A lethal dose of Cd is >350 mg.

Cadmium accumulates in the liver and kidneys, the average level in wet kidney tissue is 25-50 μg g^{-1}, a level of 200 μg g^{-1} produces irreversible kidney damage. Assessment of the cadmium body burden is difficult as levels in both blood and urine are not good indicators. Some correlations exist, for example smokers have overall higher blood cadmium levels than non-smokers, but individual variations are so great that reliable diagnosis cannot be made. The amount of a low molecular weight protein, β_2-microglobulin in urine, produced as a consequence of Cd damage of kidneys, may be used as an indicator. Neutron activation analysis *in vivo* on patients has been used successfully.

Both Zn and Ca protect against cadmium poisoning, for Zn its effect may be the reversal of reaction (13.7). A calcium deficient diet enhances Cd accumulation, therefore older people and pregnant women are most at risk. The contributing factors to Itai-Itai disease (see Chapter 12) were high cadmium intakes (>600 μg for most sufferers) low Ca diet and a lack of vitamin D. Itai-Itai disease is always accompanied by renal dysfunction. The levels of Cd were high in bones of Itai-Itai disease sufferers, 1.0 - 1.4% (ash weight). A low molecular weight protein in the liver, metallothioein (MW ~7000) approximately $\frac{1}{3}$ of which is cysteine, bonds to heavy metals (especially Cd and Hg) and protects against toxic metals. A sample of metallothioein was found to contain 4.2% Cd.

Hypertension has been attributed to cadmium though the topic is controversial. Respiratory and pulmonary damage is reported to occur from the breathing of cadmium vapour or particulates. Cadmium, unlike Hg and Pb, does not affect the central nervous system, it cannot cross the placental membrane and the mammary gland is an effective barrier. Therefore new-born babies are normally quite free of any cadmium. The average amount accumulated in the body over a life time is of the order of 30 mg.

Mercury

Mercury is the most toxic heavy metal, and many serious incidences have resulted from mercury poisoning. World production is only 8.6×10^3 tonnes a year, but the methylation of mercury and its bioamplification by marine life has magnified the pollution problem (Chapter 11).

An average daily intake of mercury is hard to estimate, because of the contribution from fish. Around 1 μg per person per day is taken in from air containing 50 ng m^{-3}. Depending on its chemical form, up to 80% of this will be absorbed into the body. Methyl mercury compounds are almost entirely absorbed, while for inorganic mercury compounds the absorption factor is around 15%. Drinking water, with a concentration of 50 ng l^{-1}, will contribute approximately 0.1 μg of Hg to the daily intake. Estimates of mercury intakes from food range from 1 to 20 μg day^{-1} $person^{-1}$, but this depends on the quantity of fish consumed. Mercury from fish is almost entirely as methyl mercury. The WHO suggested tolerable daily intake is 43 μg day^{-1} $person^{-1}$ of which no more than 29 μg be methyl mercury.

Mercury in blood and hair can be used for estimating the body burden, there being a reasonable correlation between the levels of Hg found and its effects. A blood Hg level of 20 μg 100 ml^{-1} would correspond to a daily consumption of 200 to 300 μg of Hg. Consistent intakes at these amounts could well produce mercury poisoning. The distribution of mercury in the body is; liver (55%) > kidneys > brain (12%) > heart > lungs > muscles.

Marine organisms readily concentrate mercury. Fish in contact with water containing 0.01 ppb, and sediments 30 ppb of mercury, have been found with 314 ppb in their flesh, a 31,000 bioamplification (over the water levels). At Minamata, mercury levels in some fish attained 50 μg g^{-1} (50 ppm) wet weight, while levels around 20 μg g^{-1} were common. Experiments with brook trout have shown that over a period of 9 months the fish had accumulated in their gonads, 0.9, 2.9 and 12.3 μg g^{-1} of mercury from water containing 0.09, 0.29 and 0.93 μg l^{-1} respectively.

The toxicity of mercury is related to the chemical form. Liquid mercury appears to have little effect, but mercury vapour is readily absorbed into the bloodstream producing brain damage. Mercury(I) salts are relatively non-toxic compared with Hg(II) compounds because of their low solubility. The solubility of methylmercury in lipids is a major factor regarding its toxicity, as it readily crosses the blood/brain barrier and the placental membrane. Mercury attached to membranes alters their permeability, for example the transfer of sugars is inhibited, while the transfer of K^+ ions is increased.

Arsenic

Arsenic is well known for its toxicity, because of its association with criminal and wartime use. It is widely distributed in the earth, 1.5 - 3.0 ppm in igneous rocks and 1.7 - 400 ppm in sedimentary rocks. Of the two oxidation states, As(III) and As(V), the latter is the more common in an oxidising environment, but the former is more toxic. World production of arsenic is 50 x 10^3 tonnes (1975) used mainly in agricultural chemicals (81%), in ceramics and glass (8%), chemicals (5%) and pharmaceuticals (2%). The agriculture chemicals are herbicides, insecticides, desiccants, wood preservatives and feed additives. Originally arsenic pesticides were inorganic chemicals but organo-arsenics are being increasingly used. Because of the large quantities used in agriculture the soil has become the main sink for anthropogenic additions to the environment.

Arsenic has some similar toxic properties to Pb, Hg and Cd, as regards bonding to sulphur and inhibiting enzyme action such as pyruvate dehydrogenase. The order of toxicity of arsenic compounds is; arsines (As(III)) > arsenite (As(III)) > arsenate (As(V)) and arsenic-organic acids (As(V)). Arsenic, which is found mainly in the liver, kidneys, lungs and intestinal walls, is readily absorbed if water soluble. Arsenic is also toxic by replacement of phosphorus in ATP and in coagulation of proteins. Since P and As are in the same periodic group, and since P is a major essential element, it is not surprising that arsenic interferes with phosphorus metabolism. In the biosynthesis of ATP, in the presence of arsenite, the intermediate 1,3-diphosphoglycerate is not formed, but instead 1-arseno, 3-phosphoglycerate, $(O_3PO\text{-}CH_2\text{-}CH(OH)\text{-}CO\text{-}OAsO_3)^{4-}$. This material hydrolyses to arsenate and 3-phosphoglycerate rather than undergoing the normal oxidative phosphorylation.

A lethal dose of the more toxic forms of arsenic is 1-25 mg per kg body weight, but for less toxic forms, say arsenic in drinking water, larger quantities are necessary, i.e. 100-200 mg per kg body weight. The symptom of acute arsenic poisoning is severe gastroenteritis, and in chronic poisoning, loss of weight and hair, and skin lesions. An antidote for arsenic is BAL, British Anti-Lewisite.

Copper

Since copper is essential for all forms of life, problems arise when it is deficient or in excess. In some respects intake of essential elements is more critical than for toxic elements. Mechanisms appear available, in biological systems, that protect against small excesses of essential elements.

Some sources rich in copper are, shell-fish, kidneys, liver and nuts. Poor sources are dairy products, sugar and honey. Soft water, or water with a low pH, can leach appreciable amounts of copper from pipes, while carbonated drinks, in the presence of copper, can become highly contaminated. Other sources of copper, Cu, are; wine, roofing materials, cooking utensils, coins (copper-nickel alloy), pigments (copper(II) arsenite and copper(II) acetoarsenite), insecticides (copper arsenites), fungicides ($CuSO_4 5H_2O$), algicides, molluscicides and dental materials. The Cu status of a soil is reflected in plant levels.

The daily human consumption of Cu is around 2-5 mg, while the required amount for a child is 1 to 1.6 mg day^{-1}, and for adults 2 mg day^{-1}, while less than 0.3 mg day^{-1} produces deficiency. Approximately 32% of the Cu consumed is absorbed, but this varies, for example, high Ca and Fe intakes reduce Cu absorption. Copper is mainly distributed in the liver (storage), kidneys and intestines.

Copper occurs in over 30 enzymes and proteins. Ascorbic acid oxidase is a Cu plant-enzyme which catalyses the reduction of O_2 to H_2O;

$$O_2 + 4H^+ + 4e \rightarrow 2H_2O \qquad (13.9)$$

and the oxidation of ascorbic acid to dihydroascorbate. Cytochrome-C oxidase catalyses electron transfer reactions and reduction of O_2 in some animals. The blue protein ceruloplasmin, found in the plasma of mammals, birds and reptiles, contains 7 ± 1 Cu atoms and is involved in the oxidation of Fe(II). Superoxide dismutase, containing Cu and Zn, rapidly converts superoxide, O_2^-, a destructively reactive material, to peroxide and water.

$$2O_2^- + 2H^+ \rightarrow H_2O_2 + O_2. \qquad (13.10)$$

Haemocyanins are O_2 carriers in some shell-fish, one O_2 molecule per two Cu atoms.

Copper deficiency means reduced protein and enzyme activity. Signs of this are; anaemia, loss of hair pigment, reduced growth and reduced arterial elasticity. Menkes kinky hair disease in children is due to a deficiency of cyctochrome oxidase, and is fatal by the age 3 - 5 years. Wilson's disease occurs with excess Cu where the biosynthesis of ceruloplasmin is suppressed, it is an inherited disease. The amount of serum Cu in newborn infants is used as a guide to the existence of both these diseases. Copper is readily transported across membranes since weak bonds are formed with serum albumin. The metal is used in intrauterine devices, and is said to enhance the contraceptive effectiveness.

In order to be toxic,single intakes of copper must be in gram amounts, or continual intakes of >250 mg day^{-1}. One effect of Cu toxicity is an irritation of the gastrointestinal tract.

Zinc

The element zinc occurs in over 80 proteins and enzymes, mainly peptidases and anhydrases. The metallothioneins are zinc storage proteins which are useful in combating Cd and Pb poisoning. The common problem with zinc is having a deficiency. A daily consumption of <5 mg will produce deficiency symptons, such as dwarfism, dermatitis, loss of taste, immature gonads and delayed wound healing. Sickle-back disease may be due to Zn-deficiency. Normal intakes should be around 5 - 40 mg day^{-1} (a mean of 15 mg day^{-1} for adults and 10 mg day^{-1} for children). Toxicity will occur for a daily consumption >150 mg, a cause of anaemia. A lethal dose is >6000 mg (6g) day^{-1}.

The enzyme carbonic anhydrase, found in erythocytes, is a zinc enzyme involved in the transport of CO_2 in blood, by assisting in the protonation of CO_2 to HCO_3^-. Carboxy peptidase, which contains one Zn atom per mole of protein (MW ~34600) is involved in terminal peptide bond hydrolysis, and in the binding step of ester hydrolysis. The role of zinc is bonding to the C=O group of the peptide link to be broken. The C=O bond, which is now polarized, is more susceptible to nucleophilic attach. Removal of the Zn destroys the activity of the enzyme, but it is regenerated by adding the metal back, or other divalent metals such as Co and Mn. A number of zinc enzymes are catalysts for RNA and DNA metabolism, for example thymidine kinase, RNA polymerase and DNA polymerase.

Selenium

Selenium is an essential trace element for plants and animals but it is not clear whether it is necessary for human beings. Estimates of the consumption of Se in relation to health are given in Table 13.6. Fortunately, human

TABLE 13.6 Intakes of Selenium in Relation to Health

Effect of Selenium	Human Beings (mg day^{-1})	Rats (mg day^{-1})	Plants(mg l^{-1}) of nutrient solution
Deficient	<0.006	<0.0003	<0.02
Normal	0.006 - 0.2	0.0003 - 0.004	<1
Toxic	>5	>0.004	>1 - 2
Lethal		>1 - 2	

beings do not come into contact with selenium frequently, or at high levels, unless working with it. The element is used mainly in the electronics industry, added to steel to increase machineability, in pigments and for colouring glass and glazes. Selenium is added to aminal feeds. The mean level of selenium in cereals is 0.38 $\mu g\ g^{-1}$, in egg yolk, 0.18 $\mu g\ g^{-1}$ and in sea-food, 0.53 $\mu g\ g^{-1}$.

The compounds that occur in plants and fungi are selenium analogues of cysteine and cystine, for example, methyl selenocysteine ($CySeCH_3$), i.e. CH_3-Se-CH_2-CH(NH_2)CO_2H, and selenohomocystine, $CyCH_2SeSeCH_2Cy$. Glycine reductase contains one Se atom per mole and glutathione peroxidase has 4 selenium atoms per mole. The latter enzyme catalyses the reduction of H_2O_2. The element may also be essential for the activity of the enzymes formate dehydrogenase and cyctochrome-C type enzymes found in sheep muscle. Selenium also protects against the toxic effects of Ag, Cd, Hg and Tl.

Selenium deficiency in animals affects fertility, produces muscular dystrophy, leucocyte inefficiency and liver necrosis. In excess it causes cancers, deformation of hair and nails (presumably by replacing sulphur in the keratin) giddiness, depression and nervousness. A sign of Se (and also Te) poisoning is a garlic odour of breath and sweat, due to the formation of $(CH_3)_2Se$ (and $(CH_3)_2Te$) in the body.

Cyanide

Cyanide is not a major pollutant but it has notoriety because of its use in death chambers. It is toxic to all forms of life, a lethal concentration of HCN gas, or prussic acid being 0.1 - 1.0 mg l^{-1}. Anthropogenic sources come from mineral and metal processing, the burning of polyacrylics and fumigation to kill pests.

A number of plants such as white clover, cassava and sorghum, and seeds, including apple, apricot, peach, cherry and plum, contain cyanide. The problem is removed for foods, such as cassava, by boiling in water when the cyanide is hydrolysed and lost. The cyanide is bound to a glycoside called amygdalin, which on acid hydrolysis or enzyme action (as may occur in the stomach) gives HCN.

$$C_6H_5\text{-}\underset{CN}{CH}\text{-O-}C_6H_{10}O_4\text{-O-}C_6H_{11}O_5 \rightarrow \text{HCN + glucose + benzaldehyde} \qquad (13.11)$$

Cyanide toxicity relates to its ability to bond to Fe(III). The reduction of ferricytochrome to ferrocytochrome, in the last stage of oxidative phosphorylation, is prevented by the formation of an Fe(III)-CN system. Detoxification may be achieved with thiosulphate, which in the presence of an enzyme rhodanase, found in the liver and kidneys, converts cyanide to the less toxic thiocyanate.

$$CN^- + S_2O_3^{2-} \xrightarrow[\text{rhodanase}]{} SCN^- + SO_3^{2-} \qquad (13.12)$$

Carbon Monoxide

Gaseous CO, a significant air pollutant, and a component of some fuel gases, is toxic to animals, but not plants. Both CO and O_2 have similar dimensions, but CO bonds more strongly to the iron of haemoglobin than O_2. This means O_2 transport and oxidative phosphorylation are prevented by carbon monoxide.

The coordination of CO reverses slowly if the source of CO is removed, and the oxygen concentration allowed to build up;

$$HbFeO_2 + CO \rightleftharpoons HbFeCO + O_2. \quad (13.13)$$

However, brain damage can occur if the exposure is for too long a period. A lethal concentration is 1-10 mg l^{-1}.

Nitrate and Nitrite

Large quantities of nitrate may be consumed daily, but reduction of nitrate to nitrite in living organisms;

$$NO_3^- + 2H^+ + 2e \rightarrow NO_2^- + H_2O, \quad (13.14)$$

may produce toxicity problems, as the nitrite ion is toxic to human beings, especially infants. The principal source of nitrate in food is vegetables and meats which have been treated with nitrate or nitrite to protect against bacterial growth. The main food sources and the relative amounts of NO_3^- and NO_2^- in each is given in Table 13.7. As saliva is produced internally and then swallowed, it can be considered as adding to the daily intake. Approximately 100 mg of NO_3^- and 12 mg NO_2^- are consumed daily. The high nitrite content of saliva arises from the reduction of nitrate by oral microflora, the nitrate being concentrated from plasma.

TABLE 13.7 Sources and Amounts of Nitrate and Nitrite in Food

Source	Nitrate %	Nitrite %
Vegetables	86	8
Cured meats	9	92
Water	1	-
Fruit	1	-
Other	3	-
Approximate daily intake	70 mg	3 mg
Approximate amount in saliva (daily)	30 mg	9 mg

The two principal ill-effects of NO_2^- are methemoglobinemia in infants (<4 months) and the formation of carcinogens, such as nitrosoamines. The pH of the stomach fluids of young infants is higher than for adults (>4). This is an environment in which NO_3^- reducing bacteria can exist, while for adults, because of the more acid stomach fluids, they live in the lower intestine where absorption into the bloodstream is not a problem. The NO_2^-, once in the blood, oxidises haemoglobin to methemoglobin, an Fe(III) product with no O_2 transport properties. A constant intake of water containing 10 mg (N) l^{-1} (as nitrate) or greater, may lead to methemoglobinemia in infants. Other sources are vegetables, for example micro-organisms grow in cooked spinach, which, if left standing, are able to reduce NO_3^- to NO_2^-. The reaction of NO_2^- with haemoglobin may be reversed with an enzyme such as methemoglobin diaphorase, which unfortunately is not plentiful in infants.

To form nitrosoamines from nitrate, two reactions are required to occur:

$$NO_3^- \xrightarrow{\text{bacteria}} NO_2^-, \tag{13.15}$$

and

$$NO_2^- + \begin{matrix}\text{secondary}\\\text{amines or}\\\text{amides}\end{matrix} \rightarrow \text{N-nitrosoamines} \tag{13.16}$$

Equation (13.15) requires a pH > 4 and equation (13.16) a pH < 4, this restricts both reactions occurring at the same site at the same time. However, direct ingestion of NO_2^- (from saliva) may lead to the formation of the carcinogens. The reactions may occur in the mouth or in infected bladders. In Worksop, U.K., the local water supply had a high nitrate level (90 mg l^{-1}), and it was found that there was a 25% greater incidence of gastric cancer in the town compared with control towns. The weekly use of nitrate in Worksop was around 900 mg (600 mg from water), while for the control towns it was 400 mg (100 mg from water). The results point to an influence of NO_3^- on cancer.

While potentially toxic the nitrite ion is an antidote for cyanide poisoning, and amyl nitrite is a common material, in first aid cabinets, in areas where CN^- is handled. The amyl nitrite is enhaled. The NO_2^- produces Fe(III) methemoglobin which coordinates CN^- ions and releases the CN^- from ferricytochrome oxidase.

$$\underset{\text{Haemoglobin}}{\text{HbFe(II)}} + NO_2^- \rightarrow \underset{\text{methemoglobin}}{\text{HbFe(III)}} + N_2 + NO \tag{13.17}$$

$$\text{HbFe(III)} + \underset{\substack{\text{cyanoferricytochrome}\\\text{oxidase}}}{\text{NC-Fe(III)-oxidase}} \rightarrow \underset{\substack{\text{Cyanomethe-}\\\text{moglobin}}}{\text{HbFe(III)CN}} + \underset{\substack{\text{Ferricytochrome}\\\text{oxidase}}}{\text{Fe(III)-oxidase}} \tag{13.18}$$

The cyanide may then be removed using thiosulphate.

Fluoride

The question whether fluoride is an essential element is still unresolved, but its effect in reducing dental caries is well known. A normal daily intake is 0.3 - 5 mg whereas a toxic dose is 20 mg day^{-1} and a lethal dose is 2000 mg day^{-1}. Since the difference between normal and toxic levels is not great, it is necessary to control fluoride consumption. If F^- is deficient, people can have poor bones and teeth, while an excess shows up in mottled teeth (2 to 8 times the normal dose) and bone sclerosis (20 to 40 times normal dose). The fluoride's importance is as a structural element, as it isomorphously replaces OH^- ions in hydroxy-apatite of bones and teeth to give fluoro-apatite a harder and less easily diseased material. In many places fluoride is added to public water supplies to give a concentration of 1 mg l^{-1} (1 ppm). The compounds added are NaF, or Na_2SiF_6, or $(NH_4)_2SiF_6$ or H_2SiF_6. Hydrolysis produces fluoride ions from the fluorosilicates;

$$Na_2SiF_6 + 2H_2O \rightarrow SiO_2 + 6F^- + 4H^+ + 2Na^+. \qquad (13.19)$$

From food up to 3 mg may be taken in daily, but this varies widely depending on the fluoride status of the soils.

Interelement Effects

Elements may influence each other in biological processes. The effects can be synergistic or antagonistic, or competitive and non-competitive. A simpler description can be given in terms of a positive interaction or a negative interaction.

It is possible that interelement effects are the norm in biological systems, rather than the exception. In addition to the chemical form and the duration of intake the nature and amount of other elements ingested, or present in the body can be an important factor. Hence single element studies may be misleading, and a subject may be suffering from deficiency of an element, even though the normal amount is consumed, because of an unnatural amount present of another element. For example, Cu deficiency occurs in the presence of excess Mo and SO_4^{2-}, and Cu toxicity occurs in the presence of excess Zn and Fe. Many interelement effects work both ways, i.e. toxic levels of Mo are overcome by adding Cu^{2+} and SO_4^{2-}, the reverse of copper deficiency.

A list of a few interelement interactions are given in Table 13.8. The chemical mechanism of many of these interactions are poorly understood, and

TABLE 13.8 Interelement Effects

Element		Effect of Interacting Element	Positive or Negative Effect
in toxic amounts	in deficient amounts		
Mo		corrected by Cu^{2+} & SO_4^{2-}	+
	Cu	produced by excess Mo & SO_4^{2-}	-
Cu		corrected by Zn and Fe	+
Cd		corrected by Zn	+
Se		corrected by As	+
Hg		corrected by Se	+
	Fe	produced by Mn	-
	Cu	produced by Pb	-
	Fe (utilization)	produced by Cu (deficiency)	-
	Cu (utilization)	reduced by Cd	-
	Cu	produced by Ag	-

some interactions are understood to a limited extent, especially when the two interacting elements have a similar chemistry, e.g. Zn and Cd. A number of the processes involve replacement of one element by another, at say a protein binding site. Such a process will depend on bond strengths and kinetic factors. Hence, stability constants of metal complexes can be used to suggest why certain interelement effects occur. The data given in Table 13.9 follows, to some extent, the order of stability constants for divalent

TABLE 13.9 Stability Constants of Some Transition Metal-Bio Complexes

Metal	log k	
	carboxy peptidase	glycine(k_1)
Mn^{2+}	5.6	3.4
Ni^{2+}	8.2	6.2
Cu^{2+}	10.6	8.6
Zn^{2+}	10.5	5.2

metal ions,

$$Mn^{2+} < Fe^{2+} < Co^{2+} < Ni^{2+} < Cu^{2+} > Zn^{2+}.$$

The order may be explained, in part, by the ligand field stabilization energy of the metal ion in a spin-free octahedral field. The ligand field stabilization energies increase from Mn^{2+} to Ni^{2+} and then decline. However, the stabilization energy is only one influence.

RADIATION

Biological material is susceptible to damage by ionizing radiation, that is γ and X-rays, neutron, alpha and beta particles. Controlled exposure can be beneficial in destroying unwanted tissue as in radiation therapy of cancer cells. Low doses of soft X-rays are used to take X-ray photographs of human tissues.

The potential for serious radiation damage of human beings is well known, because of the effects recorded after the atomic bombing of Hiroshima and Nagasaki. Human beings are constantly exposed to natural background ionizing radiation, namely cosmic rays, radiation from radionuclides present in rocks, soil, water, air and food, and naturally occurring nuclides in our bodies (e.g. ^{40}K). More recently radioactive decay products from nuclear explosions have added to the background levels. Man-made radiation sources include X-ray generators, nuclear power plants and the problems associated with handling nuclear wastes and reprocessing fuel, nuclear weapons, nuclear explosions, and radionuclei produced for laboratory and hospital use.

Some properties of damaging radiation are given in Table 13.10. Their damaging effect comes from either ionization or electron excitation producing ions or free radicals which are reactive in biochemical processes. Hence metabolic damage such as inhibition of enzymes, and changes to DNA and RNA can occur. Gamma and X-rays are the most damaging as they penetrate furtherest into the body (Table 13.10). Neutrons also have deep-seated effects as a

TABLE 13.10 Radiation Types

Radiation	Type	Charge	Energy	Average penetration	
				air	body
γ rays	electromagnetic		0.1 - 40 MeV	no limit	can pass through body
X rays	electromagnetic		10 - 100 keV	no limit	penetrates deeply into body
α	particle	+2	4 - 10 MeV	4 - 10cm	clothing, outer layers of skin
β^-, β^+	particle	- or +	0.025 - 2.15 MeV	several meters	few mm. into tissue
neutrons	particle	zero	several MeV	captured by particles they collide with	

result of subsequent processes, for example slow neutrons participate in $^{M}_{Z}A(n,\gamma)^{M+1}_{Z}A$ reactions, i.e. a neutron is captured producing a radionucleide (and γ rays) and which decays emitting γ-radiation. Another neutron capture reaction $^{14}_{7}N(n,p)^{14}_{6}C$ can alter amino acids. In addition to external radiation radionucleides which are ingested are a source of internal radiation.

In order to quantify the effects of radiation a unit of radiation called the *rad* (radiation absorbed dose) has been defined as the dose of radiation required to liberate 1×10^{-5} J of energy per gram of absorbing material. As this unit gives no indication of the biological effects, another unit, the *rem* (roentgen equivalent man) is defined as the amount of radiation to produce a certain biological effect. The two units are related; 1 rem = 1 rad of X-rays or γ-rays. The level of radioactivity of a material is given in Curies, where one Curie is the amount of substance undergoing 3.7×10^{10} disintegrations per second.

Radiation damage manifests itself in two ways: somatic, where the effects are experienced by the person; and genetic, where the effects may only be apparent in later generations. Somatic effects are either short term, i.e. relatively soon after the event, or long term when they become obvious months or years later. Some of the effects of whole body radiation are listed in Table 13.11. Doses in the range 200 - 700 rem that are restricted to certain areas will cause localised effects, such as X-ray burns, damage to the eyes (causing cataracts) and sterility. The radio-sensitive organs are the eyes, gastointestinal tract, bone marrow, spleen and reproductive organs. One pronounced effect is leukemia.

The impact of radionucleides on the body, that are ingested, is controlled by the radioactive ½-life and biological ½-life of the element. The effective ½-life (t_{eff}) i.e. the time to reduce the radiation in the body by half is;

TABLE 13.11 Effects of Whole Body Radiation

Dose (rems)	Effects
0 - 25	A dose around 25 rem may reduce the white blood cell count.
25 - 100	Nausea for about half those exposed, fatigue, changes to blood.
100 - 200	Nausea, vomiting, fatique, death possible, susceptible to infection (low white blood cell count).
200 - 400	A lethal dose for 50% of those exposed especially in absence of treatment. Bone marrow, spleen (blood forming organs) damaged.
>600	Fatal, probably even with treatment.

$$t_{eff} = \frac{t_{rad} t_{biol}}{t_{rad} + t_{biol}} \quad (13.20)$$

Strontium-90, a product of nuclear explosions, has t_{rad} = 28 years and t_{biol} = 35 years, hence t_{eff} = 15.5 years, a long time, especially as ^{90}Sr ends up in the bones (where t_{biol} = 50 years) and can cause continuing damage to bone marrow. Iodine-131 has t_{rad} = 8 days and t_{biol} = 138 days, therefore t_{eff} = 7.6 days, sufficient time to damage the thyroid gland where iodine accumulates. The most dangerous radionucleides are the short-lived isotopes (which have the more energetic radiation) with long biological ½-lives.

Genetic effects resulting from the effect of radiation on DNA and RNA lead to mutagens. Around 2 to 10% of natural mutations arise from exposure to background radiation for 50 years. The end product of radiation initiated genetic effects is similar to that caused by chemical mutagenic material.

The annual, whole body, dose rate due to natural radiation sources is around 0.1 rem $year^{-1}$. In addition a person may also be exposed to radiation from fall-out, medical uses of radiation, colour T.V. and air travel, increasing the dose to around 0.2 rem $year^{-1}$. People working in jobs associated with radiation are exposed to higher levels. The maximum permissable whole body dose for the general public has been set at 0.5 rem $year^{-1}$. And for reproductive organs the limit is <5 rem over 30 years.

BIO-INORGANIC SPECIES

Some of the more important bio-inorganic compounds are listed in Table 13.12. Their importance stems, in part, from the amount of research carried out on them. In discussing such materials it is easy to over-emphasize the role of the metal ion, because in a number of cases their chemistry is more readily understood. A brief mention will be made of some of the systems listed in Table 13.12.

TABLE 13.12 Some Bio-inorganic Species

Bio-inorganic species	Metal	Principal organic group	Important activity
Chlorophyll	Mg	Porphyrin	Photosynthesis.
Nitrogenase	Mo, Fe (and S)	Protein	Fixation N_2 to NH_3.
Haemoglobin	Fe	Porphyrin, protein	Binds O_2 and transfers to tissues.
Myoglobin	Fe	Porphyrin, protein	Binds O_2 and stores it.
Ferredoxins	Fe (and S)	Protein	Electron transfer reactions.
Vitamin B_{12}	Co	Corrin ring	Enzyme cofactor, catalyses atom transfer.
Haemocyanin	Cu	Protein	Binds and carries O_2 in blood of molluscs.
Zn-enzymes	Zn	Protein	Various biochemical catalytic reactions.

Haemoglobin and Myoglobin

Both haemoglobin and myoglobin bind and transfer dioxygen. They consist of an Fe(II) atom surrounded by a planar tetrapyrrole unit called a porphyrin, giving the haem group. Porphyrins are important biochemical chelating molecules and also occur in the chlorophylls (Mg), cytochrome a, b and c (Fe), chlorocruorin (Fe), protoporphyrin IX (Fe), uroporphyrin III (Cu) and modified in cobalamin (Co).

Haemoglobin (Hb) contains four haem groups as well as four polypeptide chains (150 - 160 amino acids). Myoglobin (Mb) has one haem group and one polypeptide chain. The fifth iron coordination site is a nitrogen of an imidazole side chain of a histidine group, while the sixth position (trans to the imidazole) bonds the dioxygen (or H_2O when no O_2 is present). Haemoglobin picks up O_2 from the lungs and transfers it, by means of the bloodstream, to the tissues where it is transferred to myoglobin and stored until required. The deoxy-haemoglobin returns to the lungs taking CO_2, a metabolic product of respiration, with it.

$$MbO_2 + \{CH_2O\} \rightarrow Mb + CO_2 + H_2O + \text{energy} \qquad (13.21)$$

At the partial pressure of O_2 in the lungs (~13 kPa), Hb is almost fully saturated with O_2 but at the tissues the partial pressure drops to around 5 kPa and Mb preferentially binds to the O_2. The CO_2 released lowers the pH (~6.8) which favours transfer of O_2 from Haemoglobin to Myoglobin.

Experimental evidence suggests the Fe-O-O bonding is bent, but the formal electronic configuration is still in dispute, but maybe Fe(III)-O_2^- rather than Fe(II)-O_2. The binding of one mole of O_2 to Hb increases the affinity of the next haem-group for O_2 and so on. This cooperative effect is not

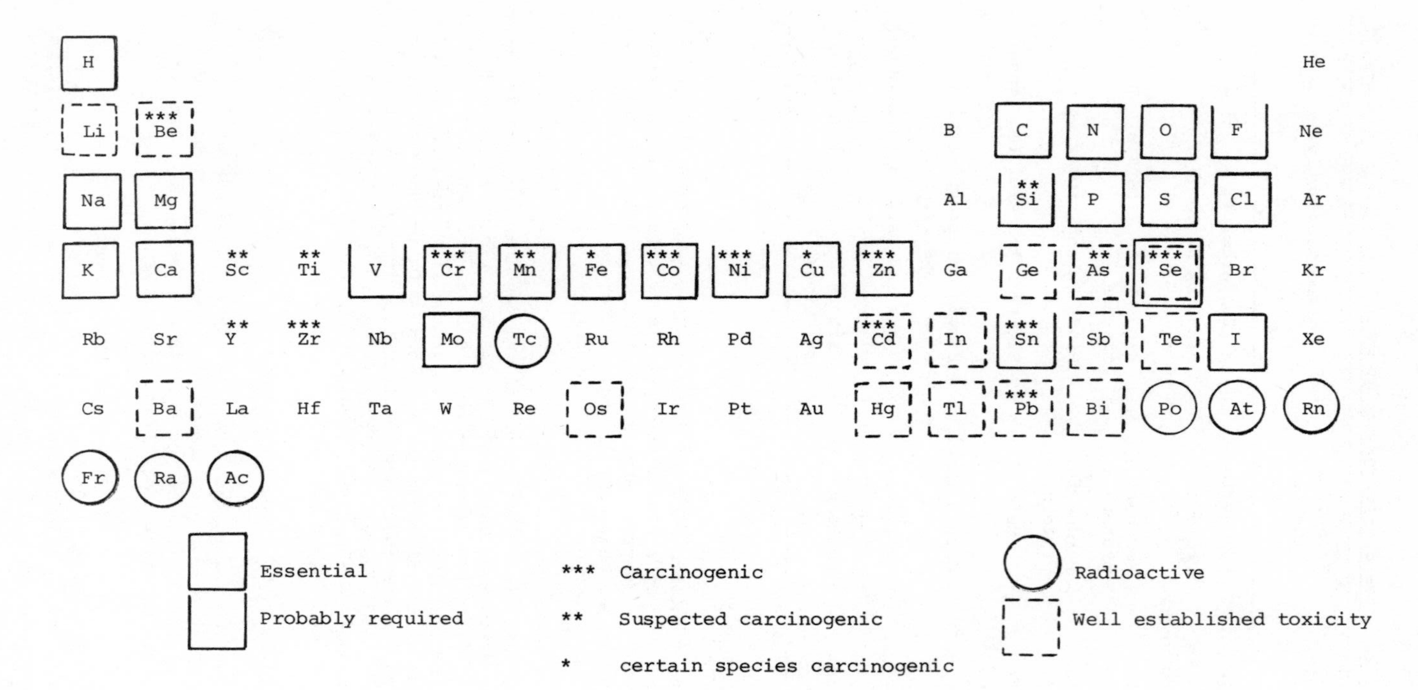

Fig. 13.6. Biological activity of the elements. Source: T.D. Luckey and B. Venugopal, Metal toxcity in Mammals, 1. Plenum Press, N.Y. 1977.

well understood. Some movement of the iron with respect to the four planar N atoms of the porphyrin, and moment of the peptide chain appear important.

Ferredoxins

Ferredoxins contain the Fe-S clusters $Fe_4S_4(SR)_4$ and $Fe_2S_2(SR)_2$. They are involved in oxidation and reduction reactions and are sinks for electrons in electron transfer reactions in sequences such as;

$$[Fe_4S_4(SR)_4]^{1-} \underset{-e}{\overset{+e}{\rightleftharpoons}} [FeS_4(SR)_4]^{2-} \underset{-e}{\overset{+e}{\rightleftharpoons}} [Fe_4S_4(SR)_4]^{3-}. \quad (13.22)$$

Such species are believed to be the source of reducing power needed to reduce N_2 to NH_3 in nitrogenase.

Vitamin B_{12}

Vitamin B_{12} is a cofactor for a number of enzymes. It occurs mainly in the liver and a deficiency gives rise to pernicious anaemia. The enzyme normally catalyses the exchange process;

$$-\overset{\overset{\displaystyle R}{|}}{C}-\overset{\overset{\displaystyle H}{|}}{C}- \rightarrow -\overset{\overset{\displaystyle H}{|}}{C}-\overset{\overset{\displaystyle R}{|}}{C}- \quad (13.23)$$

on the substrate. The reaction is complex and involves radicals, reduction of Co(III) and reoxidation may also occur. The methyl derivative of cobalamin has been discussed as a possible methylating reagent in the production of methyl mercury.

The abundant light elements of the first and second rows of the periodic table and most of first row transition metals are the essential biological elements (Fig. 13.6). Abundance is probably an important factor in the element being essential. Some of the carcinogenic elements and toxic elements are also listed in the figure. The toxic elements tend to be the heavier members of the periodic groups, these being the elements that bond strongly to soft donor atoms such as sulphur. The heavier elements also tend to form the more insoluble compounds, a factor reducing their potential toxicity, for example the barium in insoluble $BaSO_4$ is not toxic.

CHAPTER 14

Analytical Chemistry and the Environment

Chemical analysis is an integral part of environmental investigations, as the question of how much of a substance is present is important to answer, whether the problem be one of resources, pollutant levels, health or natural bio-geo-chemical processes. It is important to carry out multi-element analysis in order to investigate inter-element effects. Also it is becoming more important to know the chemical speciation of the elements being analysed. This latter area is where more effort needs to be directed in the future. Reliable information can be obtained on the quantity of material present, but less information is available on the species present. This is particularly the case for solids at the ppm level.

It is not intended to give great detail on analytical techniques; these are available in numerous texts, but rather to outline briefly some of the areas where pitfalls await the unwary analyst.

AMOUNTS TO BE ANALYSED

One aspect of analytical chemistry is estimating how much of an element is present, and its concentration in the surrounding matrix. This knowledge helps in deciding on methods of sampling, handling and analysis. Trace amounts are, 100 - 1 $\mu g\ g^{-1}$ (some use the range 1000 - 1 $\mu g\ g^{-1}$) and ultra-trace <1 $\mu g\ g^{-1}$. At these levels contamination problems are very real. Another problem is the amount of material available, anything between 0.001 - 0.01 g (1 - 10 mg) is called micro and anything <0.001 g (<1 mg) is sub-micro. In many trace element studies the analyst is interested in a range of results, rather than one particular result, i.e. is the value obtained within an expected, or required, range?

Accuracy is a measure of the closeness of a result to the true result. *Precision* is a measure of the consistency of results, i.e. how close they agree with each other, even though they may not be very accurate. An illustration of these two terms is the grouping of arrows on a target. If the arrows are on the bulls-eye the shooting has been both accurate and precise. If they, instead, group together but away from the bulls-eye, the shooting has been precise, but not accurate. Precision is estimated by the standard deviation, and is also used for estimating accuracy if the true value is known. The *detection limit* of a technique is the signal that can reliably be discerned above background, after correcting for the blank (i.e. reagents), with say a 95% certainty. Two mathematical estimates of detection limits are, d.l. $>$ $2\sqrt{\text{background}}$, or d.l. $>$ 2 or 3σ, where σ is the standard deviation of the background or blank measurements. The limits of detection can be in absolute terms, i.e. the amount of material, or in relative terms, i.e. concentration. A reported detection limit does not mean that accurate measurements can be made at these levels because of the errors involved. Normally, the lower the detection limit, the greater the error. Some detection limits (ideal) are given in Table 14.1. However,

TABLE 14.1 Detection Limits for Some Instrumental Analytical Techniques

Technique	Detection Limit	
	g	ng
Spectrophotometry[a]	10^{-10}	0.1
Fluorimetry	10^{-11}	10^{-2}
X-ray fluorescence	10^{-9}	1
Electron microprobe	10^{-14}	10^{-5}
Anode stripping voltammetry	10^{-10}	0.1
Gas chromatography	10^{-12}	10^{-3}
Atomic absorption (flameless)	10^{-13}	10^{-4}
Emission spectroscopy	10^{-10}	0.1
Neutron activation analysis[b]	10^{-11}	10^{-2}
Mass spectrometry	10^{-16}	10^{-7}

a Increased pathlength will improve detection limit

b For neutron flux 10^{14} neutrons cm^{-2} s^{-1}

the actual limit depends on the element being analysed, conditions used for the analysis, and the performance of the instruments. For example, in neutron activation analysis the detection limit for ^{56}Mn, using γ-ray spectrometry, is around 7×10^{-3} ng, whereas for ^{69}Zn it is 0.1 ng.

SAMPLING AND STORAGE

Environmental analytical studies are as good as the sampling and storage techniques used. There is little value having a sensitive and accurate analytical method, if in the process of sampling or storage the amount of analyte has altered. Sampling is important and attention must be given to it, in particular, how representative the sample is of the bulk material. The reason for concern over obtaining representative samples is because most samples are static, i.e. they have been removed from the body being studied. The better approach is of dynamic sampling, which is more difficult, and only amenable to gases and liquids.

In sampling solids, which are mostly heterogenous, e.g. soil or rocks, it is necessary to decide whether to use random sampling (e.g. spot, cross section or drilled samples) or selective sampling. It is important that the sampling instrument does not contaminate the sample. Storage of solids is not too great a problem using air-tight polyethylene containers.

Liquids could pose more of a problem over storage than sampling. Liquids are normally homogenous making it easier to obtain representative samples. Polyethylene bottles appear the most reliable for storage, but previous history and pretreatment can affect this. Two problems in the storage of liquids, are sorbtion on to the surface of the bottle, and leaching materials from the bottle into solution. The latter is averted, or reduced, by pretreatment of the bottles, for example washing in acid. Sorption is often reduced by lowering the pH of the sample down to pH = 2 (Table 14.2).

TABLE 14.2 Sorption on to Container Walls

Metal	Concentration µg l^{-1}	pH	Time of storage (days)	% Sorbtion	
				Polyethylene	Borosilicate glass
Cd	1.0	2.0	20	negligible	negligible
		6.0	20	negligible	20
Fe	trace	1.5	55	negligible	-
		8.1	55	90	70
Zn	100	2.0	60	negligible	negligible
		5.0	60	negligible	20

However, if the chemical form of the materials in the water are being studied, the pH cannot be changed. Freezing the samples, until required, can also be used for storage of liquid samples.

Sorption is also a problem with gas storage. It is necessary to use materials for the gas bottles that will not interact with the gases.

The species, and the part of a plant or animal being sampled must be noted in biological sampling. Plant material can be washed, oven dried and stored in polyethylene containers, while animal material needs to be deep-frozen until ready for analysis.

Sampling of human head hair is an example of the difficulties of obtaining representative samples. Analytical results have been reported on samples obtained from; the floor of hair-dressing salons, normal hair-cuts, close the the scalp, from a particular part of the head, or the whole fibre(s). It is necessary to be aware of the limitations of any of the sampling methods, and not to infer too much from the results. It is best to be clear on what information is required, before sampling, as each factor influences the other.

HANDLING OF SAMPLES

The major steps in preparing samples for analysis are cleaning and sorting, followed by chemical treatment if necessary. Some of the processes that may be used include; washing, drying, crushing, sieving, extraction, pre-concentration, wet or dry ashing, dissolving, filtering, precipitation, fusion, ion exchange and chromatography techniques. Opportunity exists for contamination during these processes, therefore minimal treatment is best. Reagents must be pure, and reagent blanks must be allowed for. Poor sample preparation can destroy the validity of results obtained.

For human hair, various washing techniques have been used to clean the hair. Attempts have been made to separate internally deposited elements from external contamination by washing. Different analytical results can be obtained depending on the washing procedure used and the objective of the washing, and therefore care is necessary in interpretation of the results.

ANALYSIS

Growth, and interest, in trace element analysis corresponds with developments in sensitive instrumental techniques. Normally, use of an instrument is not a great problem, but an analyst needs to be aware of difficulties in the areas of sample matrix, and interferences. Both areas relate to the specificity of the method. The matrix effect is relevant when calibrating an instrument, as often the calibrant is not in the same matrix as the samples. Interferences may be instrumental, e.g. the inability of a spectrophotometer to resolve signals from two separate species, and chemical, i.e. when one species inhibits or enhances the signal from another, because of chemical (or physical) interactions. It is necessary to be aware of these problems, if reliable analytical results are to be obtained.

The signal which is measured and correlates with the amount of material present includes; emission, absorption, scattering, diffraction, and rotation of radiation; and electrical properties such as, electric potential, current or resistance. While radiation and electrical properties are the two principal properties employed, other techniques make use of thermal, kinetic, mass-charge ratio, mass and volume properties.

The interaction of radiation with matter, and the resulting signal, is widely used in analysis. Some of the main techniques are listed in Table 14.3 in relation to the electromagnetic spectrum. Radiation ranging from γ-rays to radiowaves are all made use of.

Numerous factors need to be considered when selecting an analytical method. The information presented in Table 14.4 provides comparative data on the more common analytical techniques. However, information on the applicability of a method, and the skill necessary to carry out the analysis often only comes from experience.

Probably atomic absorption spectroscopy is the one technique that has given the greatest boost to trace analysis over the last two decades. It is one of the more widely used methods, irrespective of the fact that only one element can be measured at a time, and each element needs its own source lamp. Continuing developments increase the sensitivity, the dramatic effect of changing from flame to flameless atomic absorption is illustrated by the data in Table 14.5.

TABLE 14.3 Analytical Methods and the Electromagnetic Spectrum

Wavelength (m)	Frequency Hz	Electro-magnetic spectrum	Spectroscopy	Energy Transition
3×10^{-13} (0.3 pm)	10^{21}	γ rays	γ ray emission	Nuclear
3×10^{-11} (30 pm)	10^{19}	X-rays	X-ray emission and adsorption (XRD, XRF, Electron microprobe)	Electronic (Inner electrons)
3×10^{-9} (3 nm)	10^{17}		Vac. UV absorption	Electronic (Valence electrons)
		UV		
3×10^{-7} (0.3 μm)	10^{15}	visible	U.V.-visible emission and absorption, flame emission and absorption	
		IR	IR absorption, Raman	Molecular vibrations
3×10^{-5} (30 μm)	10^{13}			
3×10^{-3} (3 mm)	10^{11}	Micro wave	Microwave absorption	Molecular rotations
			esr	
3×10^{-1} (3 dm)	10^{9}			Magnetically induced spin states
		Radio wave	nmr	
3×10^{1} (30 m)	10^{7}			

Each analytical method has its own particular instrumentation, but most are similar in the basic requirements. Each instrument has a way of producing a signal from the sample. For example; in neutron activation analysis neutrons are absorbed by the sample producing radioactive isotopes by the

TABLE 14.4 Comparison of Some Analytical Methods

Method	Sample[a]	Instrumentation	Specificity	Sensitivity[b]	Precision[b] (%)	Applications
Gravimetric	SLG		good	100 mg-1 g 1-10 μg	0.005-0.01 0.1	major constituents
Titrimetric	SLG		good	10^{-2} M in soln. 10^{-5} M in soln. 10^{-6}-10^{-7} M in soln.	0.01 0.1 0.2-1.0	major and semi-micro constituents
Visible spectrophotometry	SL	colorimeter; spectrophotometer	fair	10-100 ppm in soln. 0.005-0.1 ppm in soln.	1-5 5-10	semi-micro and micro-constituents
Ultraviolet	SLG	UV spectrophotometer	fair	10-100 ppm in soln. 0.005-0.1 ppm in soln.	1-5 5-10	semi-micro and micro for chromophores
Flame emission spectroscopy	SL	flame photometer; spectrophotometer	good	0.1-10 ppm in soln. 0.001-0.1 ppm in soln.	0.5-3 5-10	micro; for alkali, alkaline earth, and some transition metals
Atomic absorption spectroscopy	SL	AA spectrophotometer	excellent	0.1-10 ppm in soln. 0.001-0.1 ppm in soln.	0.5-3 5-10	micro; for transition and some semi-metals
Gas chromatography	LG	gas chromatograph	excellent	0.1-2% 10-100 ppm	0.2-5 5 - ⩾10	major to micro; organic and organometallic constituents
Liquid chromatography	SL	high-pressure liquid chromatograph	good	0.001-1 ppm	2-20	micro;mainly organic constituents
Polarography	L	polarograph	good	10-100 ppm 0.1-10 ppm	1-2 3	semi-micro and micro; organic constituents and elements

Anodic stripping	L	DC-pulse polarograph	good	10-100 ppm 0.1-10 ppm 0.001 ppm (in soln.)	1-2 3 5	micro; selected trace metals; (Ag,Bi,Cd,Cu,Fe, In,Pb,Sb,Sn,Zn)
Spectro-fluorimetry	SL	spectro-fluorimeter	good	0.001-10 ppm	0.5-10	micro; for inorganic and organic constituents
Emission spectroscopy	SL	photographic/ direct-reading spectrograph	excellent	major component 0.1-100 ppm	10-15 5-10	major and micro; for many elements
X-ray fluorescence spectrometry	SL	XRF spectrometer	good	10-200 ppm generally	1-2	semi-micro; for many elements
Neutron activation	SL	neutron source, γ-ray spectrometer, β-counter	excellent	0.001 ppm-1-2%	1-5	micro, semi-micro for many elements
Mass spectrometry	SLG	mass spectrometer	good	0.003-100 ppm	1-10	micro, semi-micro for many elements

[a] S = solid, L = liquid, G = gas, [b] Sensitivity and Precision figures are approximate, and depend on the particular element being analysed.

Source: L.E. Smythe, Environmental Chemistry (1977)(Bockris, J. O'M., Ed) Plenum Press, New York.
D. Betteridge and H.E. Hallam., Modern Analytical Methods, The Chemical Society London, 1972.

TABLE 14.5 Comparison of Flame and Flameless Atomic Absorption Spectroscopy

Property	Flame AA	Flameless AA
Technique	Air/acetylene flame N_2O/acetylene flame	carbon furnace electrically heated
Detection Limit	10^{-10} g	10^{-12} g
Precision	±5%	±5-15%
Sensitivity for:		
Cd(228.8 nm)	0.001 ppm	0.00002 ppm
Cu(324.7 nm)	0.005 ppm	0.0001 ppm
Hg(253.7 nm)	0.5 ppm	0.006 ppm
Mn(279.5 nm)	0.002 ppm	0.0001 ppm
Pb(283.3 nm)	0.01 ppm	0.0002 ppm
Sample size	5-10 ml	1-20 μl (soln) 1-5 mg (solid)

process ${}^{A}_{Z}X(n,\gamma){}^{A+1}_{Z}X$, the signal being emission of γ or β^- rays. In colorimetry the signal, visible radiation with some wavelengths removed by absorption, is produced by irradiating the sample with light. Each instrument then has a way of detecting the signal, if necessary converting (transducing) it into another signal, and also if necessary amplifying the signal. The most common procedure is the conversion of the original signal into an electrical signal, which can be recorded.

It is possible to estimate quite low levels of trace elements in the environment, with good precision. Often the sensitivity obtainable is greater than necessary as frequently just an order-of-magnitude result is required. For example, in food quality studies one is mainly interested in determining the concentration of an element that lies within, or outside, some prescribed limits.

DATA HANDLING

The last step in an analytical study is deciding how to handle the data. It is at this point that mistakes can be made, and wrong interpretations given, especially if the statistical analysis of the data is in error. On the other hand, information may be missed if a statistical analysis is not carried out (or carried out incorrectly).

Many statistical tests depend on having normally distributed data, but often environmental analytical data is log-normally distributed (see Figs 3.2, 9.14). In a normal distribution the mean and median are the same, while for log-normal data the median and geometric mean (or arithmetic mean of log (data)) are the same. For example the analytical data for an element in human head-hair of a population will be (generally) log-normally distributed. This is because a number of samples, for one reason or another, have high levels. Some of the high values may be discarded if the reason is known (e.g. if the hair has been treated with a preparation containing the element being analysed), but others cannot. If the data is treated as a normal distribution, when comparing two sets of data from two populations, wrong deductions can be made, because too much weight has been given to the outlying values.

Statistical tests may be used to investigate if sets of data are different, or to correlate variables (Table 14.6). However, care is needed in interpretation of the results of the test.

TABLE 14.6 Some Statistical Tests

Test	Comment
Tests of Significance	
Null hypothesis	To test the source of samples
F	Comparison of variability or precision of two sets of data
Student t	Comparison of means of two sets of data
χ^2	Compares variation of a sample with theoretical
Relationship Between Variables	
Regression analysis	Relation between sets of data one set without errors
Correlation analysis	Testing association between two or more random variables

A necessary safeguard in analytical work, is to carry out frequent standardizations of procedures and instruments. International standards of rocks, plant and animal material, whose analyses are well known, can be used for this purpose. Also one can develop a local standard which has been compared with a primary standard.

Inter-laboratory analytical projects assist in keeping standards up in an analytical laboratory. While there is an "actual value" for a particular element in a sample, one does not expect to get exactly that value, because of limitations in the procedures. However, comparison of one's result with those of others, does give an indication of the quality of the analytical procedures, and a guide to the sort of tolerances that can be expected.

Careful analysis, and frequent cross-checks, mean it is possible to obtain reliable analytical data on environmental samples. Unfortunately some published data are in error, and wrong deductions have been made even on good data. In order to overcome these problems, so that good decisions can be made, more attention will have to be given to all the processes of chemical analysis. It is not unknown, for example, to find that the levels of trace elements in a system change (normally decrease) as time progresses. In the 1950-60 period reported values of chromium in blood were of the order of 100 $\mu g\ l^{-1}$ (100ppb) or greater, whereas in the late 1970's the value had been lowered to $<1.0\ \mu g\ l^{-1}$ (1ppb). The probable reason for the early high results was contamination of samples from stainless steel apparatus. The significance of such analytical results is that they may be used as a basis for laying down permissible levels. In a situation where results differ by more than two orders of magnitude consequences of decisions made on the incorrect data could be serious. Normally the higher the levels of the trace metals the smaller the spread of reported values. Therefore it is in the area of trace and ultra-trace analysis that considerable care is necessary.

Bibliography

The literature in the area of environmental chemistry is vast, in part, because it is an interdisciplinary subject. Published work falls into four main categories, general books, monographs, general papers and research papers. In this bibliography it is not intended to be exhaustive in coverage. Also, references are given mainly in the first three categories. However, such sources can only be produced from the details published in research papers. Each reference given will itself provide many more sources for the student. The bibliography is divided up into topic areas which, in general, follow the order of chapters in the book.

General Sources: Environmental

American Chemical Society (1978) Cleaning Our Environment, A Chemical Perspective (2nd Ed.) A.C.S. Washington D.C.

Bailey, R.A., Clark, H.M., Ferris, J.P., Krause, S. and Strong, R.L. (1978) Chemistry of the Environment, Academic Press, New York.

Bockris, J. O'M (Ed.)(1977) Environmental Chemistry, Plenum Press, New York.

Bowen, H.J.M. (1979) Environmental Chemistry of the Elements, Academic Press London.

Ehrlich, P.R., Erlich, A.H. and Holdren, J.P. (1977) Ecoscience: Population, Resources, Environment, W.H. Freeman, San Francisco.

Fairbridge, R.W. (Ed.)(1972) The Encyclopedia of Geochemistry and Environmental Science, Vol. IVA, Van Nostrand Reinhold, New York.

Garrels, R.M., MacKenzie, F.T. and Hunt, C. (1975) Chemical Cycles and the Global Environment, Kaufmann, Los Altos.

Garrels, R.M. and Perry, E.A., Cycling of carbon, sulphur and oxygen through geologic time, The Sea, Vol. 5 (Goldberg, E.D., Ed.)(1974) Wiley-Interscience, New York.

Gymer, R.G. (1973) Chemistry, an Ecological Approach, Harper and Row, New York.

Hodges, L. (1973) Environmental Pollution, Holt, Rinehart and Winston, New York.

Horne, R.A. (1978) The Chemistry of Our Environment, Wiley-Interscience, New York.

Hutzinger, O. (Ed.)(1980, 1982) The Handbook of Environmental Chemistry, Vols. 1, 2, 3, Springer-Verlag, Berlin.

Manahan, S.E. (1979) Environmental Chemistry, (3rd Ed.) Willard Grant Press, Boston.
Masters, G.M. (1974) Introduction to Environmental Science and Technology, J. Wiley, New York.
Moore, J.W. and Moore, E.A. (1976) Environmental Chemistry, Academic Press, New York.
Moore, J.W. and Moore, E.A. Resources in Environmental Chemistry, J. Chem. Ed., 52, 288 (1975), 53, 167, 240 (1976).
Pyle, J.L. (1974) Chemistry and the Technological Backlash, Prentice-Hall, New Jersey.
Raiswell, R.W., Brimblecombe, P., Dent, D.L. and Liss, P.S. (1980) Environmental Chemistry, Edward Arnold, London.
Scientific American (1973) Chemistry in the Environment - Readings, W.H. Freeman, San Francisco.
Spiro, T.G. and Stigliani, W.M. (1980) Environmental Issues in Chemical Perspective, State Univ. New York Press, Albany.
Stumm, W. (Ed.)(1977) Global Chemical Cycles and their Alteration by Man, Dahlem Konferenzen, Berlin.
Vowles, P.D. and Connell, D.W. (1980) Experiments in Environmental Chemistry, Pergamon Press, Oxford.

General Sources: General and Inorganic Chemistry

Atkins, P.W. (1978) Physical Chemistry, Oxford University Press, Oxford.
Bailer, J.C., Emeleus, H., Nyholm, R.S. and Trotman-Dickenson, A.F. (Eds.) (1973) Comprehensive Inorganic Chemistry, 5 Vols., Pergamon Press, Oxford.
Barrow, G.M. (1973) Physical Chemistry, (3rd Ed.) McGraw-Hill Kogakusha, Tokyo.
Benson, S.W., Predicting chemical reactivity, Chemtech, Feb., 121 (1980).
Campbell, J.A. (1965) Why Do Chemical Reactions Occur? Prentice-Hall, New Jersey.
Cotton, F.A. and Wilkinson, G. (1980) Advanced Inorganic Chemistry (4th Ed.) Wiley-Interscience, New York.
Cotton, F.A. and Wilkinson, G. (1978) Basic Inorganic Chemistry, J. Wiley, New York.
Dasent, W.E. (1965) Nonexistent Compounds, Edward Arnold, London.
Dasent, W.E. (1970) Inorganic Energetics, Penguin Books, Harmondsworth.
Eggins, B.R. (1972) Chemical Structure and Reactivity, Macmillan, London.
Eyring, H. and Eyring, E.M. (1963) Modern Chemical Kinetics, Reinhold, New York.
Heslop, R.B. and Jones, K. (1976) Inorganic Chemistry, Elsevier, Amsterdam.
Huheey, J.E. (1978) Inorganic Chemistry, (2nd Ed.), Harper and Row, New York.
Johnson, D.A. (1968) Some Thermodynamic Aspects of Inorganic Chemistry, Cambridge University Press, Cambridge.
Jolly, W.L. (1970) The Synthesis and Characterisation of Inorganic Compounds. Prentice-Hall, New Jersey.
Mackay, K.M. and Mackay, R.A. (1981) Modern Inorganic Chemistry, (3rd Ed.) Intertext Books, London.
Moore, W.J. (1972) Physical Chemistry, (5th Ed.) Longman, London.
Phillips, C.S.G. and Williams, R.J.P. (1965) Inorganic Chemistry, Vol. 1 and 2, Clarendon Press, Oxford.
Purcell, K.F. and Kotz, J.C. (1977) Inorganic Chemistry, W.B. Saunders, Philadelphia.
Rochow, E.G. (1977) Modern Descriptive Chemistry, W.B. Saunders, Philadelphia.
Strong, L.E. and Stratton, W.J. (1965) Chemical Energy, Reinhold, New York.

The Earth's Formation and Chemistry

Ahrens, L.H. (1965) Distribution of the Elements in our Planet, McGraw-Hill, New York.

Ahrens, L.H., The Significance of the chemical bond for controlling the geochemical distribution of the elements. In Physics and Chemistry of the Earth, Vol. 5, (1964) Macmillan, New York.

Allegre, C.J. and Michard, G. (1974) Introduction to Geochemistry, D. Reidel, Drodrecht.

Baas Becking, L.G.M., Kaplan, I.R. and Moore, D., Limits of the natural environment, J. Geol., 68, 243 (1960).

Booth, R.S., Interstellar molecules, J. Chem. Phys. Soc., Univ. College London, 4, (1976).

Broecker, W.S., Man's oxygen reserves, Science, 168, 1537 (1970).

Broecker, W.S. (1974) Chemical Oceanography, Harcourt, Brace, Jovanovich, New York.

Broecker, W.S. and Oversby, V.M. (1970) Chemical Equilibria in the Earth, McGraw-Hill, New York.

Burns, R.G. (1970) Mineralogical Applications of Crystal Field Theory, Cambridge Univ. Press, Cambridge.

Burns, R.G. and Fyfe, W.S. Site of preference energy and selective uptake of transition metal ions from a magma, Science, 144, 1001 (1964).

Butcher, S.S. and Charlson, R.J. (1972) An Introduction to Air Chemistry, Academic Press, New York.

Clayton, D.D. (1968) Principles of Stellar Evolution and Nucleosynthesis, McGraw-Hill, New York.

Cloke, P.L., The geochemical application of E_h-pH diagrams, J. Geol. Educ., XIV, 140 (1966).

Cloud, P., Evolution of ecosystems, Amer. Sci., 62, 54 (1974).

Cloud, P., Atmospheric and hydrospheric evolution on the primitive earth, Science, 160, 729 (1968).

Curtis, C.D. Chemistry of rock weathering fundamental reactions and Controls, in Geomorphology and Climate, Derbyshire, E. (Ed.),(1975), J. Wiley, London.

Dandy, A.J. and Morrison, R.J., Phase equilibria in geochemical systems, Chem 13 News, Nov. 4 (1977).

Deevey, E.S., Mineral cycles, in Chemistry in the Environment, 5 (1973), W.H. Freeman, San Francisco.

Delahay, P., Pourbaix, M. and von Russelberghe, P., Potential - pH diagrams, J. Chem. Ed., 27, 683 (1950).

Fearnsides, W.G. and Bulman, O.M.B. (1961) Geology in the Service of Man, (4th Ed.) Penguin Books, Harmondsworth.

Fergusson, J.E., Origins, 1, 2, Educ. in Chem., 15, 182 (1978), 16, 90 (1979).

Fieldes, M. and Swindale, L.D., Chemical Weathering of silicates in soil formation, N.Z. J. Sci. Techn., 36B, 140 (1954).

Fyfe, W.S. (1974) Geochemistry, Oxford Univ. Press. Oxford.

Garrels, R.M. and Christ, C.L. (1965) Solutions Minerals and Equilibria, Harper and Row, New York.

Garrels, R.M. and Mackenzie, F.T. (1971) The Evolution of Sedimentary Rocks, Norton, New York.

Garrels, R.M., Lerman, A. and Mackenzie, F.T., Controls of atmosphere O_2 and CO_2; past and future, Amer. Sci., 64, 306 (1976).

Gibson, D.T., The terrestial distribution of the elements, Quart. Revs., London, 3, 263 (1949).

Goldschmidt, V.M., The principles of distribution of chemical elements in minerals and rocks, J. Chem. Soc., 655 (1937).

Goldschmidt, V.M. (1954) Geochemistry, Clarendon Press, Oxford.

Holland, H.D. (1978) The Chemistry of the Atmosphere and Oceans, Wiley-Interscience, New York.
Huntress, W.T., Ion-molecule reactions in the evolution of simple organic molecules in interstellar clouds and planetary atmospheres, Chem. Soc. (Revs.), 16, 295 (1977).
Krauskopf, K.B. (1967) Introduction to Geochemistry, McGraw-Hill, New York.
Kroto, H., Chemistry between the stars, New Scientist, Aug., 400 (1978).
Krumbein, W.C. and Garrels, R.M., Classification of chemical sediments, J. Geol., 60, 1 (1952).
Loughman, F. (1969) Chemical Weathering of Silicate Minerals, Elsevier, Amsterdam.
Mason, B. (1966) Principles of Geochemistry (3rd Ed.), J. Wiley, New York.
Nicholls, G.D., Weathering of the earth crust, Chemical Oceanography (1976) Vol. 5, 81 (Riley, J.P. and Chester, R., Eds), Academic Press, New York.
Ollier, C.D. (1969) Weathering, Oliver and Boyd, Edinburgh.
Perrin, M.B. (1975) An Introduction to the Chemistry of Rocks and Minerals, Edward Arnold, London.
Pettijohn, E.J. (1977) Sedimentary Rocks, (3rd Ed.) Harper and Row, New York.
Rank, D.M., Townes, C.H. and Welch, W.J., Interstellar molecules and dense clouds, Science, 174, 1083 (1971).
Riley, J.P. and Skirrow, G.; Riley, J.P. and Chester, R. (Eds.)(1975- Chemical Oceanography, Vols 1-7, Academic Press, London.
Robertson, E.G. (Ed.)(1972) The Nature of the Solid Earth, McGraw-Hill, New York.
Selbin, J. The origin of the chemical elements, 1 and 2, J. Chem. Ed., 50, 306, 380 (1973).
Siegel, F.R. (1974) Applied Geochemistry, Wiley-Interscience, New York.
Siever, R., Steady state of the earth's crust, atmosphere, oceans, Sci. Amer., 230, 72 (1974).
Sillen, L.G., The ocean as a chemical system, Science, 156, 1189 (1967).
Skinner, B.J. (1976) Earth Resources (2nd Ed.) Prentice-Hall, New Jersey.
Solomon, P.M. Interstellar molecules, Phys. Today, Mar., 32 (1973).
Stumm, W. and Morgan, J.J. (1970) Aquatic Chemistry, Wiley-Interscience, New York.
Taylor, R.J. (1972) The Origins of the Chemical Elements, Wykeham, London.
Turekian, K.K. (1972) Chemistry of the Earth, Holt, Rinehart and Winston, New York.
Viola, V.E. Stellar nucleosynthesis, J. Chem. Ed., 50, 311 (1973).
Wallace, J.M. and Hobbs, P.V. (1977) Atmospheric Science, Academic Press, London.
Wheeler, J.C., After the supernova, what? Amer. Sci., 61(1), 42, 1973.
Wyllie, P.J., The earth's mantle, Sci. Amer., 232(3), 50 (1975).

Resources: Minerals

Many of the references above.
Barbour, A.K., Mineral recycling, a down-to-earth approach, Chem. in Brit., 13, 213 (1977), also in Conservation of Resources, Chem. Soc., London, Sp. Pub., 27, 205 (1976).
Bateman, A.M. (1967) The formation of Mineral Deposits, J. Wiley, New York.
Bateman, A.M. (1950) Economic Mineral Deposits, J. Wiley, New York.
Cook, E., Limits to exploitation of non-renewable resources, Science, 191, 677 (1976).
Corbett, P.F., Nature's assets, an elementary view, in Conservation of Resources, Chem. Soc., Sp. Pub. 27, 177 (1976).

Curtis, C.D., Applications of the crystal-field theory to the inclusion of trace transition elements in minerals during magmatic differentiation, Geochem. Cosmochim. Acta, 28, 389 (1964).
Garrels, R.M. (1960) Mineral Equilibria, Harper and Row, New York.
Gibson, J., Chemical Raw Materials, Chem. and Ind., 677 (1975).
Guild, P.W., Discovery of natural resources, Science, 191, 709 (1976).
Kesler, S.E. (1976) Our Finite Mineral Resources, McGraw-Hill, New York.
Keyfitz, N., World resources and the world middle class, Sci. Amer., 235, (July) 28 (1976).
Kirby, R.C. and Prokopovitsh, A.S., Technological insurance against shortage in minerals and metals, Science, 191, 713 (1976).
Lovering, T.S. (1969) Resources and Man, Committee on Resources and Man of the Division of Earth Sciences, Nat. Acad. Sci., U.S.A., Nat. Research Council, W.H. Freeman, San Francisco.
Meadows, D.H., Meadows, D.L., Randers, J. and Behrens III, W.W. (1972). The Limits to Growth, Universe Books, New York.
Sugden, T.M. and Briffa, F.E.J., The competition for resources: energy, Chem. and Ind., 669 (1975).
U.S. National Commission on Materials Policy, Towards a national policy - world perspective, 2nd interim report, Jan. 1973.
United Nations, Statistical Yearbooks.
Van Royen, W. and Bowles, O. (1952) Mineral Resources of the World, Vol II, Atlas of the World's Resources, Prentice-Hall, New Jersey.
Amer. Assoc. Adv. Sci., Various authors, Science, 191, No. 4228, (1976).
Warren, K. (1973) Mineral Resources, Penguin Books, Harmondsworth.
Williams, R.J.P., Deposition of trace elements in basic magma, Nature, 184, 44 (1959).

Resources: Energy

American Ass. Adv. Sci., Science, 184 , No. 4134, (1974).
Armstrong, G., World coal resources and their future potential, Energy in the 1980's, Roy. Soc. London, 33 (1974).
Austin, L.G., Fuel Cells, in Chemistry in the Environment, (Reading from Sci. Amer.), 188 (1973) W.H. Freeman, San Francisco.
Bainbridge, G.R., Fire, flame and energy, Conservation of Resources, Chem. Soc. London, Sp. Pub. 27, 217 (1976).
Barnea, J., Geothermal power, Sci. Amer., 226(1), 70 (1972).
Barnes, R.W., Btu's goods and the future, Chemtech, Jan., 30 (1978).
Bowie, S.H.U., Natural sources of nuclear fuel, Energy in the 1980's, Roy. Soc. London, 89 (1974).
Broadley, J.S., A chemical challenge in nuclear energy, Conservation of Resources, Chem. Soc., London Sp. Pub. 27, 59 (1976).
Broom, T. and Gow, R.S., Operational experience in nuclear power stations, Energy in the 1980's, Roy. Soc. London, 165 (1974).
Bryant, W.B., Trends in the oil industry, Chem. Eng. Prog., 70(8), 8 (1974).
Calvin, M., Solar energy by photosynthesis, Science, 184, 375 (1974).
Calvin, M., Hydrocarbons via photosynthesis, Energy and the Chemical Sciences (1978) Karcher Symp. 1977, 1, Pergamon Press, Oxford.
Christian, S.D. and Zuckerman, J.J. (Eds.)(1978) Energy and the Chemical Sciences, 1977 Karcher Symp., Pergamon Press, Oxford.
Clark, D., Energy conversion to electricity, Energy in the 1980's, Roy. Soc. London, 153 (1974).
Considine, D.M. (Ed.)(1977) Energy Technology Handbook, McGraw-Hill, New York.
Cook, E., The flow of energy in an industrial society, Chemistry in the Environment, (Readings from Sci. Amer.) 139 (1973), W.H. Freeman, San Francisco.

Coppack, C.P., Natural Gas, Energy in the 1980's, Roy. Soc., London, 57 (1974).

Culler, F.L. and Harms, W.O., Energy from breeder reactors, Phys. Today, May, 28 (1972).

Curran, S., Society's future energy needs, Conservation of Resources, Chem. Soc., London, Sp. Pub. 27, 1 (1976).

Curran, S.C. and Curran, J.S. (1979) Energy and Human Needs, Scot. Academic Press, Edinburgh.

Darmstadter, J. and Schurr, S.H., The world energy outlook to the mid-1980's, Energy in the 1980's, Roy. Soc., London, 7 (1974).

Drake, E., Oil reserves and production, Energy in the 1980's, Roy. Soc. London, 47 (1974).

Eisenbud, M., (1973) Environmental Radioactivity (2nd Ed.) Academic Press, New York.

Fells, I., Energy-transmission, storage and management, Conservation of Resources, Chem. Soc. London, Sp. Pub. 27, 33 (1976); also Chem. in Brit., 13, 222 (1977).

Foley, G. (1981) The Energy Question (2nd Ed.) Penguin Books, Harmondsworth.

Gaucher, L., Energy in perspective, Chemtech, Mar., 153 (1971).

Gough, W.C. and Eastlund, B.J., The prospects of fusion power, Chemistry in the Environment, (Reading from Sci. Amer.), 204, (1973), W.H. Freeman, San Francisco.

Grainger, L., Future trends in utilisation of coal energy conversion, Energy in the 1980's, Roy. Soc., London, 121 (1974).

Gregory, D.P., The hydrogen economy, Chemistry in the Environment, (Readings from Sci. Amer.) 219 (1973), W.H. Freeman, San Francisco.

Hainsel, V., The place of catalysis in the energy problem, Energy and the Chemical Sciences, (1978) Karcher Symp. 1977, 91, Pergamon Press, Oxford.

Hambling, J.K., Energy resources, problems and opportunities, Conservation of Resources, Chem. Soc. London, Sp. Pub. 27, 17 (1976).

Hammond, A., Metz, W. and Maugh II, T., (1973) Energy and the Future, Amer. Ass. Advan. Sci., Washington.

Hammond, G.S., Prospects for nonbiological storage of solar energy, Energy and the Chemical Sciences, (1978) Karcher Symp. 1977, 59, Pergamon Press, Oxford.

Hayes, E.T., Energy implications of materials processing, Science, 191, 661 (1976).

Hill, J., Future trends in nuclear power generation, Energy in the 1980's, Roy. Soc. London, 181 (1974).

Hore-Lacy, I. and Huberg, R., (1976) Nuclear Energy, an Australian Perspective, Australian Mining Industry Council, Australia.

Hubbert, M.K., The energy resources of the earth, Chemistry in the Environment, (Readings from Sci. Amer.) 149 (1973), W.H. Freeman, San Francisco.

Ion, D.C., General problems of collecting and understanding world energy data, Energy in the 1980's, Roy. Soc. London, 25 (1974).

Ion, D., Energy resources are not all fuel reserves, New Sci., 222 (1976).

Ion, D.C., (1980) Availability of World Energy Resources (2nd Ed.), Graham and Trotman, London.

Katz, J.J., Janson, T.R. and Wasielewski, M.R., A biomimetic approach to solar energy conversion, Energy and the Chemical Sciences, (1978) Karcher Symp. 1977, 31, Pergamon Press, Oxford.

Leardini, T., Geothermal power, Energy in the 1980's, Roy. Soc., London, 101 (1974).

Ledig, F.T. and Linzer, D.I.H., Fuel crop breeding, Chemtech, Jan., 18 (1978).

Lenihan, J. and Fletcher, W.W. (Eds)(1975) Energy Resources and the Environment, Blackie, Glasgow.

Lessing, L.P., Coal, Chemistry in the Environment, (Readings from Sci. Amer.) 159 (1973), W.H. Freeman, San Francisco.

Lechtin, N.L., Fixing sunshine abiotically, Chemtech, April, 252 (1980).

Marchette, C., The hydrogen economy and the chemist, Chem. in Brit., 13, 219 (1977).

Marsden, W. and Wainwright, M., Catalytic processes in production of liquid fuels from synthesis gas, Chem. in N.Z., 43, 188 (1979).

Maugh II, T.H., Fuel from wastes, a minor energy source, Science, 178, 599 (1972).

Mills, G.A., Gas from coal - fuel of the future, Environ. Sci. Techn., 5, 1178 (1971).

Mills, G.A. and Perry, H., Fossil fuel → power + pollution, Chemtech, Jan., 53 (1973).

Morrow, W.F., Solar energy: its time is near, Technol. Rev., 76, 30 (1973).

Nephew, E.A., The challenge and promise of coal, Technol. Rev., 76, 21 (1973).

de Nevers, N., Tar sands and oil shales, Chemistry in the Environment, (Reading from Sci. Amer.) 169 (1973), W.H. Freeman, San Francisco.

Nuckolls, J., Emmett, J. and Wood, L., Laser-induced thermonuclear fusion, Physics Today, 26, Aug., 46 (1973).

Osborn, E.F., Coal and the present energy situation, Science, 183, 477 (1974).

Penner, S.S. and Icerman, L. (Eds)(1974, 1975) Energy, Vol. I, II, Addison-Wesley, Reading, Mass.

Penner, S.S. (Ed.)(1976) Energy, Vol. III, Addison-Wesley, Reading, Mass.

Perry, H., The gasification of coal, Sci. Amer., 230(3), 19 (1974).

Pimentel, D., Hurd, L.E., Bellotti, A.C., Forster, M.J., Oka, I.N., Sholes, O.D. and Whitman, R.J., Food production and the energy crisis, Science, 182, 443 (1973) and Correspondence, Science, 187, 560 (1975).

Rand, D.A.J., Prospects for new batteries for electric vehicles, Chem. in N.Z., 43, 185 (1979).

Reece, L.E. and Kenedi, R.M., Human energy levels, limits and needs, Conservation of Resources, Chem. Soc. London, Sp. Pub. 27, 73 (1976).

Roberts, H.L., Energy-efficient processes for the chemical industry, Conservation of Resources, Chem. Soc. London, Sp. Pub. 27, 49 (1976).

Robinson, F.A. (Ed.)(1980) The Environmental Effects of Using More Coal, Roy. Soc. Chem., London.

Rooke, R.E., Future trends in gas production and transmission, Energy in the 1980's, Roy. Soc. London., 141 (1974).

Ruedisili, L.C. and Finebough, M.W. (1975) Perspectives on Energy, Oxford Univ. Press, Oxford.

Sarkanen, K.V., Renewable resources for the production of fuels and chemicals, Science, 191, 773 (1976).

Schriesheim, A., The heart of energy research, Energy and the Chemical Sciences (1978) Karcher Symp. 1977, 103, Pergamon Press, Oxford.

Seaborg, G.T. and Bloom, J.L., Fast breeder reactors, Chemistry in the Environment, (Readings from Sci. Amer.) 194 (1973), W.H. Freeman, San Francisco.

Spedding, C.R.W., Can human energy needs be met? Conservation of Resources, Chem. Soc., London, Sp. Pub. 27, 83 (1976).

Squires, A.M., Chemicals from coal, Science, 191, 689 (1976).

Squires, A.M., Clean power from dirty fuels, Chemistry in the Environment, (Readings from Sci. Amer.) 178 (1973), W.H. Freeman, San Francisco.

Steinhart, J.S. and Steinhart, C.S., Energy use in U.S. food system, Science, 184, 307 (1974).

Sugden, T.M. and Briffa, F.E.J., The competition for resources: energy, Chem. and Ind., 669 (1975).

Tatchell, J.A., Crops and fertilizers: overall energy budgets, Conservation of Resources, Chem. Soc., London, Sp. Pub. 27, 119 (1976).

Teller, E. (1979) Energy from Heaven and Earth, W.H. Freeman, San Francisco.

United Nations, Statistical Yearbooks.

Vernon, K.R., Hydro(including tidal) power, Energy in the 1980's, Roy. Soc. London, 79 (1974).

Walker III, E.B., Non-conventional hydrocarbons and future trends in oil utilization in Nth. America and their affect on world supplies, Energy in the 1980's, Roy. Soc. London, 135 (1974).

Weisz, P.B., Population and energy, Chemtech, Mar., 131 (1980).

Williams, R.J.P., Energy conservation in biology, Conservation of Resources, Chem. Soc. London, Sp. Pub. 27, 137 (1976).

Youle, P.V., Effects of total energy considerations on agriculture by the year 2000, Chemistry in Agriculture, Chem. Soc., London, Sp. Pub. 36, 275 (1979).

Inorganic Chemicals

American Chemical Society (1973) Chemistry in the Economy, A.C.S., Washington.

Barnes, R.S., Oxygen in iron and steel production, Oxygen in the Metal and Gaseous Fuel Industries, Chem. Soc., London, Sp. Pub. 32, 39 (1978).

Boulton, L.H. and Wright, G.A. (1979) Fundamentals of Metallic Corrosion and its Prevention, Australian Corrosion Ass., Australia.

Bradbury, F.R. and Jackson, J.B., Economics of the development of new materials, New Horizons for Chemistry and Industry in the 1990's, Soc. of Chem. Ind., London, 89 (1970).

Chynoweth, A.G., Electronic materials; functional substitutions, Science, 191, 725 (1976).

Cooke, G.W., Using chemicals to increase soil productivity, New Horizons for Chemistry and Industry in the 1990's, Soc. of Chem. Ind., London, 45 (1970).

Davies, D.S., Raw materials for chemicals, Chemtech, Mar., 135 (1974).

Derry, R., Oxygen in hydrometallurgical processes for winning metals, Oxygen in the Metal and Gaseous Fuel Industry, Chem. Soc., London, Sp. Pub. 32, 232 (1978).

Evans, U.R. (1972) The Rusting of Iron: Causes and Control, Edward Arnold, London.

Gilchrist, J.D. (1967) Extraction Metallurgy, Pergamon Press, Oxford.

Goeller, H.E. and Weinberg, A.M., The age of substitutability, Science, 191, 713 (1976).

Grimshaw, R.W. (1971) The Chemistry and Physics of Clays, (4th Ed.), Benn, London.

Hardy, R.W.F., Bottomley, F. and Burns, R.C. (Eds)(1979) A Treatise on Dinitrogen Fixation, Sect. I and II, Wiley-Interscience, New York.

Herington, E.F.G. (1963) Zone Melting of Organic Compounds, Blackwell, Oxford.

Hillig, W.B., New materials and composites, Science, 191, 733 (1976).

Ives, D.J.G. (1969) Principles of the Extraction of Metals, Monographs for Teachers, 3, Chem. Soc., London.

Jones, D.G. (Ed.)(1967) Chemistry and Industry, Clarendon Press, Oxford.

Leigh, J., A chemical fix for nitrogen? New Sci., 385 (1976).

Michaelis, E., The use of oxygen in the LD process for steel production, Oxygen in the Metal and Gaseous Fuel Industries, Chem. Soc., London, Sp. Pub. 32, 61 (1978).

Moore, W.J.(1967) Seven Solid States, W.A. Benjamin, New York.
Notholt, A.J.G., Phosphate rock; world production, trade and resources. Proc. First Indust. Minerals Inst. Congress, London, 104 (1974).
Packer, J.E. (Ed.)(1978) Chemical Processes in New Zealand. N.Z. Inst. Chem.
Parker, R.H. (1967) An Introduction to Chemical Metallurgy, Pergamon Press, Oxford.
Pearce, C.A. (1972) Silicon Chemistry and Applications, Chem. Soc., London, Monographs for teachers, No. 20.
Radcliffe, S.M., World changes and chances: some new perspectives for materials, Science, 191, 700 (1976).
Reuben, B.G. and Burstall, M.L. (1973) The Chemical Economy, Longmans, London.
Richards, R.L., Nitrogen fixation, Chemistry and Agriculture, Chem. Soc. London, Sp. Pub. 36, 258 (1979).
Robinson, F.A. Chemistry and the new industrial revolution, Chem. Revs. (Lond.) 317, (1975).
Rosenbaum, J.B., Minerals, extraction and processing: new developments, Science, 191, 720 (1976).
Rosenqvist, T., (1974) Principles of Extractive Metallurgy, McGraw-Hill, New York.
Schmalzried, H., (1974) Solid State Reactions, Verlag Chemie, Weinheim.
Selinger, B. (1978) Chemistry in the Market Place, Australian National Univ. Press, Canberra.
Society of Chemical Industry (1970) New Horizons for Chemistry and Industry in the 1990's, Soc. of Chemical Industry, London.
Thompson, R. (Ed.)(1977) The Modern Chemicals Industry, Chem. Soc., London, Sp. Pub. 31.
U.S. Department of Agriculture and TVA (1964) Superphosphate.
Warner, N.A., Developments in oxygen smelting, Oxygen in the Metal and Gaseous Fuel Industries, Chem. Soc., London, Sp. Pub. 31, 209 (1978).
Wynne, M.D. (1970) Chemical Processing in Industry, Roy. Inst. Chem., London, Monograph for teachers, 16.

Air Chemistry and Pollution

American Chemical Society (1972) Photochemical Smog and Ozone Reactions, A.C.S. 113, Washington.
Bach, W. (1972) Atmospheric Pollution, McGraw-Hill, New York.
Baker, B.G., Control of noxious emissions from internal combustion engines, Environmental Chemistry, (1977) 243 (Bockris, J. O'M., Ed.), Plenum Press, New York.
Biswas, A.K. (Ed)(1979) The Ozone Layer, Pergamon Press, Oxford.
Bockris, J. O'M., Environmentally clean fuels for transportation, Environmental Chemistry, (1977) 583 (Bockris, J. O'M., Ed.) Plenum Press, New York.
Bond, R.G. and Straub, C.P. (Eds)(1972) Handbook of Environmental Control, Vol. I, Air Pollution, Chem. Rubber Co. Press, Cleveland.
Briscard, J., Aerosol production in the atmosphere, Environmental Chemistry, (1977) 313 (Bockris, J.O'M., Ed.) Plenum Press, New York.
Butler, J.D., Controlling emissions from motor behicle exhausts, Chem. in Brit., 8, 258 (1972).
Cadle, R.D. and Allen, E.R., Atmospheric photochemistry, Science, 167, 243. (1970).
Haagen-Smit, A., The control of air pollution, Chemistry in the Environment, (Readings from Sci. Amer.), 247 (1973).

Hall, H.G. and Bartok, W., NO_x control from stationary sources, Environ. Sci. and Techn., 5, 320 (1971).
Hall, S.K., Sulphur compounds in the atmosphere, Chemistry, 45, 16 (1972).
Hecht, T.A. and Seinfeld, J.H., Development and validation of a generalized mechanism for photochemical smog, Environ. Sci. and Techn., 6, 47 (1947).
Heicklen, J. (1976) Atmospheric Chemistry, Academic Press, New York.
Hobbs, P.V., Harrison, H. and Robinson, E., Atmospheric effects of pollutants, Science, 183, 909 (1974).
Kellogg, W.W., Cadle, R.D., Allen, E.R., Lazrus, A.L. and Martell, E.A., The sulphur cycle, Science, 175, 587 (1972).
Kerr, J.A., Calvert, J.G. and Demerjian, K.L., Mechanism of photochemical smog formation, Chem. in Brit., 8, 252 (1972).
Lee, R.E., The size of suspended particulate matter in air, Science, 178, 567 (1972).
Leh, F. and Chan, K.M., Sulphur compounds, pollution, health effects, and biological function, J. Chem. Ed., 50, 246 (1973).
Leighton, P.A. (1961) Photochemistry of Air Pollution, Academic Press, New York.
Lynn, D.A. (1976) Air Pollution: Threat and Response, Addison-Wesley, Reading, Mass.
McEwan, M.J., The ozone controversy, Chem. in N.Z., 42, 64 (1978).
McEwan, M.J. and Phillips, L.F. (1975) Chemistry of the Atmosphere, Edward Arnold, London.
Molina, M. and Rowlands, F.S., Stratospheric sink for chlorofluoromethanes: chlorine atom-catalysed destruction of ozone, Nature, 249, 810 (1974).
Newell, R.E., The global circulation of atmospheric pollutants, Sci. Amer., 224, 32 (1971).
Noyes Data Corp., (1975) Handbook of Environmental Sources and Emissions, Noyes Data Corp., New Jersey.
Perkins, H.C. (1974) Air Pollution, McGraw-Hill, New York.
Pitts, J.N., Keys to photochemical smog control, Environ. Sci. and Techn., 11, 456 (1977).
Rasool, S.I. (Ed.)(1973) Chemistry of the Lower Atmosphere, Plenum Press, New York.
Robinson, E. and Robbins, R.C., Gaseous nitrogen compound pollutants from urban and natural sources, J. Air Poll. Control Ass., 20, 303 (1970).
Sawyer, J.C., Man-made carbon dioxide and the greenhouse effect, Nature, 239, 23 (1972).
Schaefer, V.J., Auto exhaust, pollution and weather patterns, Bull. Atom. Sci., Oct., 31 (1970).
Schneider, S.H. (1976) The Genesis Strategy Climate and Global Survival, Plenum Press, New York.
Schuck, E.H. and Stephens, E.E., Oxides of Nitrogen, Adv. in Environ. Sci. and Techn., 1, 73 (1969) Wiley-Interscience, New York.
Seinfeld, J.H. (1975) Air Pollution: Physical and Chemical Fundamentals, McGraw-Hill, New York.
Sekihara, K., Possible climate changes from CO_2 increase in the atmosphere, Environmental Chemistry (1977)(Bockris J. O'M., Ed.) 285, Plenum Press, New York.
Singer, S.F. (Ed.)(1975) Global Effects of Environmental Pollution, D. Reidel Pub. Co., Boston.
Spedding, D.J. (1974) Air Pollution, Oxford Univ. Press, Oxford.
Spedding, D.J., The interaction of gaseous pollutants with materials at the surface of the earth, Environmental Chemistry, (1977)(Bockris, J.O'M., Ed.) 213, Plenum Press, New York.

Stephens, E.R., The formation, reactions and properties of peroxyacyl nitrates (PANS) in photochemical air pollution, Adv. in Environ. Sci. and Techn., 1, 119 (1969) Wiley-Interscience, New York.

Stern, A.C.(Ed.)(1976-77) Air Pollution, Vols. 1-5 (3rd Ed.), Academic Press, New York.

Stern, A.C., Wohlers, H.C., Boubel, R.W. and Lowry, W.P. (1973) Fundamentals of Air Pollution, Academic Press, New York.

Strauss, W., Formation and control of air pollutants, Environmental Chemistry (1977)(Bockris, J.O'M., Ed.) 179, Plenum Press, New York.

Stuiver, M., Atmospheric carbon dioxide and carbon reservoir changes, Science, 199, 253 (1978).

Suffett, I.H. (Ed.)(1977) Fate of Pollutants in the Air and Water Environments, Part 1 and 2, Wiley-Interscience, New York.

Williams, J. (Ed.)(1978) Carbon Dioxide, Climate and Environment, Pergamon Press, Oxford.

Wilson, D.G., Alternative automobile engines, Sci. Amer., 239, 39 (1978).

Woodwell, G.M., The carbon dioxide question, Sci. Amer., 238, 38 (1978).

World Health Organisation (1977) Environmental Health Criteria, 4, Oxides of Nitrogen, WHO, Geneva.

World Health Organisation (1979) Environmental Health Criteria, 8, Sulphur Oxides and Suspended Particulate Matter, WHO, Geneva.

Water Chemistry and Pollution

Many of the references given above.

Agino, E., Wixson, B. and Smith, I., Drinking water quality and chronic disease, Environ. Sci. and Techn., 11, 660 (1977).

American Chemical Society (1968) Trace Inorganics in Water, Adv. in Chem. Series, 73, A.C.S., Washington.

American Chemical Society (1975), Marine Chemistry in the Coastal Environment, A.C.S. Symposium Series, 18, A.C.S., Washington.

American Chemical Society (1971) Non-equilibrium Systems in Natural Water, Adv. in Chem. Series, 106, A.C.S., Washington.

Bascom, W., The disposal of waste in the ocean, Sci. Amer., 231(2), 16 (1974).

Berner, R.A. (1971) Principles of Chemical Sedimentology, McGraw-Hill, New York.

Bond, R.G. and Straub, C.P. (1973) Handbook of Environmental Control, Vol. III, Water Supply Treatment, Chemical Rubber Co., Cleveland.

Bond, R.G. and Straub, C.P. (1974) Handbook of Environmental Control, Vol. IV, Waste Water Treatment and Disposal, Chemical Rubber Co., Cleveland.

Culp, R.L. and Culp, G.L. (1971) Advanced Wastewater Treatment, Van Nostrand-Reinhold, New York.

Diamant, R.M.E., The chemistry of sewage purification, Environmental Chemistry (1977)(Bockris, J.O'M., Ed.) 95, Plenum Press, New York.

Environmental Science and Technology, Zeroing in on water pollution, Environ. Sci. and Techn., 8, (10)(1974).

Fair, G.M., Geyer, J.C. and Okun, D.A. (1971) Elements of Water Supply and Waste Water Disposal (2nd Ed.), J. Wiley, New York.

Franks, F. (Ed.)(1973) Water - a Comprehensive Treatise, Plenum Press, New York.

Gibbs, R.J., Mechanism of trace metal transport in rivers, Science, 180, 71 (1973).

Goldberg, E.D. (1976) The Health of the Oceans, Unesco, Paris.

Goldberg, E.D. (Ed.)(1974) The Sea, Vol. 5, Marine Chemistry, Wiley-Interscience, New York.
Goldberg, E.D. (Ed.)(1975) The Nature of Seawater, Dahlem Konferenzen, Berlin.
Hart, B.T. and Davies, S.H.R., (1978) A Study of the Physico Chemical Forms of Trace Metals in Natural Waters and Wastewaters, Water Resource Council Tech. Rep. 35 Australian Govt. Pub., Canberra.
Higgins, I.R., Ion exchange - its present and future use, Environ. Sci. and Techn., 7, 1110 (1973).
Howe, E.D., The desalination of water, Environmental Chemistry (1977) (Bockris, J.O'M., Ed.) 617, Plenum Press, New York.
Hutchinson, G.E., Eutrophication, Amer. Sci., 61, 269 (1973).
Hyde, H.C., Utilization of wastewater sludge for agriculture soil enrichment, J. Water Pollut. Control Fed., 48, 77 (1976).
Keith, L.H. and Telliard, W.A., Priority pollutants - a perspective view, Environ. Sci. and Techn., 13, 416 (1976).
Lee, C., Rast, W. and Jones, R., Eutrophication of water bodies, insights for an age-old problem, Environ. Sci. and Techn., 12, 900 (1978).
MacIntyre, F., Why the sea is salt, Sci. Amer., 223, 104 (1970).
Mackenzie, F.T. and Garrels, R.M., Chemical mass balance between rivers and oceans, Amer. J. Sci., 264, 507 (1966).
McCaull, J. and Crossland, J. (1974) Water Pollution, Harcourt Brace Jovanovich, New York.
Mulligan, T.J. and Fox, R.D., Treatment of industrial wastewaters, Chem. Eng., 83, 49 (1976).
Mullins, T., The chemistry of water pollution, Environmental Chemistry, (1977)(Bockris, J.O'M., Ed.) Plenum Press, New York.
National Academy Science, (1977) Drinking Water and Health, Vol. 1. Nat. Acad. Sci., Washington.
National Academy Science, (1980) Drinking Water and Health, Vols. 2, 3, Nat. Acad. Sci., Washington.
Pagenkopf, G.K. (1978) Introduction to Natural Water Chemistry, Marcel Dekker, New York.
Penman, H.L., The water cycle, Chemistry in the Environment (1973)(Readings from Sci. Amer.) 23, W.H. Freeman, San Francisco.
Riley, J.P. and Chester, R. (Eds.)(1976, 1978) Chemical Oceanography, Vols. 5, 6, 7, Academic Press, New York.
Riley, J.P. and Skirrow, G. (Eds.)(1974, 1975) Chemical Oceanography, Vols. 1, 2, 3, 4, Academic Press, New York.
Rubin, A.J. (Ed.)(1976) Aqueous Environmental Chemistry of Metals, Ann Arbor Science, Ann Arbor.
Singer, P.C. (Ed.)(1973) Trace Metals and Metal-Organic Interactions in Natural Waters, Ann Arbor Science, Ann Arbor.
U.S. Environmental Protection Agency (1974) National Water Quality Inventory (2 vols.) U.S. Govt., Washington.
U.S. Environmental Protection Agency, Interim primary drinking water standards, Fed. Registrar, 40, 11990 (1975).

The Lithosphere and its Pollution

Black, C.A. (1968) Soil-Plant Relationships (2nd Ed.), John Wiley, New York.
Blum, S.L., Tapping resources in municipal solid waste, Science, 191, 669 (1976).
Bond, R.G. and Straub, C.P., (1973) Handbook of Environmental Control; Solid Wastes, Chemical Rubber Co., Cleveland.

Bowen, D.H.M. (Ed.) Solid wastes, Environ. Sci. and Techn., collection of articles (1967-1971).

Brady, N.C. (1974) The Nature and Properties of Soils, (8th Ed.), Macmillan, New York.

Brooks, R.R., Pollution through trace elements, Environmental Chemistry, (1977)(Bockris, J.O'M., Ed.) Plenum Press, New York.

Brooks, R.R. (1972) Geobotany and Biogeochemistry in Mineral Exploration, Harper and Row, New York.

Davies, B.E.(Ed.)(1980) Applied Soil Trace Elements, John Wiley, New York.

Ermolenko, N.F. (1972) Trace Elements and Colloids in Soils (2nd Ed.) Israel Programme for Scientific Translations, Jerusalem.

Environmental Science and Technology, Recycling sludge and sewage effluent by land disposal, Environ. Sci. and Techn., 6, 871 (1972).

Evans, J.O., The soil as a resource renovator, Environ. Sci. and Techn., 4, 733 (1970).

Fenchel, T. and Blackburn, T.H. (1979) Bacteria and Mineral Cycling, Academic Press, London.

Greenland, D.J. and Hayes, M.H.B.,(1978) The Chemistry of Soil Constituents, John Wiley, New York.

Griffen, M.F. and Thomas, B., The safe use of chemicals in agriculture, Chemistry and Agriculture, Chem. Soc. London, Sp. Pub. 36, 237 (1979).

Hunt, C.B. (1972) Geology of Soils, W.H. Freeman, San Francisco.

Jackson, F.R. (1975) Recycling and Reclaiming of Municipal Solid Wastes, Noyes Data Corp., New Jersey.

Jones, M.J.(Ed.)(1975) Minerals in the Environment, Inst. of Mining and Metallurgy, London.

Kenahan, C.B., Solid waste, Environ. Sci. and Techn., 5, 594 (1972).

Kothng, E.L. (1973) Trace Elements in the Environment, Amer. Chem. Soc. No. 123, Washington.

Leet, L.D., Judson, S. and Kauffman, M.E. (1978) Physical Geology, (5th Ed.) Prentice-Hall, New Jersey.

Millot, G., Clay, Sci. Amer., 240 (April) 77 (1979).

National Academy of Sciences (1972) Soils of the Humid Tropics, N.A.S., Washington.

Nicholas, D.J.D. and Egan, A.R. (Eds.)(1975) Trace Elements in Soil-Plant-Animal Systems, Academic Press, New York.

Plumb, R.C., Trace elements make Australia fertile, J. Chem. Ed., 51, 675 (1974).

Postgate, J.R. (1979) The Sulphate-Reducing Bacteria, Cambridge Univ. Press, Cambridge.

Purves, D., (1977) Trace Element Contamination of the Environment, Elsevier, Amsterdam.

Rose, D.J., Gibbons, J.H. and Fulkerson, W., Physics looks at waste management, Phys. Today, 25, Feb., 32 (1972).

Russell, E.W. (1973) Soil Conditions and Plant Growth, (10th Ed.) Longman, London.

Skitt, J., (1972) Disposal of Refuse and Other Waste, Charles Knight and Co., London.

Trudinger, P.D. and Swaine, D.J. (1979) Biogeochemical Cycling of Mineral-Forming Elements, Elsevier, Amsterdam.

Young, A., (1976) Tropical Soils and Soil Survey, Cambridge Univ. Press, Cambridge.

Inorganic Species in the Environment

Many of the above references.

Bryce-Smith, D., Behavioural effects of lead and other heavy metal pollutants, Chem. in Brit., 8, 240 (1972).

Chappell, W.R. and Peterson, K.K. (1976, Vol. 1, 1977, Vol. 2) Molybdenum in the Environment, Marcel Dekker, New York.

Chamberlain, A.C., Heard, M.J., Little, P., Newton, D., Wells, A.C. and Wiffen, R.D. (1978) Investigation into Lead from Motor Vehicles, AERE Report - 9198 Harwell.

Chow, T.J., Our daily lead, Chem. in Brit., 9, 258 (1973).

D'Itri, F.M. (1972) The Environmental Mercury Problem, Chemical Rubber Co., Cleveland.

D'Itri, F.M. and D'Itri, P.A. (1977) Mercury Contamination, a Human Tragedy, J. Wiley, New York.

Ewing, B.B. and Pearson, J.E., Lead in the environment, Adv. in Environ. Sci. and Techn., Vol. 3 (1974), Wiley-Interscience, New York.

Gavis, J. and Ferguson, J.F., The cycling of mercury through the environment, Water Research, 6, 989 (1972).

Goldwater, L.J., Mercury in the environment, Chemistry in the Environment, (1973)(Readings from Sci. Amer.) 328, W.H. Freeman, San Francisco.

Griffen, T.B. and Knelson, J.H. (Eds.)(1975) Lead, G. Thieme and Academic Press, Stuttgart.

Griffith, E.J., Beeton, A., Spencer, J.M. and Mitchell, D.T. (Eds.)(1973) Environmental Phosphorus Handbook, J. Wiley, New York.

Grimstone, G., Mercury in British fish, Chem. in Brit., 8, 244 (1972).

Harrison, R.M. and Laxen, D.P.H., Metal in the Environment, 1-chemistry, Chem. in Brit., 16, 316 (1980).

Krenkel, P.A., Mercury in the environment, CRC Critical Reviews in Environmental Control, 3, 303 (1973); and Mercury: environmental considerations, 4, 251 (1974).

National Science Foundation, Lead in the Environment, N.S.F., Washington.

National Academy of Science (1973) Manganese, Nat. Res. Council, Washington.

National Academy of Science (1977) Arsenic, Nat. Res. Council, Washington.

Newland, L.W. and Daum, K.H., Lead, Handbook of Environmental Chemistry, Anthropogenic Compounds, (Hutzinger, O., Ed.)(1982) 1, Springer-Verlag, Berlin.

Newland, L.W., Arsenic, beryllium, selenium and vanadium, Handbook of Environmental Chemistry, Anthropogenic Compounds, (Hutzinger, O.,Ed.) (1982) 27, Springer-Verlag, Berlin.

Nriagu, J.O. (Ed.)(1978) The Biogeochemistry of Lead in the Environment, Part A, Ecological Cycles, Part B, Biological Effects, Nth. Holland, Amsterdam.

Nriagu, J.O. (Ed.)(1979) Copper in the Environment, Part I, Ecological Cycling, Part II, Health Effects, J. Wiley, New York.

Nriagu, J.O. (Ed.)(1979) The Biogeochemistry of Mercury in the Environment, Elsevier - Nth. Holland, Amsterdam.

Nriagu, J.O. (Ed.)(1980) Zinc in the Environment, Part I, Ecological Cycling, Part II, Health Effects, Wiley-Interscience, New York.

Inorganic Species in Biological Systems

Brodine, V., (1975) Radioactive Contamination, Jovanovich Harcourt Bruce, New York.

Bryce-Smith, D., (1975) Heavy Metals as Contaminants of the Human Environment, Chemistry Cassettes, Chemical Society, London.

Chesters, J.K., Nutritional chemistry of inorganic trace constituents in the diet, Chem. Soc. Revs., 10, 270 (1981).

Chisolm, J.J., Lead poisoning, Chemistry in the Environment, (1973)(Readings from Sci. Amer.) 335, W.H. Freeman, San Francisco.

Clarkson, J.J., Recent advances in the toxicology of mercury with emphasis on alkylmercurials, CRC Crit. Revs. Toxicol., 1, 203 (1972).

Department of the Environment (1974) Lead in the Environment and its Significance to Man, H.M. Stationery Office, London.

Department of Health and Social Security, (1980) Lead and Health (Report of a DHSS Working Party on Lead in the Environment, Lawther, P.J.), H.M. Stationery Office, London.

Dhar, S.K. (Ed.)(1973) Metal Ions in Biological Systems, Plenum Press, New York.

Di Ferrante, E. (Ed.)(1979) Trace Metals, Exposure and Health Effects, Pergamon Press, Oxford.

Fells, G.S., Metals in the environment, 2-Health effects, Chem. in Brit., 16, 323 (1980).

Fiabane, A.M. and Williams, D.R., (1977) The Principles of Bio-Inorganic Chemistry, Chemical Society, Monographs for Teachers No. 31, Chem. Soc., London.

Friberg, L. Piscator, M. and Nordberg, G. (1971) Cadmium in the Environment, Chemical Rubber Co., Cleveland.

Garnys, V.P., Freeman, R. and Smythe, L.E. (1979) Lead Burden of Sydney School Children, (Parts 1 and 2), Univ. of New South Wales, Sydney.

Goyer, R.A. and Mehlman, M.A. (Eds.)(1977) Toxicology of Trace Elements, Halstead Press, New York.

Higgins, I.J. and Burns, R.G. (1975) The Chemistry and Microbiology of Pollution, Academic Press, New York.

Houtman, J.P.W. and van den Hamer, C.J.A. (Eds.)(1975) Physiological and Biochemical Aspects of Heavy Elements in Our Environment, Delft Univ. Press, Delft.

Lave, L.B. and Seaken, E.P. (1977) Air Pollution and Human Health, Johns Hopkins Univ. Press, Baltimore.

Luckey, T.D. and Venugopal, B., (1977) Metal Toxicity in Mammals, Vol. 1, Plenum Press, New York.

Luckey, T.D. and Venugopal, B., (1978) Metal Toxicity in Mammals, Vol. 2, Plenum Press, New York.

McIntyre, A.D. and Mills, C.F. (Eds.)(1975) Ecological Toxicology Research; Effects of Heavy Metal and Organo-halogen Compounds, Plenum Press, New York.

Meyer, E. (1977) The Chemistry of Hazardous Materials, Prentice-Hall, New Jersey.

National Academy of Science (1975) Principles for Evaluating Chemicals in the Environment, N.A.S., Washington.

National Academy of Science (1972) Lead: Airborne Lead in Perspective, N.A.S., Washington.

Nordberg, G.F. (Ed.)(1976) Effects and Dose-Response Relationships of Toxic Metals, Elsevier, Amsterdam.

Ochiai, E.I., Environmental bioinorganic chemistry, J. Chem. Ed., 51, 235 (1974).

Peisach, J., Aisen, P. and Blumberg, W.E. (Eds.)(1966) The Biochemistry of Copper, Academic Press, New York.

Phillips, D.J.H. (1980) Quantitative Aquatic Biological Indicators, Applied Science, London.

Phipps, D.A. (1976) Metals and Metabolism, Clarendon Press, Oxford.

Repco, J.D., Corum, C.R., Jones, P.D. and Garcia, L.S. (1978) The Effects of Inorganic Lead on Behavioural and Neurologic Function, NIOSH Report, U.S. Dept. Health, Education and Welfare, Washington.

Ridley, W.P., Dizikes, L.J. and Wood, J.M., Biomethylation of toxic elements in the environment, Science, 197, 329 (1977).

Saito, N., Use of mercury and its compounds in industry and medicine, Mercury Contamination in Man and His Environment, (1972) Int. Atomic Energy Agency, Tech. Rept. No. 137.

Schubert, J., Beryllium and berylliosis, Chemistry in the Environment, (1973) (Readings from Sci. Amer.) 321, W.H. Freeman, San Francisco.

Shakman, R.A., Nutritional influences on the toxicity of environmental pollutants, Arch. Environ. Health, 28, 105 (1974).

Siegel, H. (Ed.)(1974-79) Metal Ions in Biological Systems, Vols. 1-8, Marcel Dekker, New York.

Singhal, R.L. and Thomas, J.A. (Eds.)(1980) Lead Toxicity, Urban and Schwarzenberg, Baltimore-Munich.

Valee, B.L. and Ulmer, D.D., Biochemical effects of mercury, cadmium and lead, Ann. Rev. Biochem., 41, 91 (1972).

Waldbott, G.L. (1973) Health Effects of Environmental Pollutants, C.V. Mosby.

Williams, R.J.P. and da Silva, J.R.R.F. (1978) New Trends in Bio-inorganic Chemistry, Academic Press, New York.

Wood, J.M., Biological cycles of toxic elements in the environment, Science, 183, 1049 (1974).

World Health Organisation (1976) Environmental Health Criteria 1, Mercury, W.H.O., Geneva.

World Health Organisation (1977) Environmental Health Criteria 3, Lead, W.H.O., Geneva.

Environmental Analytical Chemistry

Allen, S.E., Grimshaw, H.M., Parkinson, J.A. and Quarmby, C. (Eds.)(1974) Chemical Analysis of Ecological Materials, Blackwell, Oxford.

Betteridge, D. and Hallam, H.E. (1972) Modern Analytical Methods, Monographs for Teachers, No. 21, Chemical Society, London.

Black, C.A. (Ed.)(1965) Methods of Soil Analysis (2 Vols.), Amer. Soc. Agronomy, Madison.

Ewing, G.W. (1975) Instrumental Methods of Chemical Analysis, (4th Ed.) McGraw-Hill, Kogakusha, New York, Tokyo.

Farrer, K.T.H., Tangled in the traces, Food Techn. in Australia, Aug., 312 (1978).

Gibb, T.R.P. (Ed.)(1975) Analytical Methods in Oceanography, Amer. Chem. Soc., Adv. Chem. Series, 147, Washington.

Horwitz, W., The problems of utilizing trace analysis in regulatory analytical chemistry, Chem. in Aust., 49, 56 (1982).

Jackson, M.L. (1958) Soil Chemical Analysis, Prentice-Hall, New Jersey.

Lenihan, J. and Fletcher, W.W. (Eds.)(1978) Measuring and Monitoring in the Environment, Blackie, Glasgow.

Meinke, W.W. and Taylor, J.K. (Eds.)(1972) Analytical Chemistry, Key to Progress on National Problems, Nat. Bureau Standards, Washington.

Mertz, W., Trace element nutrition in health and disease, contributions and problems of analysis, Clin. Chem., 21, 468 (1975).

Moreton, J. and Falla, N.A.R. (1980) Analysis of Airborne Pollutants in Working Atmospheres, Anal. Sciences Monograph No. 7, Chemical Society, London.

Pickford, C.J., Sources of, and analytical advances in trace inorganic constituents of food, Chem. Soc. Revs., 10, 245 (1981).

Reeves, R.D. and Brooks, R.R. (1978) Trace Element Analysis of Geological Materials, J. Wiley, New York.

Risby, T.H. (Ed.)(1979) Ultratrace Metal Analysis in Biological Sciences and Environment, Adv. in Chem. Series, No. 172, Amer. Chem. Soc., Washington.

Rodier, J., (1975) Analysis of Water, J. Wiley, New York.

Ryan, D.E., Stuart, D.C. and Chattopadhyay, A., Rapid multielement neutron activation analysis with a SLOWPOKE reactor, Anal. Chim. Acta, 100, 87 (1978).

Skoog, D.A. and West, D.M. (1980) Principles of Instrumental Analysis, (2nd Ed.), W.B. Saunders, Philadelphia.

Smythe, L.E., Analytical chemistry of pollutants, Environmental Chemistry (Bockris, J.O'M., Ed.), 677, Plenum Press, New York.

Svehla, G. (Ed.)(1975-) Wilson and Wilson's Comprehensive Analytical Chemistry, Elsevier, Amsterdam.

Toras, M.J., Greenberg, A.E., Hook, R.D. and Rand, M.C. (Editorial Bd.)(1971) Standard Methods for the Examination of Water and Wastewater, (13th Ed.) Amer. Public Health Ass., Washington.

Toribara, T.Y., Coleman, J.R., Dahneke, B.E. and Feldman, I. (Eds.)(1978) Environmental Pollutants, Detection and Measurement, Plenum Press, New York.

Wilson, A.L. (1974) The Chemical Analysis of Water, Anal. Sci. Monograph No. 2., Soc. for Anal. Chemistry, London.

Warner, P.O. (1976) Analysis of Air Pollutants, Wiley-Interscience, New York.

Zief, M. and Mitchell, J.W. (1976) Contamination Control in Trace Element Analysis, Wiley-Interscience, New York.

Index